PHYSICS OF LUMINOUS BLUE VARIABLES

ASTROPHYSICS AND SPACE SCIENCE LIBRARY

A SERIES OF BOOKS ON THE RECENT DEVELOPMENTS
OF SPACE SCIENCE AND OF GENERAL GEOPHYSICS AND ASTROPHYSICS
PUBLISHED IN CONNECTION WITH THE JOURNAL
SPACE SCIENCE REVIEWS

PROCEEDINGS
VOLUME 157

PHYSICS OF LUMINOUS BLUE VARIABLES

PROCEEDINGS OF THE 113TH COLLOQUIUM
OF THE INTERNATIONAL ASTRONOMICAL UNION,
HELD AT VAL MORIN, QUEBEC PROVINCE, CANADA,
AUGUST 15-18, 1988

Edited by

KRIS DAVIDSON

Department of Astronomy, University of Minnesota,
Minneapolis, U.S.A.

A. F. J. MOFFAT

Département de Physique,
Université de Montréal, Canada

and

H. J. G. L. M. LAMERS

SRON, Laboratory for Space Research,
Utrecht, The Netherlands

KLUWER ACADEMIC PUBLISHERS
DORDRECHT / BOSTON / LONDON

Library of Congress Cataloging in Publication Data

International Astronomical Union. Colloquium (113th : 1988 : Val Morin, Québec)
Physics of luminous blue variables : proceedings of the 113th Colloquium of the International Astronomical Union, held at Val Morin, Québec Province, Canada, August 15-18, 1988 / [edited by] Kris Davidson, A.F.J. Moffat & H.J.G.L.M. Lamers.
p. cm. -- (Astrophysics and space science library ; v. 157)
ISBN-13: 978-94-010-6955-7
1. Luminous blue variables--Congresses. I. Davidson, Kris. II. Moffat, A. F. J. III. Lamers, Henny J. G. L. M., 1941- IV. Title. V. Series.
QB843.L85I57 1988
523.8'44--dc20 89-19905
CIP

ISBN-13: 978-94-010-6955-7 e-ISBN-13: 978-94-009-1031-7
DOI: 10.1007/978-94-009-1031-7

Published by Kluwer Academic Publishers,
P.O. Box 17, 3300 AA Dordrecht, The Netherlands.

Kluwer Academic Publishers incorporates
the publishing programmes of
D. Reidel, Martinus Nijhoff, Dr W. Junk and MTP Press.

Sold and distributed in the U.S.A. and Canada
by Kluwer Academic Publishers,
101 Philip Drive, Norwell, MA 02061, U.S.A.

In all other countries, sold and distributed
by Kluwer Academic Publishers Group,
P.O. Box 322, 3300 AH Dordrecht, The Netherlands.

Softcover reprint of the hardcover 1st edition 1989

TABLE OF CONTENTS

LIST OF PARTICIPANTS -- IAU COLLOQUIUM 113

(Numbers refer to the group photograph; there may be some errors here!)

I. Appenzeller, *Heidelberg (16)*
L.A. Balona, *S.Africa (62)*
G.B. Baratta, *Rome (58)*
B. Bohannan, *Colorado (22)*
J.P. Cassinelli, *Wisconsin (21)*
C. Chavarria-K., *Mexico City (47)*
P.S. Conti, *Colorado (33)*
M. Corcoran, *NASA/Goddard (68)*
K. Davidson, *Minnesota (63)*
M. de Groot, *Armagh (49)*
C. de Jager, *Utrecht (10)*
L. Drissen, *Montréal (56)*
E.L. Fitzpatrick, *Princeton*
M. Friedjung, *Paris (61)*
A. Fullerton, *Toronto*
R. Gabler, *Munich (4?)*
J.S. Gallagher, *Flagstaff (36)*
C. Garmany, *Colorado*
E. Gosset, *ESO (35)*
H.G. Groth, *Munich (2)*
B. Hauck, *Lausanne (15)*
S. Heap, *NASA/Goddard (46)*
A.G. Hearn, *Utrecht (12)*
H. Henrichs, *Amsterdam*
J. Hillier, *Harvard-SAO (28)*
D.G. Hummer, *Colorado (26)*
R.M. Humphreys, *Minnesota (24)*
R.H. Johnson, *Oxford (51)*
F.D. Kahn, *Manchester (3)*
J. Klapp, *Mexico City (1)*
G. Koenigsberger, *Mexico City (67)*
R.P. Kudritzki, *Munich (39)*
H.J.G.L.M. Lamers, *Utrecht (34)*
N. Langer, *Göttingen (31)*
C. Leitherer, *Colorado (7)*
M.-C. Lortet, *Meudon (40)*
A. Maeder, *Geneva (32)*
J.M. Marlborough, *Ontario (6)*
S.R. McCandliss, *Colorado (42)*
P.J. McGregor, *Mt. Stromlo (25)*
A.F.J. Moffat, *Montréal (65)*
V. Niemela, *Buenos Aires (27)*
H. Nieuwenhuijzen, *Utrecht (48)*
S. Owocki, *Delaware (20)*
F. Paresce, *STScI, Baltimore*
A. Pauldrach, *Munich (13?)*
R.K. Prinja, *London (14)*
J. Puls, *Munich*
C. Robert, *Montréal (54)*
C. Rossi, *Rome*
D.D. Sasselov, *Toronto*
W. Schmutz, *Colorado (45)*
R.E. Schulte-Ladbeck, *Wisconsin (8)*
M. Shara, *STScI, Baltimore (23)*
S.N. Shore, *Socorro (37)*
L.J. Smith, *London (53)*
J. Smolinski, *Torun (29)*
S.R. Sreenivasan, *Calgary (55)*
N. St.-Louis, *London (17)*
O. Stahl, *Heidelberg (66)*
C. Sterken, *Brussels (50)*
K. van der Hucht, *Utrecht (52)*
D. Vanbeveren, *Brussels (57)*
A. Wagner-Gabler, *Munich (5)*
N.R. Walborn, *STScI, Baltimore (64)*
P. Williams, *Edinburgh (18)*
A.J. Willis, *London (41)*
B. Wolf, *Heidelberg (44)*
F.-J. Zickgraf, *Heidelberg (19)*

PREFACE

A strange field of speculation is opened by this phenomenon...here we have a star fitfully variable to an astonishing extent, and whose fluctuations are spread over centuries, apparently in no settled period, and with no regularity of progression. What origin can we ascribe to these sudden flashes and relapses? What conclusions are we to draw as to the comfort or habitability of a system depending for its supply of light and heat on so uncertain a source?

-- J.F.W. Herschel[1]

We can imagine, at least, that the study of Luminous Blue Variable stars began 5000 years ago as the official court astrophysicists of Eridu speculated about a new first-magnitude star which persisted in their southern sky for several years.[2] After that the topic languished for a while, but now it has been renewed following recent observations by Blaeu (P Cygni, *c.*1600) and Herschel (η Argus, *c.*1840)...

-- More soberly, it is only within the past few years that we have begun to understand LBV's and to recognize their coherence as a class. Thus only a minority of astronomers are familiar yet with this relatively new topic, whose potential implications are of great interest (see below). One purpose of this book, we hope, will be to provide an accessible overview of the subject for interested astronomers and astrophysicists in general.

What is an LBV? One answer is that '**Very** Luminous Blue Variable' would be more accurate and that S Doradus, η Carinae, and P Cygni are now thought to be examples. Their luminosities are of the order of $10^6 L_\odot$ and they lose mass sporadically in giant eruptions. They flirt with disaster near the Eddington Limit, and in so doing they exhibit a remarkable richness of physical behavior. They are near the observed Upper Luminosity Limit of stars in the Hertzsprung-Russell diagram, and their outbursts may be clues to why this Upper Limit exists. Opinions on the proper definition of 'LBV' are expressed by various people in this book, and Voltaire's quip about the Holy Roman Empire may apply: at times some LBV's may appear neither Luminous, nor Blue, nor Variable. Some authors prefer simply to call them S Doradus stars. But 'LBV' is pragmatically convenient here, just as a label.

Why are we interested in LBV's? The reasons embrace topics from spectral classification to galaxy evolution. Their dramatic eruptions are only beginning to be understood but apparently involve non-linear chaotic systems of hydrodynamics with radiative transfer. Their large, unsteady mass-loss rates appear to be crucial for evolution at the top of the H-R diagram. They may provide clues to interior processes relevant, for instance, to the progenitor of SN 1987A (whose blueward evolution was

---

[1]*Results of Astronomical Observations...at the Cape of Good Hope*(London, 1847), p. 36.

[2]This is based on thoroughly unreliable but fascinating passages by P.C.A. Jensen (1890; *Die Kosmologie der Babylonier*, pp. 24–28) and Shklovsky & Sagan (1966; *Intelligent Life in the Universe*, ch. 33). Modern experts agree that the cuneiform texts do not really support Jensen's idea that η Car was seen and identified with *Ía;* but the story is so pertinent to LBV's that it **should** have been true!

acknowledged most readily by those astronomers who were most acquainted with very luminous stars). LBV's are thought to be crucial for understanding the formation of some types of Wolf-Rayet stars. They can be observed in other galaxies, even beyond the Local Group. They are suspected to be related, indirectly or perhaps directly, to any physical justification for using red supergiants as extragalactic distance indicators. They may be prominent in nearly-primordial dwarf galaxies, according to some speculations. And finally, they are sufficiently ill-understood, as yet, to provide scope for interesting surprises and reversals of consensus!

This book contains papers presented at IAU Colloquium 113, held at the Far Hills Inn near Val Morin, Quebec, on 15–18 August 1988. LBV's had played leading roles at the de Jager workshop in Lunteren, the Netherlands, in 1986 (see *Instabilities in Luminous Early-Type Stars*, ed. by Lamers and de Loore, 1987) and to a lesser extent they were discussed on other occasions such as IAU Symposia 105 and 116; but Val Morin was the first meeting devoted specifically to LBV's *per se*. Since other classes of very luminous stars are relevant to LBV's (*e.g.*, B[e] stars, not to be confused with Be stars!), many or even most of the stars mentioned at Val Morin were strictly speaking not real LBV's; we have relied on the authors' own self-discipline to keep resulting centrifugal tendencies within reasonable bounds. The discussions following the papers **have** been edited; we have tried to make these clear and potentially useful, rather than aiming for a transcript of precisely what words each participant used.

* * *

A note concerning multiple alternative star names and catalogs containing LBV's -- When one astronomer mentions HDE 326823 and another refers to He 3 - 1330, it is useful to know that these are the same star! Therefore *we have provided many cross-references in the object index at the end of this book.* We hope that these references to alternative star names will have a utility beyond this book. Moreover, preceding the index *we have listed a few catalogs* whence most commonly-cited LBV designations come; these may be useful if one must find the actual location of a given star.

* * *

We are sincerely grateful to many people who made the colloquium and this book possible -- especially the local organizing committee who helped to make Val Morin such a superb location: Madeleine Bergevin (coordinator), Louise Cloutier, Marie-Claude Beaudet, Jean-Eudes Samuel, Laurent Drissen, Carmelle Robert, Normand Lapierre, Andre Grandchamps, and Luc Turbide. Hans Nieuwenhuijzen and J.-E. Samuel provided most of the photographs used here, Louise Cramer efficiently prepared the first version of the discussions, and M.-C. Lortet, S.N. Shore, B. Bohannan, and R.M. Humphreys provided information on catalogs and alternative star names that we have used in the index pages as mentioned above. We thank M. de Groot, K. van der Hucht, and A.J. Willis for giving talks summarizing the many poster papers at the meeting. In addition to ourselves (Davidson, Moffat, Lamers), the Scientific Organizing Committee included P. Benvenuti, M. Feast, B. Hauck, L. Luud, O. Stahl, R. Viotti, and N.R. Walborn; while financial support was generously provided by the Canadian NSERC, the IAU, the Canadian Institute for Theoretical Astrophysics, and the Université de Montréal.

SECTION I:
MAIN PAPERS,
GENERAL DISCUSSIONS,
AND
TWO PANEL DISCUSSIONS

WHAT ARE LBV'S? -- THEIR CHARACTERISTICS AND ROLE IN THE UPPER H-R DIAGRAM

Roberta M. Humphreys
Landessternwarte Heidelberg, and
University of Minnesota

1. INTRODUCTION

Only ten years ago a symposium on the special class of stars we now call Luminous Blue Variables (LBV's) would not have been possible. At that time we were only just beginning to recognize the similarities among the stars we call S Doradus variables, P Cygni stars, and Hubble-Sandage variables, and their importance in the evolution of the most massive stars.

About a decade ago it became clear that the observed H-R diagram has an upper luminosity limit and that η Car, P Cyg, and S Dor are all near the temperature-dependent boundary for hot stars. With hindsight the luminosity boundary can be seen in an H-R diagram for hot stars published by Hutchings (1976), but the full temperature range is necessary to show that it really exists. When more data became available in 1977--1978, including the evolved and cool supergiants, the upper luminosity limit became obvious; and we proposed that an instability causing **rapid** and **unsteady** mass loss (*i.e.*, LBV's) is the basic reason for this boundary (Humphreys and Davidson 1979). We emphasized the temperature-dependence of the boundary for the most luminous hot stars, the lack of cooler counterparts at similar luminosities, the nearly temperature-independent limit to the luminosities of cool hypergiants, and the critical role that objects like η Car, P Cyg, and S Dor may play in the evolution of the most massive stars.

The location of these LBV's near the luminosity limit was crucial to understanding its possible cause. The empirical boundary could in principle be explained by steady mass loss, but only at rates much higher than normally observed; and LBV's are notorious for their **sporadic** mass-ejection events, their "eruptions". A stability limit, probably caused by radiation pressure, is a possibility, but the classical Eddington limit due to electron scattering did not show the observed dependence on temperature for the hot stars. However, as temperatures decrease below 30000 K, opacity tends to increase as low-level ions such as H I, Fe II, *et al.*, begin to appear; so we can imagine a **modified** Eddington luminosity limit which decreases with decreasing temperature, like the observed limit. Unfortunately this is not as simple an explanation as it might seem at first sight. Opacities depend on density as well as on temperature, and as the limit is approached a classical static atmosphere would puff up and its opacity would tend to decrease to the electron-scattering value after all. Nevertheless, an instability may occur under these circumstances and several people have written about the hypothetical modified Eddington limit (Humphreys and Davidson 1984; Appenzeller 1986, 1987; Lamers 1986; Davidson 1987; Lamers and Fitzpatrick 1988). Other processes such

K. Davidson et al. (eds.), Physics of Luminous Blue Variables, 3–14.

as interior evolution or atmospheric turbulence may contribute to the instability (de Jager 1980, 1984; Maeder 1983).

It should be emphasized that there was no theoretical prediction of the luminosity boundary or stability limit in the upper H-R diagram. Eta Car, P Cyg, and S Dor have been famous for many decades, but it is only recently that we have learned enough about these unstable, massive stars to begin to discuss "the Physics of LBV's." There has actually been very little theoretical work on the mechanism of LBV eruptions and the origin of the instability (see Stothers and Chin 1983, Appenzeller 1986, and *Instabilities in Luminous Early Type Stars*, edited by Lamers and de Loore, 1987). In this field, theory has followed observation.

2. WHAT IS AN 'LBV'?

The term Luminous Blue Variable or LBV was first proposed by Peter Conti (1984) to describe the S Dor variables, Hubble-Sandage variables, and P Cyg stars in recognition of the many properties they share. Luminous Blue Variables are characterized by:

1) *Variability* — Photometric variations are observed on different timescales and to vastly different degrees:

a) Eruptions or outbursts of ≥ 3 mag refer to sudden ejections of large amounts of mass. The best example is of course η Car's famous outburst from 1837 to 1860 when it became the second brightest star in the sky. The timescales for these giant eruptions are of course very uncertain, but because we see so few, hundreds to thousands of years are reasonable estimates of the frequencies of these events.

Other possible examples of these violent eruptions are P Cyg's behavior in the 1600's (see de Jager 1980 and Lamers 1986 for reviews), when it brightened from invisibility to about third magnitude, and V 12 in NGC 2403 (Tammann and Sandage 1968). After such major outbursts, these stars remain relatively quiescent. Davidson's description of 'Plinian' outbursts in this volume gives more information on these violent outbursts.

b) Moderate variations of 1 to 2 mag are often observed on timescales of 10 to 40 years. From the minimum or quiescent phase, the star may brighten by up to two magnitudes, often within a few months. Both the minima and the maxima may last several years, although smaller variations may be superposed. At visual maximum, the star's atmosphere is greatly expanded to form an optically thick pseudo-photosphere. S Dor and R 71 in the LMC, plus several of the Hubble-Sandage variables (Hubble and Sandage 1953) are good examples of this type of variability which may even be semi-regular. Wolf will discuss moderate LBV outbursts (these proceedings).

c) Smaller oscillations of about half a magnitude are often observed on timescales of months to a few years, on top of the longer-term moderate variations.

d) Microvariations of ≤ 0.1 mag have been described by van Genderen *et al.* (1985) for R71, but these variations have also been reported in normal supergiants.

2) Their *spectra* typically show prominent emission lines of H, He I, Fe II, and [Fe II] often with P Cygni profiles when observed at sufficiently high resolution. The spectra are variable, corresponding to the photometric variations.

At visual minimum or quiescence an LBV has the spectrum of a hot supergiant with H and He I emission; the Fe II and [Fe II] emission is strongest at minimum. At visual maximum, the optically thick pseudo-photosphere resembles a much cooler supergiant of spectral type A or F. At maximum the Fe II emission is weaker. At high resolution, P Cyg profiles are still observed in hydrogen and other lines.

3) The *temperature variations* correspond to the spectral and photometric variations. At visual minimum when the LBV is hottest, its temperature is ≥ 15000--20000 K and at maximum the LBV's all seem to have minimum temperatures near 8000 K. Davidson (1987) showed that the expanded pseudo-photospheres asymptotically approach a minimum value near 7500 K no matter how high the mass-loss rate.

4) As their name implies, the LBV's are stars of *high intrinsic luminosity*. Their bolometric magnitudes remain essentially constant during visual light variations of up to 2 mag (see Appenzeller and Wolf 1982 and Wolf *et al.* 1981 concerning R 71; Stahl and Wolf 1986 for R 127; and Leitherer *et al.* 1985 for S Dor.) The well-studied LBV's have absolute bolometric magnitudes brighter than -9 mag ($L \geq 10^{5.5}\ L_{\odot}$).

5) The active or shell ejection (pseudo-photosphere) phase of an LBV coincides with *high mass-loss rates*, typically 10^{-5} to $10^{-4}\ M_{\odot}$/yr. These rates are 10 to 100 times those of normal supergiants. In quiescence their mass-loss rates are more like those of normal supergiants of the same temperatures and luminosities. The observed mass-loss rates as well as most of our information on temperatures have resulted from extensive moderate- and high-resolution ground-based and UV studies of these stars in our galaxy and the LMC by many different groups.

6) Most LBV's show some evidence for an excess of *infrared radiation* and *circumstellar ejecta*. A small excess in the near-IR (1--3 μm), due to free-free and free-bound emission, is common (Humphreys *et al.* 1984; Leitherer *et al.* 1985). Many of these stars also have longer-wavelength radiation due to circumstellar dust (McGregor *et al.* 1988). In the case of η Car the dust is thick enough to obscure the star (Westphal and Neugebauer 1969). These circumstellar shells and ejecta are produced by the high mass loss and ejection of shells from the LBV's.

Some of this ejecta is clearly visible, as in the 'homunculus' of η Car, although this case is perhaps exceptional. More common is the presence of a ring nebula as in AG Car (Thackeray 1977; Stahl 1987). Papers in this volume by Stahl, Paresce, and McGregor will review observations of circumstellar shells and ejecta. If LBV's shed a few × 10^{-5} $M_{\odot}$/yr, then during the LBV phase, usually estimated as 10^{4}--10^{5} yr, they may lose up to a solar mass or slightly more.

But there is considerable circumstantial or *circumstellar* evidence that the total mass lost in the LBV phase may be much more than inferred from the nominal few × 10^{-5} $M_{\odot}$/yr. Eta Car probably lost 2--3 $M_{\odot}$ during its famous 1840's outburst and its current mass-loss rate is estimated at 10^{-4} to $10^{-3}\ M_{\odot}$/yr.

More typical examples are P Cyg, AG Car, and R 127. In quiescence P Cygni's mass-loss rate is ~ $1.5 \times 10^{-5}\ M_{\odot}$/yr. If it is like other LBV's, then the rate during its outburst was 10--100 times greater or 10^{-4} to $10^{-3} M_{\odot}$/yr. From direct imaging of circumstellar shells around AG Car, R 127, and others, Stahl (1987) estimated their kinematic ages and masses and concluded that their average mass-loss rate must be > $10^{-4} M_{\odot}$/yr. In our poster paper on the LBV's in M31 and M33 we show that a shell of photoionized gas around AE And contains > $6 \times 10^{-3} M_{\odot}$. This is probably the material ejected during its last maximum, which lasted ~ 20 years, implying > $3 \times 10^{-3} M_{\odot}$/yr during the ejection event. Thus, a massive star could lose several solar masses during its LBV stage.

Quantitative analyses of the ejecta, as in η Car, and the atmospheres and circumstellar envelopes of these stars and related objects show that they are nitrogen- and helium-rich (Davidson *et al.* 1982, 1986; Allen *et al.* 1985; and review by Walborn 1988). This is presumably CNO-processed material brought to the surface by mixing and mass loss.

The observed characteristics are consistent with a scenario for the LBV's as *evolved, very luminous, unstable* hot supergiants which suffer *irregular ejections*. The cause of their instability is very likely radiation pressure (see refs. cited above), resulting in a greatly enhanced average mass outflow which leads to the formation of a pseudo-photosphere (Leitherer et al. 1985) at visual maximum. At this stage the slowly expanding (100--200 km/s) envelope is cool (8000---9000 K) and dense (N ~ 10^{11} cm^{-3}), and the star resembles a very luminous A-type supergiant. At minimum light, or the quiescent state, the LBV is at its 'normal' high temperature (> 15000--20000 K) and the mass-loss rate is lower. During these variations the bolometric luminosity remains essentially constant. The visual light variations are caused by the apparent shift in the star's energy distribution driven by the instability.

With these remarkable properties, it is not surprising that the LBV's have come to play a critical role in our current thinking about the evolution of the most massive stars. With their high luminosities, enriched ejecta, and high mass-loss rates, similar to WR stars, they are commonly considered to be evolved objects in transition to the WR stage (Maeder 1983). Thus the LBV's may be a relatively short, highly unstable, but important phase in the lives of the most massive stars.

An alternative possibility, first proposed by Bath (1979) and further developed by Gallagher *et al.* (1981) and Kenyon and Gallagher (1985), is that LBV's are contact binaries (or at least close binaries) of lower mass (~ 20 $M_{\odot}$) and their peculiarities are then caused by mass flow onto an accretion disk. There is no evidence for mass exchange in the LBV's, but R81 in the LMC is an eclipsing binary (Stahl et al. 1987). Gallagher will discuss the possible role of binaries later in this volume.

Much of our information about LBV's has derived from studies of only a few well-known examples, especially P Cyg, S Dor, η Car, and more recently R127 and AG Car. There are actually very few known or confirmed LBV's. The known LBV's in nearby galaxies are listed in Table 1. Although there are many luminous emission-line stars, most are not known to be significantly variable. Because long quiescent periods are possible, some of those stars may also eventually be found to be LBV's. Although I expect more LBV's to be identified, only careful analyses will show whether they are LBV's or other types of luminous emission-line stars.

Table 1. CONFIRMED LBV'S IN NEARBY GALAXIES*

Milky Way:	η Car, AG Car, HR Car, P Cyg.
LMC:	S Dor, R71, R127, HDE 269582.
M31 (1,2,3,4):	AE And, AF And, Var A-1, Var 15.
M33 (1,2,3):	Var A (?), Var B, Var C, Var 2, Var 83.
NGC 2403 (5):	V12, V22, V35, V37, V38.
M81 (6):	I1, I2, I3.
M101 (7):	V1, V2, V10.

*The underlined stars are included in Table 2 as relatively well-studied LBV's.
Refs.: (1) Hubble and Sandage 1953; (2,3) Humphreys 1975, 1978; (4) Rosino and Bianchini 1973; (5) Tammann and Sandage 1968; (6,7) Sandage 1983, 1984.

Table 2. OBSERVED PROPERTIES OF WELL-STUDIED LBV'S

Galaxy	*Star*	*Observed temperature* (K) 'at min'	'at max'	M_{bol}	$\dot{M}$ ($M_\odot$/y)
Milky Way	η Car	27000:	---	-11.3	10^{-3} to 10^{-1}:
	P Cyg	19000	---	-9.9	2×10^{-5}
	AG Car	25000:	9000	-10.1	3×10^{-5}
LMC	S Dor	20000--25000:	8000	-9.8	5×10^{-5}
	R 71	13600	9000:	-8.8	5×10^{-5}
	R 127	30000	8500	-10.5	6×10^{-5}
M 33	Var C	20000--25000	7500--8000	-9.8	4×10^{-5}
	Var A	3500:	8000:	-9.5	2×10^{-4}

Except for the LMC, the galaxies with confirmed LBV's are all spirals. There are no known LBV's in the SMC and as far as I know none have been confirmed yet in IC 1613 or NGC 6822, two less luminous irregular galaxies. This may be primarily a statistical effect, because these smaller galaxies have fewer of the most massive stars; they also contain few WR stars. The lack of LBV's is consistent with the ideas that LBV's are in a relatively short stage in the evolution of massive stars ($\geq 40\ M_\odot$) and that they are possible progenitors of WR stars.

3. THE ROLE OF LBV'S IN THE UPPER H-R DIAGRAM AND THEIR RELATIONSHIP TO OTHER EVOLVED MASSIVE STARS

As I mentioned above, the upper H-R diagram is characterized by an obvious upper envelope to stellar luminosities. The temperature dependence of the observed boundary for the hot stars and the relatively well-defined upper luminosity limit for the late-type hypergiants suggest that this is a critical location in the diagram (Humphreys and Davidson 1979, 1984). Garmany and Fitzpatrick will discuss the observed luminosity/stability limit in these proceedings.

Figure 1 shows a schematic H-R diagram with the well-studied LBV's from Table 2. The LBV's maintain essentially constant luminosity between their quiescent (minimum visual light) and shell-ejection (visual maximum) phases. The dashed lines in Fig. 1 show their variations in photospheric temperature between these phases. The solid lines are the upper luminosity boundary for the hot stars and for the cool hypergiants, although the limit for the hot stars is not yet precisely defined.

Very massive stars ($\geq 40\ M_\odot$ in their post-main sequence evolution eventually encounter the stability limit as they attempt to evolve to cooler temperatures, leading to enhanced mass loss, the pseudo-photosphere stages. Apparently the usual result is unsteady behavior (outbursts) rather than an enhanced long-term steady wind. After an LBV has lost enough mass in an eruption, the instability is temporarily relieved and the star returns to its quiescent state, presumably until the instability develops again. Appenzeller (1986) and Lamers (1986) suggested that the LBV instability is caused by radiation pressure in numerous ultraviolet Fe lines. Then the modified Eddington luminosity should depend on metallicity. The observed luminosity boundary and the

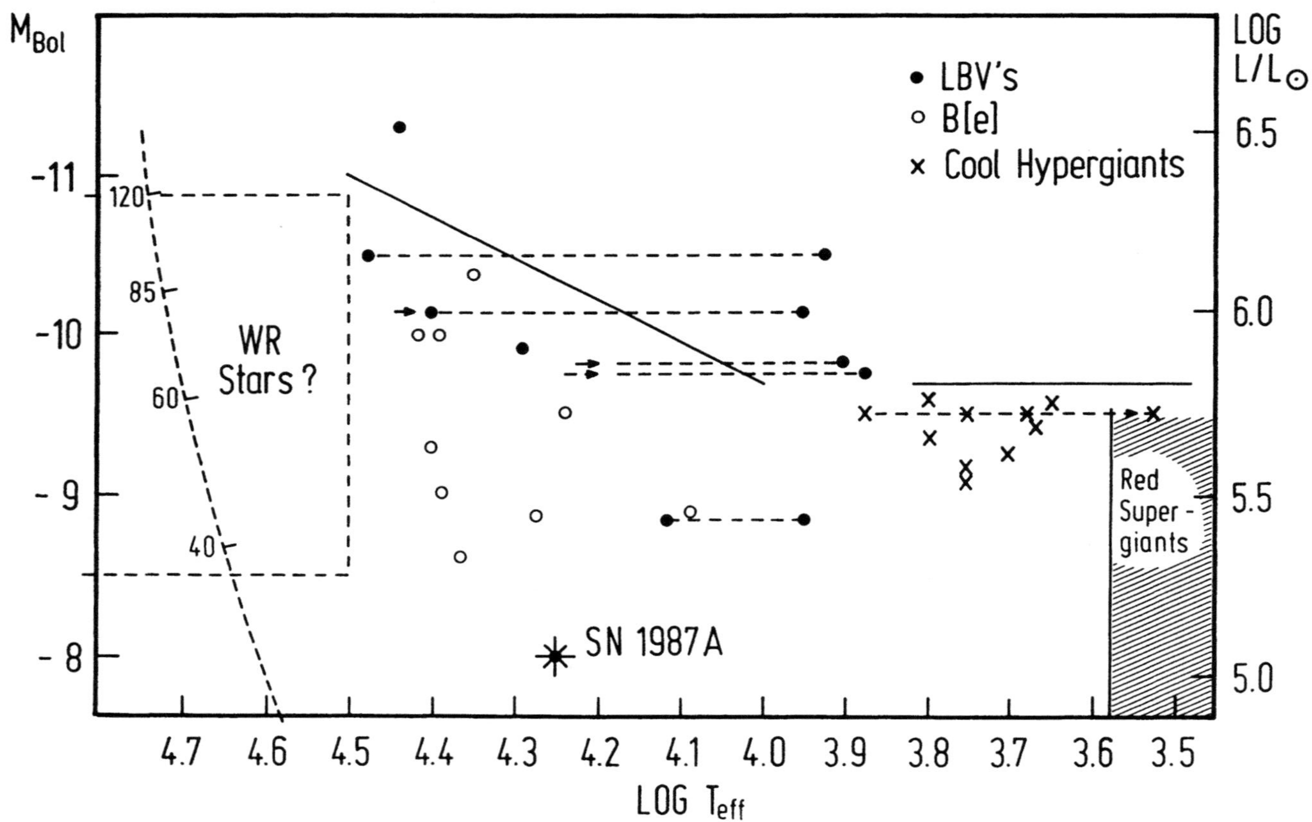

Figure 1. The upper HR diagram showing the location of the LBV's in Table 2, the B[e] supergiants and the cool hypergiants. The solid lines mark the temperature-dependent luminosity boundary for the hot stars and luminosity limit for the cool supergiants.

luminosities of LBV's should be higher in low-metallicity systems. This is not discernible in the available data for massive stars in different galaxies, but other effects such as population statistics may dominate.

Several of the stars in Figure 1 deserve special comment:

R 127 is especially important because it was one of the few Of/WN stars originally recognized by Walborn (1977) before its recent eruption (Stahl *et al.* 1983). It is now at maximum light and its spectrum is nearly identical with that of S Dor at maximum.

Var C in M33, after nearly 40 years of relative quiescence, is now in its false-photosphere stage (Humphreys *et al.* 1988). In 1985 it was at visual maximum with a spectrum like a very luminous F-type supergiant, and it is still bright.

Var A, another of the original Hubble-Sandage variables in M33, is perhaps the most peculiar. Its spectrum at maximum in the early 1950's was that of a very luminous F-type supergiant. It then declined rapidly by nearly 4 mag, becoming red. Recent observations reveal a very large infrared excess (Humphreys, Jones, and Gehrz 1987) and its total luminosity today is equal to that at maximum light. What is peculiar about Var A is its current spectrum; rather than being a hot star within a dust shell, it has the TiO bands of an M supergiant. Our best explanation for its remarkable behavior is that Var A is not a true LBV but rather is an evolved very luminous hypergiant, very close to its stability limit, which ejected an optically thick shell (*cf.* papers by Smolinski and by de Jager, this volume, on cool supergiants). Whether it will actually evolve to the red supergiant stage is uncertain. Its location on the HR diagram at maximum and minimum is shown in Figure 1 together with the location of cool hypergiants. I think Var A should be considered with the cool hypergiants, many of which are unstable, such as ρ Cas and HR 8752.

R 71, with absolute bolometric magnitude ≈ -9.0, is below the observed luminosity limit and is the faintest of the well-studied LBV's. Since it is in the LMC, there is little uncertainty in its distance. It has a lower temperature at minimum light than other LBV's (Wolf *et al.* 1981). It certainly deserves further attention.

AG Car has usually been assigned a distance of ~ 2500 pc corresponding to the large young associations in the Carina region. At that distance its luminosity would be only $M_{bol} \approx -8.7$ mag, far below the stability limit. For this reason I recently looked for other clues to its distance. Its systemic velocity of +18 km/s, measured by Wolf and Stahl (1982) from the center of the P Cygni profiles of the strongest Fe II lines, or +21 km/s, measured from Fabry-Perot interferometry of the nebula (Johnson 1976), implies a distance far beyond 2500 pc. If this velocity is due to differential galactic rotation, then its kinematic distance is 7 kpc and M_{bol} is around −11.2 mag, a luminosity more appropriate to its spectral characteristics. In their poster paper at this meeting, Lamers and his collaborators present additional evidence for a larger distance based on its extinction. Stahl (1986) has noted that the spectrum of AG Car at minimum is very similar to the Of/WN9 stars.

The close relationship between the Of/WN9 stars and LBV's such as R 127, AG Car, and HDE 269582 = MWC 112 provides important support for suggestions that LBV's are the predecessors of WR stars. There are now seven known Ofpe/WN9 stars in the LMC (Walborn 1972, 1982, 1985). R 127 is one of the Of/WN9 stars and of course is a known LBV. Two others, R 84 and R 99, are also variable with smaller amplitudes (Stahl *et al.* 1984) and are probably related to the LBV's. Obviously this is a group of stars that should be closely monitored and more effort should be made to find their counterparts in our Galaxy (*e.g.*, He 3 - 519 -- Stahl 1986).

Many luminous emission-line stars share spectroscopic characteristics, such as H and He I emission with P Cygni profiles and Fe II and [Fe II] emission, with LBV's. Among them are the B[e] ("B-bracket" or "B-bracket-e") supergiants. Zickgraf *et al.* (1985, 1986) have identified several important distinctions between LBV's and B[e] supergiants. The B[e] stars are not variable; their emission-line spectra are hybrid with narrow Fe and [Fe II lines but broad absorption components in H, He I, and UV resonance lines with terminal velocities typical of B supergiant winds. These hybird characteristics are explained by a two-component model with a normal B supergiant wind from the poles and a denser, slower-moving equatorial disk. Zickgraf gives more information about the B[e]'s in this volume. Their location in the H-R diagram is shown in Figure 1. They overlap the LBV region but tend to be somewhat less luminous on the average. Is there a possible evolutionary connection between the B[e] supergiants and LBV's? Can the latter recover from their instability, shed their angular momentum, and evolve to the red supergiant stage, or are they binaries?

The H-R diagram in Figure 1 illustrates the variety of objects and the diversity of behavior near the luminosity/stability limit. We have made great progess in our understanding of the structure and evolution of the most massive stars, but as we learn more about the different stars that populate the upper H-R diagram, the situation appears not to be as straightforward as we might have assumed a few years ago.

4. QUESTIONS AND FUTURE PROBLEMS

New observations of LBV's and other objects in the upper H-R diagram continually reveal new and important clues to their evolution and therefore to physical processes inside them. Spectroscopic and photometric monitoring of these unstable stars must continue. An important observational problem is how to identify more LBV's for better statistics on their relationships to other objects.

The scenario in which LBV's have evolved primarily from very massive stars and are progenitors of WR stars is increasingly clear. A possible sequence is

$$\mathrm{O} \rightarrow \mathrm{Of} \rightarrow \mathrm{LBV} \longleftrightarrow \mathrm{Of/WN} \rightarrow \mathrm{WR}.$$

What about their relationship to other luminous massive stars? In the upper H-R diagram we also observe the B[e] supergiants, unstable cool hypergiants, and luminous red supergiants with high mass-loss rates, some with thick circumstellar shells. What are the relationships among the different kinds of massive stars near the luminosity/ stability limit? Is the LBV phenomenon indirectly responsible for the red-supergiant upper limit which is so useful as an extragalactic distance indicator? And precisely what are the roles of initial mass, age, evolutionary status, rotation, internal angular momentum distribution, and binarity?

The fundamental question concerns the cause or causes of the LBV instability. The star's evolution has brought it close to the limit of its stability. We talk about radiation pressure and turbulence, but what is the actual mechanism or trigger for the moderate ejections like that of S Dor and the violent outbursts like that of η Car? Is the physics the same for both types of event? Why and how does an LBV recover from its pseudophotosphere eruptive state? Shedding a lot of mass apparently relieves the instability, but why? Perhaps the most pressing questions about LBV's are more theoretical than observational, though there are enough observational questions! In this colloquium on the 'Physics of Luminous Blue Variables' we may make some progress toward understanding their structure and behavior.

References

Allen, D.A., Jones, T.J., & Hyland, A.R. 1985, *Astrophys.J.* **291**, 280.

Appenzeller, I. 1986, *IAU Symposium 116, Luminous Stars and Associations in Galaxies,* p. 139.

Appenzeller, I., and Wolf, B. 1982, *ESO Workshop on The Most Massive Stars,* p. 131.

Bath, G.T. 1979, *Nature* **282**, 173.

Conti, P.S. 1984, *IAU Symposium 105, Observational Tests of Stellar Evolution Theory,* p. 233.

Davidson, K. 1987, *Astrophys.J.* **317**, 760.

Davidson, K., Dufour, R.J., Walborn, N.R., & Gull, T.R. 1986, *Astrophys.J.* **305**, 867.

Davidson, K., Walborn, N.R., & Gull, T.R. 1982, *Astrophys.J.(Letters)* **254**, L47.

de Jager, C. 1980, *The Brightest Stars* (Reidel: Dordrecht)

de Jager, C. 1984, *Astron.Astrophys.* **138**, 246.

Gallagher, J.S., Kenyon, S.J., & Hege, E.K. 1981, *Astrophys.J.* **249**, 83.

Hubble, E.P., & Sandage, A. 1953, *Astrophys.J.* **118**, 353.

Humphreys, R.M. 1975, *Astrophys.J.* **200**, 426.

Humphreys, R.M. 1978, *Astrophys.J.* **219**, 445.

Humphreys, R.M., Blaha, C., D'Odorico, S., Gull, T.R., & Benvenuti, P. 1984, *Astrophys.J.* **278**, 124.

Humphreys, R.M., & Davidson, K. 1979, *Astrophys.J.* **232**, 409.

Humphreys, R.M., & Davidson, K. 1984, *Science* **223**, 243.

Humphreys, R.M., Jones, T.J., & Gehrz, R.D. 1987, *Astron.J.* **94**, 315.

Humphreys, R.M., Leitherer, C., Stahl, O., Wolf, B., & Zickgraf, F.J. 1988, *Astron.Astrophys.* **203**, 306.

Hutchings, J.B. 1976, *Astrophys.J.* **203**, 438.

Johnson, H.M. 1976, *Astrophys.J.* **206**, 469.

Kenyon, S.J., & Gallagher, J.S. 1985, *Astrophys.J.* **290**, 542.

Lamers, H.J.G.L.M. 1986, *IAU Symposium 116, Luminous Stars and Associations in Galaxies,* p. 157.

Lamers, H.J.G.L.M., & Fitzpatrick, E. 1988, *Astrophys.J.* **324**, 279.

Leitherer, C., Appenzeller, I., Klare, G., Lamers, H.J.G.L.M., Stahl, O., Waters, L.B.F.M., & Wolf, B. 1985 *Astron.Astrophys.* **153**, 168.

Maeder, A. 1983, *Astron.Astrophys.* **120**, 113.

McGregor, P.J., Hyland, A.R., & Hillier, D.J. 1988, *Astrophys.J.* **324**, 1071.

Rosino, L., & Bianchini, A. 1973, *Astron.Astrophys.* **22**, 953.

Sandage, A. 1983, *Astron.J.* **88**, 1569.

Sandage, A. 1984, *Astron.J.* **89**, 621.

Stahl, O. 1986, *Astron.Astrophys.* **164**, 321.

Stahl, O. 1987, *Astron.Astrophys.* **182**, 229.

Stahl, O., & Wolf, B. 1986, *Astron.Astrophys.* **154**, 243.

Stahl, O., Wolf, B., Leitherer, C., Zickgraf, F.-J., Krautter, J., & de Groot, M. 1984, *Astron.Astrophys.* **140**, 459.

Stahl, O., Wolf, B., & Zickgraf, F.-J. 1987, *Astron.Astrophys.* **184**, 193.

Stothers, R., & Chin, C. 1983, *Astrophys.J.* **264**, 583.

Tammann, G.A., & Sandage, A. 1968, *Astrophys.J.* **151**, 831.
Thackeray, A.D. 1977, *Mon.Not.Roy.Astron.Soc.* **180**, 95.
van Genderen, A.M., Steemers, W.J.G., Feldbrugge, P.T.M., de Groot, M., Damin, E., & van den Boogaart, A.K. 1985, *Astron.Astrophys.* **153**, 163.
Walborn, N.R. 1977, *Astrophys.J.* **215**, 53.
Walborn, N.R. 1982, *Astrophys.J.* **256**, 452.
Walborn, N.R. 1985, *IAU Symposium 116, Luminous Stars and Associations in Galaxies*, p. 185.
Walborn, N.R. 1988, *IAU Colloquium 108, Atmospheric Diagnostics of Stellar Evolution*, p. 70.
Westphal, J.A., & Neugebauer, G. 1969, *Astrophys.J.(Letters)* **156**, L45.
Wolf, B., Appenzeller, I., & Stahl, O. 1981, *Astron.Astrophys.* **103**, 94.
Wolf, B., & Stahl, O. 1982, *Astron.Astrophys.* **112**, 111.
Zickgraf, F.-J., Wolf, B., Stahl, O., Leitherer, C., & Appenzeller, I. 1986, *Astron. Astrophys.* **163**, 119.
Zickgraf, F.-J., Wolf, B., Stahl, O., Leitherer, C., & Klare, G. 1985, *Astron. Astrophys.* **143**, 421.

Roberta Humphreys

DISCUSSION

De Jager: Sometimes it has been proposed that LBV's be called hypergiants, but this would not be correct in terms of spectroscopic classification. Hypergiants are stars that show the supergiant spectral characteristics still more pronouncedly than normal supergiants, but have only moderate emission lines or P Cygni profiles, not at all like LBV's. We should note that some hypergiants may be closer to the upper limit than LBV's, but they do not show the large episodical eruptions.

Wolf: I like your new distance to AG Carinae. As I will point out in my talk, there seems to be a correlation between the amplitude of S Dor-type eruptions and luminosity. Accordingly I would expect a much higher luminosity for AG Car than previously thought.

Walborn: The possibility of an amplitude-luminosity relation (later types at visual minimum for less luminous LBV's) is interesting. However, the true minimum states of P Cyg and S Dor may not yet have been observed spectroscopically, so it is possible that at minimum they could also turn out to be Ofpe/WN9 objects.

Humphreys: On this H-R diagram P Cyg is shown in its present quiescence and an arrow indicates that S Dor has probably come from higher temperatures.

Gallagher: Can the radial velocity of AG Car be influenced by the presence of circumstellar dust? For example, optical depth effects could produce an apparent blue shift.

Humphreys: Its radial velocity has been determined in two ways, one of which uses nebular lines. If significant circumstellar optical depth effects occur, the two determinations would presumably differ.

McGregor: The optical depth of the dust in the AG Car shell is low; the extinction is small.

Sreenivasan: You talked about the expansion of the pseudo-photosphere. Do you believe that the observations suggest that the expanding material actually falls back on the star, or do they suggest that the expanding material leaves the star and thereby reveals a hotter object?

Humphreys: The pseudo-photosphere is the apparent surface, representing optical depths of order unity, and material can move through it. After an outburst the pseudo-photosphere presumably contracts back to its former state which may be a photosphere in the normal, static sense.

Appenzeller: At least in one case (S Dor, see Leitherer *et al.* 1985, *Astron. Astrophys.* **153**, 168), the radial velocities of absorption lines formed deep in the pseudo-photosphere indicate that the pseudo-photosphere actually contracts when the visual brightness decreases at the end of an outburst.

Friedjung: I think that we should clarify what is meant by a pseudo-photosphere. In the case of a nova the optically thick region produced by the initial explosion does not last very long. Then one sees an optically thick wind. Does one see optically thick winds in the case of LBV's?

Humphreys: Yes, when they are not in their quiescent states.

Shore: About 100 years after Herschel's remarks about η Car there was a possibly analogous explosion near Alamagordo. In a nuclear explosion, the photosphere initially expands and then stalls, the outer shock breaks away and the pseudo-photosphere recontracts. The effective temperature of this 'photosphere' remains roughly constant, the shock advances and is optically thin, while the photosphere cools and contracts.

Baratta: We show in our work that not only Var C in M33, but also η Car in the actual phase, displays the spectral type of an F supergiant.

Lortet: Where is R81 in the H-R diagram? -- This one may be a non-variable star.

Humphreys: I did not include R81, an eclipsing binary, in this diagram, because its LBV characteristics are somewhat uncertain.

Moffat: Polarization studies are often neglected by astronomers. Particularly, the ejection phenomenon and its geometry (with asymmetries!) can be effectively, even uniquely(?) probed.

Humphreys: That is a good idea. Polarimetry has scarcely been done on the LBV's, except for Hayes' work on P Cyg (1985, *Astrophys. J.* **289**, 726).

ON THE EVOLUTIONARY STATUS AND INSTABILITY MECHANISM OF THE LUMINOUS BLUE VARIABLES (LBV)

André Maeder
Geneva Observatory
CH-1290 Sauverny, Switzerland

ABSTRACT. Various evolutionary sequences leading to LBV are examined. The sequence O-Of-LBV-WR-SN is well supported by the models; some LBV with relatively lower luminosity may turn into OH/IR sources. The overall duration of the LBV phase depends mainly on the average mass loss rate; for $< \dot{M} > = 10^{-3} M_{\odot} y^{-1}$, it lasts about 10^4 y.

Very massive stars undergo, when they reach log $T_{eff} = 3.9$, strong departure from hydrostatic equilibrium due to supra-Eddington luminosities at some depth in the outer layers. This results in heavy mass loss, as the growth rate of the instability is very fast. We suggest that the amount of mass ejected in a shell episode is mainly determined by the mass of such a layer that its thermal adjustment timescale is within an order of magnitude of the stellar dynamical timescale. Simulations of B-light curves due to shell ejections by LBV are performed and some sensitive properties are identified.

1 INTRODUCTION

LBVs are extreme supergiants which have experienced one or several violent and large shell ejections; in the Galaxy, such typical objects are η Car, P Cyg, AG Car and HR Car. We shall examine here what stellar models may teach us about the evolutionary sequences leading to and from the LBV stage. The structural properties will also be examined in order to find indications about the driving instability mechanism leading to the observed outbursts and about the characteristic timescales of these objects.

2 EVOLUTION TOWARDS AND FROM THE LBV STAGE

Several observational properties of the LBV are to be matched by the evolutionary models:

1. Their location in the HR diagram (cf. Humphreys, 1987).
2. Their He, C/N and O/N ratios (cf. Davidson et al., 1984).
3. Their high average mass loss rates $\dot{M}$ and their episodic outbursts (cf. Lamers, 1987).
4. The existence of transition objects, like R 127 and AG Car (cf. Stahl et al. 1983; Stahl 1986).
5. Their light curve and in particular the constancy of L during the variations (cf. Appenzeller and Wolf 1981).
6. Possibly, their association to young stellar groups (cf. Lortet, e.g. this meeting).

K. Davidson et al. (eds.), Physics of Luminous Blue Variables, 15–26.

This list is certainly non-exhaustive. These observational constraints have generally been considered to support the view that LBV are massive post- main sequence supergiants undergoing large shell ejections (cf. Humphreys, Davidson 1979; de Jager 1984; Stothers and Chin 1983; Maeder 1983; Appenzeller 1986, 1987; Lamers 1986, 1987; Humphreys 1987). The observed surface abundances in LBV are particularly decisive arguments in this context.

We have now to figure out more closely what might be the evolutionary sequence leading towards and then from the LBV, the status of which is still somehow controversial. For purpose of clarity, let us classify the evolutionary sequences of massive stars by the stage as a result of which a bare core (i.e. a WR star) is formed (cf. Chiosi and Maeder, 1986). Starting by the most massive stars, the first evolutionary sequence to be identified is the following one:

2.1 Bare core formation during the H-burning phase

It is possible that, as a result of heavy mass loss during the O and Of stages, the very massive stars loose most of their original envelope during the H-burning phase. As a consequence, the evolutionary sequence would be

$$\text{O} - \text{Of} - \text{WR} - \text{SN} \qquad M > 120\, M_\odot$$

This is indeed just the Conti scenario (1975) for the formation of WR stars; illustrations of it can be found in Maeder (1980). In view of the currently observed mass loss rates, this scenario appears feasible for initial masses above about 120 $M_\odot$. Thus, it is possible that the most massive stars due to heavy mass loss on the MS avoid the LBV phase, a suggestion also consistent with Lortet (cf. this meeting). Any mixing mechanisms, such as overshooting, rotational or tidal diffusion, would also favour this scenario.

2.2 Bare core after the LBV stage

This sequence illustrated by several models with M $\geq$ 60 $M_\odot$ (cf. Maeder 1983; Maeder and Meynet 1987) is likely the typical one leading to the formation of LBV

$$\text{O} - \text{Of} - (\text{BSG}) - \text{LBV} - \text{WR} - \text{SN} \qquad \sim 50 - 120\, M_\odot$$

Here, after central H-exhaustion, the star undergoes a fast increase of its radius and a rapid (in $10^3 - 10^4\, y$) redwards evolution in the HR diagram (cf. Fig. 1). Then it reaches an instability limit, the nature of which is discussed below (cf. § 3) and is likely to be related to the supra- Eddington luminosity reached in some of its outer layers. As a consequence, mass loss occurs at instantaneous rates $\dot{M} \geq 10^{-2}\, M_\odot y^{-1}$. Some first results on the physical mechanism driving this ejection, the $\dot{M}$-rates, the involved timescales, the stellar models and simulations of light curves are examined in § 3 below. After an ejection, the models show a rapid blue shift, in a few years or tens of years depending on the mass loss rates in the shell ejection (cf. Fig. 2). Then the internal evolution again brings the star to the red with a recovery time, which for a given stellar mass is function of the amount of mass ejected in the shell. Keeping nearly at constant L, the star again reaches the instability limit corresponding to its new mass and composition and is ready for a new violent episode.

We emphasize that *the properties and duration of the LBV phase are essentially determined by the mass loss rates before and during the LBV phase.* For a typical $< \dot{M} > = 10^{-3}\, M_\odot y^{-1}$ over all the LBV phase (cf. Lamers, 1986), the LBV phase may be estimated

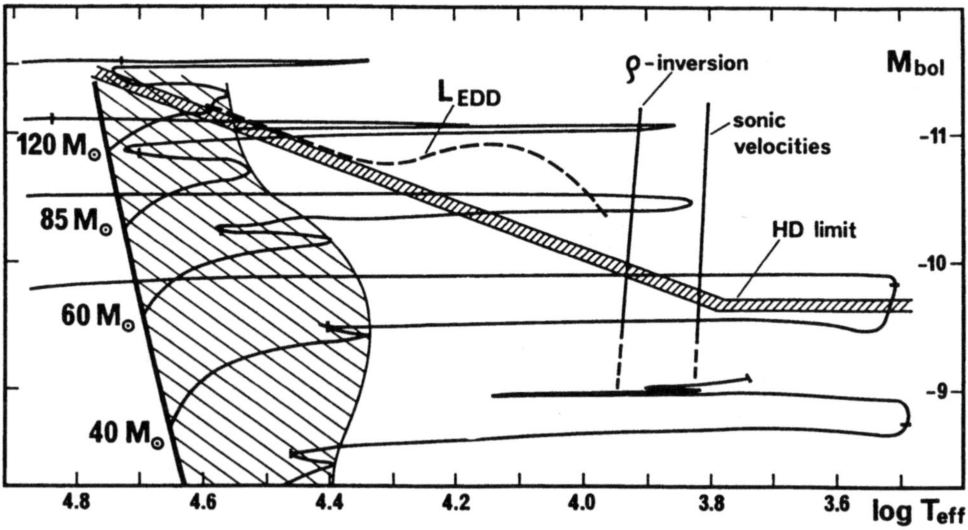

Figure 1: Evolutionary tracks in the HR diagram from Maeder and Meynet (1987), with indications of the Humphreys-Davidson limit (1979), of the limit where $L = L_{EDD}$ (cf. Lamers and Fitzpatrick, 1988), and of the locations where ρ-inversions and sonic velocities are met in the models.

to have a duration $t_{LBV} \leq 10^4 y$. The argument is the following one. The amount of mass located between the surface left at the end of the MS phase and the H-burning shell is less than or of the order of 10 $M_\odot$ (cf. Fig. 3 by Maeder and Meynet, 1987). This amount gives an upper bound to the mass to be removed in the LBV phase. Depending on $< \dot{M} >$ in LBV, different durations t_{LBV} may therefore be obtained. Of course, the star may run into several LBV episodes. When the surface H-content is lower than about $X = 0.3$, which corresponds to different core mass fractions according to the initial mass, the star makes its last blueward motion likely towards the WNL stage. We notice that for a given initial mass, the lifetime ratio t_{LBV}/t_{WR} is of the order of 0.02, within a factor of two at least.

Observational evidences for the sequence Of - LBV - WR are provided by the objects R 127 (Stahl et al. 1983) and AG Car (cf. Stahl 1986).

2.3 Bare core after the stage of OH/IR source

The following sequence through the stage of OH/IR source should in principle be possible :

$$\text{O - Of - (BSG) - LBV - OH/IR - WR - SN} \qquad \sim 40 \text{ or } 50\ M_\odot$$

OH/IR sources are near the upper L limit of red supergiants, with IR excess, OH maser and optically thick dust shell (cf. Humphreys, 1987).

Indeed, if $\dot{M} < 10^{-3}\, M_\odot y^{-1}$ (depending on mass), the redwards motion through the HR diagram is not stopped and the star reaches the red stage. We may speak of a "transparency of the instability limit".

The opposite sequence of OH/IR source evolving to LBV is not supported by our models. The stellar mass is smaller, but the opacity is also smaller, due to the much lower H-content, so that the Eddington luminosity may not be smaller. The observations of Var. A in M33 and IRC+104020 by Humphreys (1987) support the above scenario.

Mass domains for lower initial masses may of course be distinguished, but their relevance to the LBV is uncertain, although the case of SN1987A has shown that further progress is also needed in this range.

3 PHYSICAL CONSIDERATIONS ON LBV OUTBURSTS

3.1 Generalities

We shall not discuss here all the possible and impossible instabilities considered for explaining the LBV variability, since most have been ruled out (cf. Stothers and Chin 1983). Let us just mention that the recent claim that vibrational instability by the ϵ-mechanism (cf. Doom et al. 1986) is responsible for the LBV mechanism cannot be supported. As stars become supergiants the ratio of central to average densities $\rho_c/\overline{\rho}$ grows and the stability with respect to vibrational instability increases very much (cf. Maeder 1985).

Appenzeller (1986), Lamers (1986), Lamers and Fitzpatrick (1988) have convincingly proposed that radiation pressure effects on the atmospheres cause the LBV instability. According to them the contribution of metal line opacities in the Balmer continuum considerably reduces the Eddington luminosity L_{EDD}. The limit they obtain for L_{EDD} is drawn in Fig. 1 and it more or less coincides with the blue part of the Humphreys- Davidson limit.

3.2 The occurrence of a catastrophe during redwards evolution

What can be learned about the LBV instability from evolutionary models? Let us examine here some first preliminary results.

The most conspicuous effect in models of very massive stars evolving, after MS evolution, towards the red side of the HR diagram is the occurrence of a *density inversion* (cf. also Bisnovatyi-Kogan and Nadyozhin 1972). In our models (cf. Maeder 1987), this density inversion typically appears for stars with log $T_{eff} \simeq 3.9$ (cf. Fig. 1) and it grows very fastly at lower T_{eff}. The ρ-inversion occurs at a distance corresponding to 10^{-4} to 10^{-6} in mass fraction from the surface; it lies in the layers where is located the opacity peak due to partial H-ionisation and where L happens to be larger than the local Eddington luminosity (account being properly given to the various opacity sources of the models). Physically, the origin of the ρ-inversion is the following one:

As the star expands

$$(\rho, T) \downarrow \Rightarrow \kappa \nearrow \Rightarrow \frac{L}{L_{EDD}} > 1 \Rightarrow \rho - \text{inversion}$$

since $F_{conv} \simeq 0$.

Let us clearly emphasize that the occurrence of ρ-inversions in the models does not necessarily mean that ρ-inversions effectively occur in real stars. Rather, the occurrence of a strong ρ-inversion, which is a consequence of the local supra-Eddington luminosity, must be considered as the sign that *hydrostatic models are no longer valid in the upper right corner of the HR diagram* and that massive outflows occur. An opposite point of view was supported by Stothers and Chin (1973). However, in our opinion the absence of hydrostatic solutions in the upper right corner of the HR diagram is well indicated by the fact that the matching of the Henyey interior solutions with the outer solutions can progressively no longer be achieved as the star moves to the right. We may also point out that ρ-inversions occur in the subsonic regime of convective velocities, i.e. for $V_{conv}/V_{sound} \lesssim 0.1$; thus, the instability revealed by the ρ-inversion is not the same as that due to turbulent pressure and studied by de Jager (1984). Indeed, on a redwards track the ρ-inversions are met before sonic velocities are reached (cf. Fig. 1). The breakdown of hydrostatic solutions is due to supra-Eddington luminosities in the region of the opacity peak, and is thus the continuation for stars of lower T_{eff} of the instability proposed by Appenzeller and Lamers (above reference) from atmospheric studies of hotter stars.

The departure of hydrostatic equilibrium revealed by the ρ-inversions which can reach a factor of 10 near log $T_{eff} \simeq 3.9$, is quite large and therefore heavy mass loss is to be expected. Rates of mass loss as high as $0.5\, M_{\odot} y^{-1}$ have been obtained from non-static solutions (cf. Bisnovatyi-Kogan and Nadyozhin 1972). We can roughly estimate the above figure by simple analytical arguments. Let us call $\tau_{\dot{M}} = M/\dot{M}$ the characteristic time of mass loss and τ_{ρ} the characteristic time for the growth of the ρ-inversion in the models. If $\tau_{\rho} < \tau_{\dot{M}}$, the non-contraried ρ-inversion would grow without any limit. At the opposite, $\tau_{\dot{M}}$ cannot be much shorter than τ_{ρ}, since the mass loss is driven by the instability announced by the ρ-inversion. Thus $\tau_{\dot{M}}$ and τ_{ρ} must be of the same order of magnitude, which means

$$\dot{M} \simeq \frac{M}{\tau_{\rho}}$$

For an initial 60 $M_{\odot}$ model in the LBV stage, we have typically $M \simeq 40\ M_{\odot}$ and $\tau_{\rho} \simeq 10^2\, y$, thus we get $\dot{M}$-rates of the order of magnitude of several $0.1\, M_{\odot} y^{-1}$. The above value is only a rough estimate of the order of magnitude, but it tells us that *the strong departure from hydrostatic equilibrium in the upper right corner of the HR diagram must result in phases of intense mass loss* since the growth rate of the instability is extremely short.

3.3 Why shell ejections?

Central questions about LBV are why matter is ejected in violent outbursts rather than in the form of a steady outflow, and what determines the amount of mass in the ejected shell. This question is a critical one, which could by the way also be asked for other kinds of variable stars. Indeed, if only the matter in and above the ρ-inversion would be expelled in one episode, the amount of matter in the ejected shell would be at most $\sim 10^{-4}$ of the stellar mass; this is far less than observed.

Some understanding may come from considerations about the timescales involved in the problem. The timescale for thermal adjustment is very short in the outer layers of an LBV (i.e. about 70 days in an external layer of 0.1 $M_{\odot}$ with the numerical data given below). Thus, during the dynamical process of ejection itself which lasts t_{ej} (which typically is some multiple χ of the dynamical timescale t_{ff} since matter is progressively accelerated outwards by radiation effects) the opacity peak will be able to go down within the star by as much as the local thermal timescale t_{KH} allows it. For a layer with mass Δm, this local timescale is of the order of

$$t_{KH} \simeq \frac{G M \Delta m}{R L}$$

The κ-peak may move down during the ejection by a distance encompassing the mass Δm fixed by the condition

$$\frac{G\,M\Delta\,m}{R\,L} \simeq t_{ej} \simeq \chi\,t_{ff} \simeq \chi\left(\frac{R^3}{G\,M}\right)^{1/2}$$

which gives

$$\frac{\Delta\,m}{M} \simeq \frac{\chi\,R^{5/2}L}{G^{3/2}\,M^{5/2}} \tag{1}$$

$$\simeq 1.61\cdot 10^{-12}\,\frac{\chi\,(R/R_\odot)^{5/2}\,(L/L_\odot)}{(M/M_\odot)^{5/2}} \tag{2}$$

This gives an estimate of the order of magnitude of the maximum possible mass ejected by the κ-peak during some dynamical time scales.

As a numerical example let us consider a model of initially 60 $M_\odot$ having at an age of $3.71\cdot 10^6 y$ an actual mass of 41.2 $M_\odot$, log $L/L_\odot = 6.05$, log $T_{eff} = 3.89$ and $R = 586\,R_\odot$. This model has just a growing ρ-inversion with $\tau_\rho \simeq 10^2 y$. For the above values we get

$$\Delta\,m \simeq 0.06\,\chi\,M_\odot$$

If we take $\chi = 1$ to 10, we get $\Delta\,m = 0.06$ to 0.6 $M_\odot$ This amount is ejected during the time t_{ej} and this implies that the mass loss rate during the ejection amounts to about the critical value of

$$\dot{M} \simeq \frac{\Delta\,m}{t_{ej}} = \frac{LR}{GM}$$

i.e. 0.75 $M_\odot\,y^{-1}$ for the above model; this rate is of the same order as those previously obtained. The above discussion shows that the breakdown of the hydrostatic equilibrium in LBV may give rise to ejections of several tenths of $M_\odot$ at rates of about 0.1 to 1 $M_\odot\,y^{-1}$. *The physical reason why a large amount of matter is ejected in an outburst despite the fact that the κ-peak is very close to the surface resides in the fact that the blasting front (i.e. the κ-peak) has time to deeply move inwards during the explosion event itself (some dynamical timescales).* As a consequence of the heavy mass loss the star undergoes rapid changes of physical conditions in the outer layers and it rapidly moves bluewards in the HR diagram on timescales which depend on the parameters of the ejection as illustrated by the light curves given in § 3.4 below. The opacity decreases in the critical layers which are then no longer characterized by large supra-Eddington luminosities, thus the catastrophic mass loss rates must soon cease. The hydrostatic solutions are again an acceptable approximation until secular evolution once more brings the star to the instability limit after a certain recovery time, as shown below.

3.4 Response of an LBV to a large shell ejection. Theoretical light curves

Let us now examine some consequences of massive shell ejections with parameters in the ranges suggested above. Many simulations have been performed for various choices of $\Delta\,m$, the mass of the ejected shell and for various $\dot{M}$-rates during the outburst. Only the consequences of the thermal response of the LBV to a mass ejection are considered here, and not

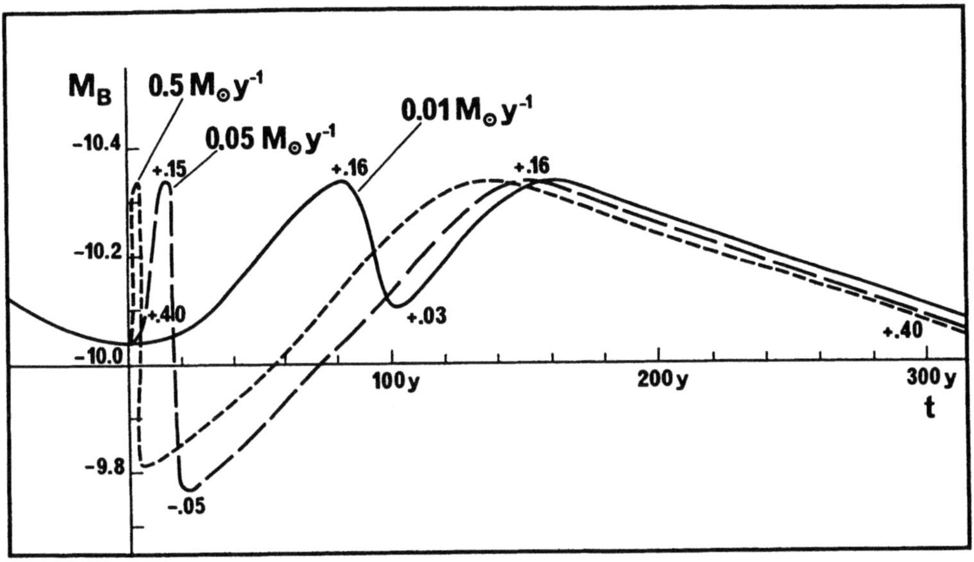

Figure 2: Example of the light curves in B-magnitude resulting from shell ejections of 1 $M_\odot$ at various mass loss rates. Before the ejection the model had 46.9 $M_\odot$, $\log L/L_\odot = 6.04$ and log $T_{eff} = 3.82$; the initial mass was 60 $M_\odot$.

the effects of the wind and dust on the light curve; for this last aspect the reader should refer to Davidson (1987).

For ejections made at rates larger than a few $10^{-3}\, M_\odot\, y^{-1}$ (the exact value depends on the stellar mass), the thermal adjustment overpasses the effects of secular evolution, and the net result is a shrinkage of the stellar radius and thus a bluewards evolution in the HR diagram. The overall luminosity undergoes only limited changes. The main effect for the light curves in B or V magnitudes results from the change of the appropriate bolometric corrections as T_{eff} varies during the horizontal oscillations in the HR diagram.

As an example and subject to the above mentioned restrictions, Fig. 2 shows three light curves for a mass ejection of $\Delta\, m = 1\, M_\odot$ in $10^2 y$, in $20\, y$ and in $2\, y$, i.e. with $\dot{M} = 0.01,\ 0.02$ and $0.5\, M_\odot y^{-1}$ respectively. From the many such examples calculated we can derive a few properties of LBV light curves:

1. The peak-to-peak amplitude of the B or V light curve is mainly a function of the total amount $\Delta\, m$ lost in the shell ejection: the larger $\Delta\, m$, the larger the amplitude (cf. Table below). From 10^{-2} to $1\, M_\odot\, y^{-1}$, $\dot{M}$-rates have only a small effect ($<$ factor 2).

2. The $\dot{M}$-rates during the outburst essentially determine the initial slope of the variations, i.e. the rapidity of the initial changes: the higher $\dot{M}$, the faster the initial variation (cf. Fig. 2).

3. The recovery time after an ejection, i.e. the time to again reach the same T_{eff} as

time determines the slope of the final part of the light curve after an ejection. As an example, for a 46 $M_\odot$ star with log $L/L_\odot = 6.05$ and log $T_{eff} = 3.82$, one has for various cases of Δm in the ejected shell:

Δm	Amplitude (B-magn)	Recovery time
0.3 $M_\odot$	0.06	125 y
1.0 $M_\odot$	0.51	350 y
3.0 $M_\odot$	2.16	746 y

4. The light curves of LBV may have very different shapes depending on Δm and also on the T_{eff} at the beginning of the ejection.

   This is a first simple approach of the LBV phenomenon, but from this we see that *a proper analysis of the LBV light curve might give an insight on the parameters of the shell ejections.* Ideally, future models should be non static and should also incorporate the effect of the optically thick wind and of the dust, in order to describe more closely the totality of LBV phenomena.

4 FINAL REMARKS ON ABUNDANCES AND MODEL CONSTRUCTION

Theoretical abundances in LBV have been given by Maeder (1983, 1987), Maeder and Meynet (1987). Essentially, LBV are predicted to have He-enhancements with $Y = 0.4 - 0.7$ instead of the classical value $Y = 0.28$. The C/N ratios should be equilibrium values around 0.02, while O/N is close to but not necessarily at equilibrium with values of about 0.1 to 0.9 (see above reference). In the case of η *Car* (cf. Davidson et al. 1984, 1986) the observations agree with the above values.

Since the occurrence of large density inversions suggest non static solutions, does it mean that in calculations of stellar models we cannot go through the LBV phase in the framework of hydrostatic code? If no special care is taken, the models certainly break down as many have experienced it. However, an appropriate and physically consistent treatment of the outer layers (cf. Maeder 1987) may be devised to properly handle these stages even in the framework of classical stellar codes.

5 REFERENCES

Appenzeller, I.: 1986, in *Luminous stars and associations in galaxies,* IAU Symp. 116, Ed. C. de Loore and A.J. Willis, p. 139

Appenzeller, I.: 1987, in *Instabilities in luminous early type stars,* Ed. H. Lamers and C. de Loore, Reidel Publ. Co., p. 55

Appenzeller, I., Wolf, B.: 1981, in *The most massive stars,* ESO workshop, Ed. S. D'Odorico, D. Baade and K. Kjär, p. 131

Bisnovatyi-Kogan, G.S., Nadyozhin, D.K.: 1972, *Astrophys. Space Sci.* **15**, 353

Chiosi, C., Maeder, A.: 1986, *Ann. Rev. Astron. Astrophys.* **24**, 329
Conti, P.S.: 1975, *Mém. Soc. Roy. Sci. Liège*, 6ème série, tome IX, p. 193
Davidson, K.: 1987, *Astrophys. J.* **317**, 760
Davidson, K., Dufour, R.J., Walborn, N.R., Gull, T.R.: 1984, in *Observational tests of the stellar evolution theory*, Ed. A. Maeder and A. Renzini, Reidel Publ. Co. p. 261
Davidson, K., Dufour, R.J., Walborn, N.R., Gull, T.R.: 1986, *Astrophys. J.* **305**, 867
Doom, C., de Greve, J.P., de Loore, C.: 1986, *Astrophys. J.* **303**, 136
Humphreys, R.M.: 1987, in *Instabilities in luminous early type stars*, Ed. H. Lamers and C. de Loore, Reidel Publ. Co. p. 1
Humphreys, R., Davidson, M.K.: 1979, *Astrophys. J.* **232**, 409
de Jager, C.: 1984, *Astron. Astrophys.* **138**, 246
Lamers, H.: 1986, in *Luminous stars and associations in galaxies*, IAU Symp. 116, Ed. C. de Loore and A.J. Willis, Reidel Publ. Co. p. 157
Lamers, H.: 1987, in *Instabilities in luminous early type stars*, Ed. H. Lamers and C. de Loore, Reidel Publ. Co. p. 99
Lamers, H., Fitzpatrick, E.L.: 1988, *Astrophys. J.* **324**, 279
Maeder, A.: 1980, *Astron. Astrophys.***92**, 101
Maeder, A.: 1983, *Astron. Astrophys.***120**, 113
Maeder, A.: 1985, *Astron. Astrophys.***147**, 300
Maeder, A.: 1987, *Astron. Astrophys.***173**, 287
Maeder, A., Meynet, G.: 1987, *Astron. Astrophys.***182**, 243
Stothers, R., Chin, C.W.: 1973, *Astrophys. J.***179**, 555
Stothers, R.N., Chin, C.W.: 1983, *Astrophys. J.***264**, 583
Stahl, O.: 1986, *Astron. Astrophys.***164**, 321
Stahl, O., Wolf, B., Klare, G., Cassatella, J., Krautter, J., Persi, P., Ferrari-Toniolo, M.: 1983, *Astron. Astrophys.***127**, 49

Andre Maeder

DISCUSSION

Owocki: Is it correct to view the density inversion as arising when the luminosity exceeds the Eddington limit, yielding a net outward effective gravity, and thereby requiring a positive pressure (and hence density) gradient? Have you attempted to model the time-dependent mass loss that this density inversion implies? What are the expected flow speeds of ejecta? Can they be supersonic? What determines how much mass is ejected?

Maeder: The answer to your first question is yes. The actual luminosity exceeds the Eddington limit in the region of the opacity peak and this causes the density inversion in hydrostatic calculations. However, I do not expect the same in non-static models, where very likely an outward motion would occur instead. What I have modeled is essentially the star's thermal response to shell ejections having a variety of parameters, in order to see how the resulting light curves can give information on the masses of ejected shells and on the rates of mass loss during ejections.

The depth of the layers where the density inversion occurs is certainly a critical parameter determining the amount of mass ejected. However, this critical level may move downward during the shell ejection itself, by an amount which is fixed by the condition that the local thermal timescale should not become longer than the duration of the shell ejection. Thus, I expect the amount of mass in a shell ejection to be a multiple of the mass above the density inversion.

Langer: Some comments on your proposition that the density inversion in supergiant envelopes may be responsible for violent shell ejections:
(1) The density inversion may be avoided by taking the density scale height instead of the pressure scale height in the mixing length theory. (2) Even if a density inversion develops, the Rayleigh-Taylor instability will sweep it out. (3) How can a turbulent process like the Rayleigh-Taylor instability lead to outward-directed forces and subsequent shell ejections?

Maeder: The density inversion can be avoided by some artifices, such as multiplying the pressure scale height by 10 or more. Although I consider that static models are not appropriate for very luminous red supergiants, what I showed is that by a proper analysis of the Rayleigh-Taylor instability we can find a better way of handling the problem in the framework of static models than by just artificially increasing the mixing-length ratio.

Sreenivasan: Semi-convection zones tend to appear in our models for $log\ T_{eff}$ between 3.9 and 3.7 or so. These can effectively smooth out any such density inversions. The semi-convection zones appear far above the core just as you find for the density inversion. There is not much mass loss between these points on the evolution track. The CAK-type mass loss has switched itself off and there is not yet an extensive convection zone to produce a real giant-type wind driven by mechanisms analogous to the solar wind. A Rayleigh-Taylor instability should also smooth out density inversions but it does not, so far as I know, produce abrupt mass ejections. So I really do not understand what you are saying. Could you explain it further please?

Maeder: As is well known (*cf.* Chiosi and Maeder in *Ann. Revs. Astron. Astrophys.*, 1986), mass loss and convective overshooting can make the semi-convection zones essentially disappear in massive stars (except for initial masses around 12 to 15 $M_{\odot}$, well below the LBV range). The statement I made is that the density inversion frequently found in static models of stars evolving toward extremely luminous red supergiants (above and near the H-D limit) is just the sign that static models are no longer valid for these stars and that an instability with heavy mass loss is encountered. Thus there is no question of the Rayleigh-Taylor instability itself driving the ejection.

Lamers: Your mechanism related to density inversion occurs around $T_{eff} \approx 9000$ K; but the large LBV outbursts are observed to occur when $T_{eff} \approx 20000$ to 30000 K.

Maeder: The instability effectively occurs at about $log\ T_{eff} \approx 3.9$; then the star very rapidly moves to a much higher T_{eff}, the value of which depends on the amount of gas ejected. The star model spends relatively much more time at this higher T_{eff}. I agree that whether this behavior corresponds to the observed one, or not, has to be carefully verified.

Kahn: Do similar effects occur also for stars with higher surface temperatures, where the density inversion demanded by a static model is smaller, and if so do they lead to a less violent dynamic response?

Maeder: Of course, before being large, the density inversion starts small. Its growth rate is quite short, though (a few times 10 y), which implies that the initially small instability is likely to become a larger one soon.

De Groot: You showed a viewgraph with light curves for three different mass-loss rates during an outburst, and identified one of these with η Carinae. However, while the light curve looks very much like that of η Car, you also said that dust was not included in your calculation. Since dust is crucial for the light curve of η Car, I do not see the relevance of the curve that you showed. On the other hand, your next viewgraph showed a light curve that very much resembles the historical light curve of P Cygni; so maybe you do have a point in all this.

Maeder: The simulated light curves are just illustrations of the thermal responses of LBV stars for various choices of the mass of the ejected shell and for various choices of the mass-loss rates during the episode of shell ejection.

Baratta: What is the mass-loss rate in the case of the more massive stars, which may avoid the LBV phase?

Maeder: I have given some numerical values for quasi-homogeneous evolution in a 1980 paper, *Astr.&Astrophys.* 92, 101. Essentially, the observed rates allow this scheme for masses above 100 $M_\odot$.

Van der Hucht: In their original paper (1972), Bisnovatji-Kogan and Nadyazkin related the occurence of the density-inversion-induced shell ejection to the presence of ring nebulae around some WR stars. Can it be that WR stars with "ejecta-type" ring nebulae are post-LBV objects?

Maeder: This is an interesting suggestion. In order to verify it, more data on the masses and compositions of the ring nebulae are necessary.

Conti: With O-type stars we have to keep in mind that the main sequence, the core H-burning phase, may well include stars that are classified as supergiants. The total mass loss during the main sequence phase is a product of the mass-loss rate(s) and the total lifetime. Could LBV's be H-burning objects?

Maeder: I would not expect LBV stars to be in the core H-burning phase but more likely in the shell H-burning or possibly at the very beginning of the central He-burning phase.

De Jager: Van Genderen monitored some hundred super- and hyper-giants in a recent survey. Among a considerable number of LBV's, he found that their short-time-scale variations have longer "quasi-periods" (or "semi-periods") than non-LBV supergiants in the same part of the H-R diagram. With *period* $\sim \rho^{-1/2}$, this means that LBV's have smaller masses than supergiants of the same L and T_{eff}. Hence, LBV's are on the blueward track of evolution.

Maeder: LBV's are likely on tracks evolving back and forth in the H-R diagram and whether they are moving blueward or redward is difficult to determine. But I agree with your conclusion that the longer periods may be consequences of lower actual mass for the LBV stars, in or between successive shell ejections -- provided, of course, that the nature of the pulsations and the pulsation modes are the same for both categories of stars observed.

EVOLUTIONARY DIAGNOSTICS OF LBV SPECTRA AND SYSTEMS IN THE LMC

NOLAN R. WALBORN
Space Telescope Science Institute
3700 San Martin Drive
Baltimore, Maryland 21218, USA

ABSTRACT. The characteristics of the Ofpe/WN9 class in the LMC, including two newly discovered members, are reviewed; a strong relationship between this class and the LBVs is established by HDE 269858 = Radcliffe 127, currently in outburst with a spectrum identical to that of S Doradus. The evidence for CNO-cycle products, *i.e.*, enhanced N and He, on the surfaces and in the circumstellar nebulae of Ofpe/WN9 objects and LBVs is discussed, with emphasis on recent near-IR observations of He/H. Finally, the occurrence of several Ofpe/WN9 and LBV stars in compact multiple systems will enable valuable information about the evolutionary status of the peculiar objects to be derived. Strong threads unite these three topics: the same objects frequently recur in each.

1. The Ofpe/WN9 Class

1.1. DEFINITION AND MEMBERSHIP

A group of four unusual, hot emission-line spectra was isolated among the brightest OB supergiants in the Large Magellanic Cloud by Walborn (1977); their outstanding peculiarity is the combination of high-ionization Of (He II, N III) and low-ionization (He I, N II) emission features with comparable strengths. Despite their overall similarity, when examined in detail they divide into three subgroups, particularly in terms of the rich spectral morphology in the vicinity of Hδ: HDE 269858 and Sanduleak (Sk) $-67°266$ show exceptionally narrow absorption lines of Si IV λ4089 and N III λ4097; HDE 269227 displays a striking configuration of partially overlapping P Cygni profiles; while HDE 269445 has no absorption features except for a possible diffuse N III λ4097.

K. Davidson et al. (eds.), Physics of Luminous Blue Variables, 27–34.

Subsequent discoveries have expanded the membership of the class, but not the number of subgroups. Sk−66°40 (Walborn 1986) and HDE 269582 (Stahl 1986) belong to the HDE 269858/Sk−67°266 subclass; HDE 269927C (Azzopardi and Breysacher 1979) and Bohannan - Epps (BE) 381 (Bohannan 1979) are similar to HDE 269227 (see also Walborn 1982); while no further spectra exactly like that of HDE 269445 have been found to date. Very recently, two new examples of the HDE 269858/Sk−67°266 subclass have been identified by Bohannan and Walborn (1988); they are HDE 269687 and Sk−69°297, which bring the membership of the Ofpe/WN9 class in the LMC to ten. This last reference also illustrates the spectra of most members of the class in terms of modern digital data, as a basis for future work.

There is some indication that the envelopes of the HDE 269227 subgroup have higher ionization than those of the HDE 269858/ Sk−67°266 type, in the sense that the He II/He I and N III/N II emission-line ratios are larger in the former, despite similar ranges in line strengths. The differing appearances near Hδ may then be due to relatively stronger Si IV and N III emissions there in the HDE 269227 subclass.

1.2. VARIATIONS

In or about 1980, the Ofpe/WN9 prototype HDE 269858 = Radcliffe 127 unexpectedly entered an outburst phase (Stahl *et al.* 1983, Walborn 1984), during which its spectrum has evolved through an intermediate B-type to a peculiar supergiant A-type identical to that of S Doradus, while the bolometric luminosity has remained constant (Stahl and Wolf 1986a, Wolf *et al.* 1988). Thus HDE 269858 has firmly established its own LBV nature, as well as a connection between the Ofpe/WN9 and LBV classes. A connection in the opposite direction has been made by the recent discovery of Ofpe/WN9 characteristics during a light minimum of the galactic LBV AG Carinae (Stahl 1986), which at other times displays a B-type spectrum identical to that of P Cygni (Hutsemékers and Kohoutek 1988), and an A-type similar to S Doradus (Wolf and Stahl 1982).

These remarkable observations indicate previously unsuspected relationships among the apparently disparate objects in question. Moreover, strong hypotheses are suggested that all Ofpe/WN9 objects may be quiescent LBVs, and conversely that many or perhaps all LBVs are of Ofpe/WN9 type at minimum. A corollary of the latter would be that the true minimum states of P Cygni and S Doradus have not yet been observed, and that they are in extended intermediate or maximum states.

A different kind of variability is shown by HDE 269582, which evidently oscillates between high-ionization (Ofpe/WN9) and low-ionization (Bpe) states (Shore and Sanduleak 1984, Stahl 1986, Bohannan and Walborn 1988). The relation between these spectrum variations and the poorly observed light variations of this object has not yet been established. The recent reappearance of its Ofpe/WN9 spectrum is delimited to sometime between November 1983 and December 1984 by observations shown by Stahl (1986) and Bohannan and Walborn (1988). This behavior also raises a question whether other similar Bpe objects in the LMC might be members of the Ofpe/WN9 class in transitory low-ionization disguise.

2. Evolutionary Abundance Anomalies

The evidence for He and CNO anomalies in Ofpe/WN9 and LBV objects was recently reviewed by Walborn (1988) in the context of such effects in a wider array of massive stellar atmospheres and winds, so that only the salient points and more recent developments will be summarized here.

2.1. CIRCUMSTELLAR NEBULAE

Another important property of several Ofpe/WN9 objects in the LMC, including HDE 269858, is their associated nitrogen-rich, nonspherical circumstellar nebulae, which were first detected spectroscopically by Walborn (1982). Stahl and Wolf (1986b) found evidence for similar structures in the spectra of a wider range of LMC emission-line supergiants, including the LBVs S Doradus and HDE 269006 = R71. In the case of the galactic LBV AG Carinae, a spectacular ring nebula can be observed directly in detail, and the first abundance results indicating nitrogen enhancement and oxygen depletion in this object have recently been reported by Dufour and Mitra (1987).

Eta Carinae may be regarded as the most extreme LBV. The large N, He overabundances and C, O underabundances in its spectacular ejected nebulosity have been discussed by Davidson, Walborn, and Gull (1982); Allen, Jones, and Hyland (1985); and Davidson, Dufour, Walborn, and Gull (1986).

2.2. LBV B-TYPE SPECTRA

The λ4600 region in early-B supergiant spectra provides a sensitive diagnostic of CNO anomalies, as discussed by Walborn (1971, 1972, 1976); when N dominates over C and O, striking departures from the normal line-strength patterns occur. In every well-observed LBV B-type phase, such dominance by N II is seen. The examples include P Cygni (Luud 1967), AG Carinae (Hutsemékers and Kohoutek 1988), AF Andromedae and M33 Variable B (Kenyon and Gallagher 1985), HDE 269006 = R71 (Wolf, Appenzeller, and Stahl 1981), and HDE 269858 = R127 (Stahl *et al.* 1983). No quantitative abundance analyses yet exist for such spectra, but it is virtually certain that when they are done large nitrogen overabundances will be found.

2.3. HE/H IN THE IR

The unprecedented quality of the recent 2μ spectroscopic survey of LMC emission-line supergiants by McGregor, Hillier, and Hyland (1988 - see also these Proceedings) has led to several significant new results, including the identification of a category in which the He I 2.058μ line is stronger than H Brackett-γ, which they have reasonably suggested to indicate an abundance anomaly. Of the eight objects in this category, five are members of the Ofpe/WN9 class (HDE 269445, 269582, 269687; Sk$-66°40$,$-67°266$), while the other three have been suggested to be related to it by Stahl and Wolf (1987) and Bohannan and Walborn (1988). This is a remarkable degree of *a posteriori* correlation! Evidently the morphological categories are well defined.

Two Ofpe/WN9 objects observed by McGregor *et al.* did not show He>H, but plausible explanations are available in both cases. One is HDE 269858 observed in its A-type outburst state, and the 2μ spectrum is also identical to that of S Doradus; the He I is most likely not seen because the ionization is too low, and it is predicted that if these objects can be observed at 2μ in Ofpe/WN9 states, they will show He>H. The other case is HDE 269227, in which a red supergiant companion dominates the 2μ region and also the ionization is *higher* than in the other observed objects as noted above. Moreover, from a quantitative analysis of the optical spectrum of HDE 269227 = R84, W. Schmutz, C. Leitherer *et al.* (these Proceedings) have in fact derived a substantial He overabundance, as well as a surprisingly low initial mass of 25 $M_\odot$, a point which will recur in the next section.

All of these results provide strong evidence that both the Ofpe/WN9 class and the LBVs are highly evolved objects, which have CNO-cycle products at their surfaces and are ejecting them into their circumstellar environments.

3. Multiple Systems

A number of Ofpe/WN9 and LBV objects in the LMC are located in compact multiple systems which are most likely physically associated and coeval. The only case which has been investigated in detail so far is that of HDE 269227 by Schild (1987), who found that the associated stars are predominantly of type B and derived a likely initial mass for HDE 269227 of 25 $M_\odot$—in perfect agreement with the independent determination from the atmospheric analysis by Schmutz *et al.* cited above. Other outstanding multiple systems about which little or nothing is known are those including HDE 269858; HDE 269927C (Azzopardi and Breysacher 1979, Walborn 1982); and S Doradus (Leitherer *et al.* 1985). Another spectacular case involving a different kind of emission-line supergiant is that of HDE 269599 (Appenzeller *et al.* 1984). Detailed spectroscopic and photometric studies of these systems will provide unique information about the initial mass, age, and evolutionary status of the peculiar objects.

REFERENCES

Allen, D.A., Jones, T.J., and Hyland, A.R. 1985, *Ap. J.*, **291**, 280.

Appenzeller, I., Klare, G., Stahl, O., Wolf, B., and Zickgraf, F.-J. 1984, *ESO Messenger*, No. 38, p.28.

Azzopardi, M., and Breysacher, J. 1979, *Astr. Ap.*, **75**, 243.

Bohannan, B.E. 1979, in *IAU Symposium 83, Mass Loss and Evolution of O-Type Stars*, ed. P.S. Conti and C.W.H. de Loore (Dordrecht: Reidel), p. 479.

Bohannan, B.E., and Walborn, N.R. 1988, in press.

Davidson, K., Dufour, R.J., Walborn, N.R., and Gull, T.R. 1986, *Ap. J.*, **305**, 867.

Davidson, K., Walborn, N.R., and Gull, T.R. 1982, *Ap. J. (Letters)*, **254**, L47.

Dufour, R.J., and Mitra, P. 1987, *B.A.A.S.*, **19**, 1090.

Hutsemékers, D., and Kohoutek, L. 1988, *Astr. Ap. Suppl.*, **73**, 217.

Kenyon, S.J., and Gallagher, J.S. III 1985, *Ap. J.*, **290**, 542.

Leitherer, C., Appenzeller, I., Klare, G., Lamers, H.J.G.L.M., Stahl, O., Waters, L.B.F.M., and Wolf, B. 1985, *Astr. Ap.*, **153**, 168.

Luud, L.S. 1967, *Soviet Astr.*, **11**, 211.

McGregor, P.J., Hillier, D.J., and Hyland, A.R. 1988, preprint.

Schild, H. 1987, *Astr. Ap.*, **173**, 405.

Shore, S.N., and Sanduleak, N. 1984, *Ap. J. Suppl.*, **55**, 1.

Stahl, O. 1986, *Astr. Ap.*, **164**, 321; err. **170**, 197.

Stahl, O., and Wolf, B. 1986a, *Astr. Ap.*, **154**, 243.

____________. 1986b, *Astr. Ap.*, **158**, 371.

____________. 1987, *Astr. Ap.*, **181**, 293.

Stahl, O., Wolf, B., Klare, G., Cassatella, A., Krautter, J., Persi, P., and Ferrari-Toniolo, M. 1983, *Astr. Ap.*, **127**, 49.

Walborn, N.R. 1971, *Ap. J. (Letters)*, **164**, L67.

____________. 1972, *Ap. J. (Letters)*, **176**, L119.

____________. 1976, *Ap. J.*, **205**, 419.

____________. 1977, *Ap. J.*, **215**, 53.

____________. 1982, *Ap. J.*, **256**, 452.

____________. 1984, In *IAU Symposium 108, Structure and Evolution of the Magellanic Clouds*, ed. S. van den Bergh and K.S. de Boer (Dordrecht: Reidel), p. 239.

____________. 1986, in *IAU Symposium 116, Luminous Stars and Associations in Galaxies*, ed. C. W. H. de Loore, A. J. Willis, and P. Laskarides (Dordrecht: Reidel), p. 185.

____________. 1988, in *IAU Colloquium 108, Atmospheric Diagnostics of Stellar Evolution*, ed. K. Nomoto, Lecture Notes in Physics, **305** (Berlin: Springer-Verlag), p. 70.

Wolf, B., Appenzeller, I., and Stahl, O. 1981, *Astr. Ap.*, **103**, 94.

Wolf, B., and Stahl, O. 1982, *Astr. Ap.*,**112**, 111.

Wolf, B., Stahl, O., Smolinski, J., and Cassatella, A. 1988, *Astr. Ap. Suppl.*, **74**, 239.

Nolan Walborn

DISCUSSION

De Groot: A spectrum of P Cygni is sometimes classified as OBN, Ofpe, Of/WN, or otherwise. Don't we have a problem of nomenclature?

Walborn: I would not classify P Cyg as any of those, only as B1 Iape. He II λ4686 emission would be required for an Ofpe/WN9 classification, and this has never been seen in P Cyg so far as I know. Your interesting light curve shows that P Cyg has remained at an intermediate level since the 17th century, consistent with its relatively constant intermediate-B-type LBV spectrum. If we could observe it spectroscopically at the extreme minimum and maximum levels as reached during the 17th century, we might see Ofpe/WN9 and A-type spectra, respectively.

Heap: What is the difference between late OIaf spectra and Ofpe/WN spectra? What percentage of stars is in each class?

Walborn: The original defining characteristic was that the Ofpe/WN9 spectra show strong low-ionization emission features (N II, He I) along with the normal high-ionization Of-features (N III, He II). In the LMC, ten Ofpe/WN9 objects are now known, which is comparable to the number of normal Of stars known there. However, the statistics are most likely incomplete; the situation in our Galaxy is much worse, though improving.

Maeder: For the interesting star HDE 269227 studied by Schmutz and by Schild, do you have any CNO-ratio data in addition to the helium abundance?

Walborn: No, the analysis includes only He and H; but I believe the study of additional elements is planned.

Langer: Is any LBV known **not** to have CNO-cycled surface material? Such objects might exist, according to standard evolutionary calculations.

Walborn: None to my knowledge. All cases for which there is relevant evidence have nitrogen enhancements.

Kudritzki: Just related to your question, Norbert -- we have quantitatively studied nine B Ia -type supergiants (*poster paper, this meeting*) and none of them has a normal helium abundance. Interestingly, these stars are classified as apparently normal. Atmospheric CNO-cycle enrichment is more common in the post-main-sequence stage than previously thought.

Walborn: This point was addressed in my review at IAU Colloquium 108 in Tokyo. There is growing evidence that the morphologically normal majority of OB supergiants have CNO-cycle products in their atmospheres, while the rarer morphologically nitrogen-deficient ones (*e.g.* OBC) have normal (*i.e.* main sequence) abundances.

Sreenivasan: Is there any evidence of radial motion in the nebula around AG Carinae? Is there inward or outward motion?

Walborn: Thackeray found that it is expanding at about 50 km/s. Paresce (*these proceedings*) will show us evidence for more complex kinematics.

Conti: Line widths and expansion velocities of some of these stars imply relatively slow outflow, not all that different from that of supergiants.

Humphreys: Did you say that R 71 is a multiple system? If so this is new.

Walborn: No, this was a misunderstanding. R 71 is not known to be multiple.

Leitherer: This was why we did our first fine-analysis of an LBV with R 71 and not with R 127. Unfortunately R 127 has a close companion which may affect the observations.

Wolf: I should like to note that S 111 in the LMC has recently been found to be a B[e] star (see *poster paper* by Stahl et al.).

Shore: HD 38489 (= S 134/LMC) is a member of a "normal" populous blue cluster and its place in the H-R diagram is consistent with a high mass (~ 50 $M_{\odot}$). HDE 269445 (= S 30/LMC) is one star with He II λ4686 and may be helium-enhanced. It is nitrogen-rich and is one of the highest-excitation objects in the LMC sample.

Walborn: It is worth noting that you found nitrogen enhanced in HD 38489.

THE DISTRIBUTION OF TYPES OF LUMINOUS BLUE VARIABLES

Bruce Bohannan
Astrophysical, Planetary and Atmospheric Sciences
University of Colorado, Boulder, Colorado, U.S.A.

Abstract: If Luminous Blue Variables (*LBVs*) are not each unique types, three broad groups can be characterized depending on the luminosity and location of *LBVs* in the Hertzsprung-Russell diagram. To assist in defining the evolutionary nature of *LBVs*, connections are made to stars that have similar spectral character with the suggestion that some of these objects that may someday become *LBVs*.

1. Introduction

To be designated a Luminous Blue Variable (*LBV*), a star has to be luminous, blue in color and variable in light and spectrum. The variability typically is a magnitude or so with an accompanying change in spectral type. Since the number of *LBVs* known in any one galaxy can be counted on the fingers of one hand, there is a possibility that each *LBV* is an individual expression of the evolution of massive stars. If *LBVs* are, however, to be better understood, the identification of broad groups of *LBVs* with similar character will provide clearer insight to the origin of the—as yet unknown—physical processes which drive their variability. In order to look at the distribution of types, a rather liberal approach will be taken to group *LBVs* into categories by their luminosity and location in the Hertzsprung-Russell (HR) diagram. We will also seek stars of similar spectroscopic character to *LBVs* which may have the potential to become *LBVs* at some time in the future.

2. The Distribution of *LBVs* in the Hertzsprung-Russell Diagram

The term Luminous Blue Variables was coined by Conti (1984) to include stars which are known by a variety of variable star names—P Cygni, S Doradus, Hubble-Sandage—but that have remarkably similar spectroscopic characteristics and variability. Known *LBVs* are among the brightest stars in a galaxy with spectral types varying in the range between O and A, and with variations in light output of more than 0.2 mag. The best known and most studied of the *LBVs* are P Cygni, AG Carinae and η Carinae in the Galaxy, and HDE 269858 (R 127) and S Doradus in the LMC. Table 1 lists the *LBVs* that are known at this time. References to studies of individual *LBVs* are given by R. Humphreys in this volume.

Lamers (1987) has categorized the different types of spectroscopic and photometric variability exhibited by these most peculiar objects. In general, *LBVs* show evidence for variations that are both large and associated with extensive mass loss; they change their spectroscopic appearance on time scales of years to that of generations of

K. Davidson et al. (eds.), Physics of Luminous Blue Variables, 35–43.

astronomers. Spectroscopic changes from as early as type O to as late as type A have been observed in one star. In general, *LBVs* vary at constant bolometric brightness with minimum light at the highest temperatures and maximum light at the coolest, a variation that implies a change in radius. The mass loss rate is greatest at maximum light.

Table 1:
Known Luminous Blue Variables

The Galaxy	P Cyg	AG Car	HR Car	η Car	
LMC (HDE)	S Dor	269582	269858	269006	
M 31	AE And	AF And	Var A-1	Var 15	
M33	Var A	Var B	Var C	Var 2	Var 83
M81	I1	I2	I3		
M101	V1	V2	V3		
NGC 2403	V12	V22	V35	V37	V38

The true nature of Luminous Blue Variables is most often cloaked by a pseudo-photosphere arising from a dense stellar wind. The visible spectrum only crudely reflects the true parameters of the underlying star. In an HR diagram of *LBVs* (Figure 1), most have maximum light conditions that are relatively close to the upper luminosity limit first drawn by Humphreys and Davidson (1979). Minimum light is characterized by a B or late-O type spectrum.

From their luminosity and location in the HR diagram, one can break LBVs into 3 broad groups:

- The most luminous group contains only one star, η Car.
- The second group contains LBVs in the range $-11 > M_{bol} < -9.9$. HDE 269858 (R 127) is the prototype of this group which includes most of the well known LBVs, such as S Dor, P Cyg and AG Car. Var C in M 33 would be near the lower luminosity range of this group.
- The lowest luminosity *LBVs* are found below the Humphreys-Davidson limit and are in a group characterized by HDE 269006 (R 71) in the LMC. M 33 Var A would be at the upper luminosity boundary of this group.

The underlying basis for the differing character of these groups may rest in the available range for radius variation before the star runs into whatever physical process is behind the Humphreys-Davidson limit. η Car has a very limited range in radius before reaching this physical limit and consequently undergoes very large mass ejections with attendant variations in brightness of over 3 mag. in V. The 1837-1860 event, which produced the presently observed extensive nebulosity around η Car, is an example of such a major eruption. The *LBVs* in the second group are able to vary over a wider range in radius with accompanying change in spectral type from mid-B to A or F, or from late O to A. Here the instability usually produces only moderate variations in light, on the order of a mag. or so in V; these are the most commonly observed variations of *LBVs*. For example, S Dor and AG Car have gone through several of these active stages in the past decades (*c.f.* Lamers 1987). The variations are not continuous. Often, there are long times of relative quiescence, for example P Cyg in this century.

The Humphreys-Davidson limit likely maps those conditions at which massive stars return to higher temperatures after post-hydrogen burning expansion due to rapid high mass loss (*e.g.* Maeder 1983). *LBVs* in the two most luminous groups would not then have evolved through a red supergiant phase. While stars in the second group may undergo large eruptions which may leave behind nebulae that are lower in mass than that of η Car, it is not necessary for all LBVs of this group to do so since they can use a number of mass ejections of smaller scale in order to loose sufficient mass and to return to higher temperatures in the HR diagram as Wolf-Rayet stars. By contrast, those *LBVs* in the lowest luminosity group most likely have been red supergiants. HDE 269006, the prototype of the low luminosity group, was characterized as an S Dor variable by Thackeray (1974). This outburst ended in 1977 (*e.g.* Wolf *et al.* 1981). Noteworthy of the probable evolution through a red supergiant phase for this star is the observation of Glass (1984) of a large excess at 10 μm which he interprets as thermal emission from cool dust.

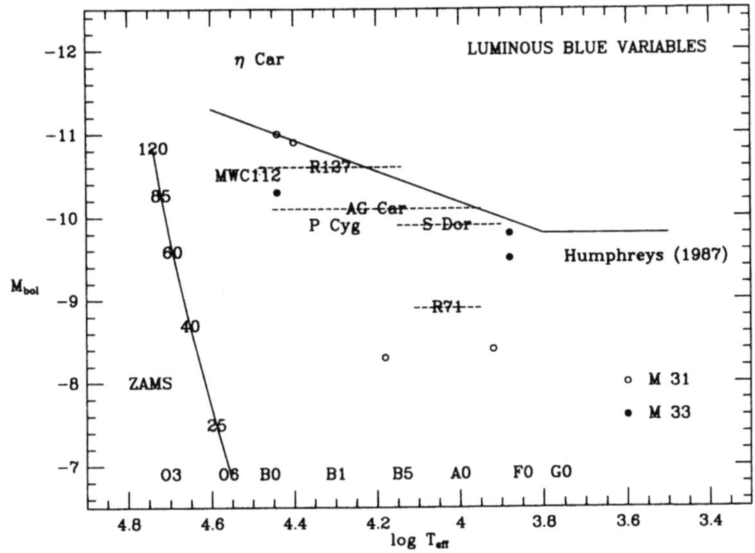

Figure 1. The Distribution of Luminous Blue Variables in the Hertzsprung-Russell Diagram. Temperatures and luminosities for this plot are based upon investigations cited by R. Humphreys in this Colloquium. The luminosity limit is that drawn by Humphreys (1987).

3. The Relationship of LBVs to Other Luminous Emission-line Stars

A clearer understanding of the physical nature of Luminous Blue Variables and of the role that they play in the evolution of massive stars can be gained by finding more *LBVs*, relating *LBVs* to other stars which also characterize the evolution of massive stars, and identifying stars which may become *LBVs* or that perhaps are quiescent *LBVs*.

If this Colloquium had been held before 1982, the number of known *LBVs* would have been one less. In January 1982, Stahl observed that HDE 269858 (R 127) was 0.75 mag. brighter than previously and that its spectrum had undergone a remarkable transformation (Stahl *et al.* 1983). Walborn (1977) had classified this star as OIape extr or alternately as WN9. HDE 269858 had shown essentially the same spectroscopic appearance from the epochs of Feast, Thackeray and Wesselink (1960) to that of Walborn. However, in early 1982 Stahl found that the characteristic Of emission features had disappeared and that no He II absorption lines were visible. No photospheric absorption lines were present at all, and the spectrum was dominated by strong P Cygni features of H I, He I, N II, and Fe III. Stahl *et al.* suggest a mid-B type spectral equivalent. Most recently, the spectrum of HDE 269858 has changed to be essentially identical to S Doradus, a prototype *LBV* (Wolf *et al.* 1988). HDE 269858 became an *LBV* while we weren't looking!

Could it be that the Ofpe/WN9 spectrum represents the minimum light stage of many of the second class of Luminous Blue Variables? Recent behavior shown by AG Carinae suggests that this is an idea worth pursuing. During a recent deep light minimum, AG Car exhibited a previously unseen Of-type spectrum that was remarkably similar to three LMC stars that are classified as Ofpe/WN9 (Stahl 1986). If the Ofpe/WN9 classification does represent the minimum light stage of this class of *LBVs*, then the true minimum *LBV* spectrum of P Cyg and S Dor has not yet been observed; these objects have for our lifetimes been in a long-term intermediate state.

Ten stars in the LMC are now designated as Ofpe/WN9 (Bohannan and Walborn 1989). Most have not been observed to vary in light or spectrum; only two are designated as *LBVs*. Besides HDE 269858, HDE 269582 (MWC 112) is a known *LBV*. In this second object, the Of features disappear and the spectrum consists of H I and He I emission lines only; in the simplest terms, a classification of B pe. The variability of HDE 269582 can even include the absence of Hα (Henize 1956). Several other LMC supergiants have similar B pe spectra to HDE 269582; for example, HD 37836 and HDE 268840 in 1983-84. Stahl and Wolf (1987) find that HD 37836 has a surface temperature equivalent to an O-type supergiant. Other LMC stars that they relate to HD 37836 are HDE 269445 and Sk -69° 240, both of which have He II λ4686Å in emission. The infrared observations of LMC supergiants by McGregor, Hillier and Hyland (1989, see also McGregor in this volume) suggest that there is an evolutionary connection between these B pe objects, the Ofpe/WN9 stars, and *LBVs*. For eight supergiants in the LMC, they find that He I λ2.058 μm emission is stronger than H I Brackett-γ, an observation that they link to a surface abundance anomaly arising from stellar evolution. Five of these eight are of the Ofpe/WN9 class, the others are the previously mentioned HD 37836, HDE 268840, and Sk -69° 240.

4. Looking for Potential LBVs

From the discussion above, it appears that a large number of emission-line stars in the LMC seem to be in the post-hydrogen burning stages of stellar evolution, just as has been proposed for *LBVs*. Will some become Luminous Blue Variables? Are some quiescent *LBVs*? What characteristic makes an *LBV* an *LBV* that might point us to potential *LBVs*? *LBVs* are luminous, blue and variable. However, all stars are variable at some level. It is just a question of how hard one wants to look and when one is looking. Recall the situation for HDE 269858. The *LBVs* in M 31 and M 33 were discovered by Hubble and Sandage (1953) in a survey of these galaxies that was very similar in intent to the Harvard program in the Large Magellanic Cloud. Yet, only 2 of the 4 known LMC *LBVs* are Harvard variables, S Dor = HV 90 and HD 269582 = HV 5495. Detected variability may not be the best way to seek potential *LBVs*!

While the famous *LBVs* exhibit a range of spectral character, the common element is the presence of emission lines much stronger than normal for their temperature and luminosity. I will use this characteristic to point to stars that might be considered as *potential LBVs*; not true *LBVs*, but ones that might someday blossom into *LBVs*.

In the Galaxy, the constellation of Carina would be good place to seek potential *LBVs* as 3 of the 4 famous Galactic *LBVs* are found there. Henize and Carlson (1979) studied 20 of the southern hemisphere emission-line stars surveyed by Henize (1976); 13 have P Cygni-like or B ep spectra very similar to those shown by the well-known *LBVs*. Moreover, one of these stars, He 3-519, has a ring nebula just like the *LBVs* HDE 269858 and AG Car (Stahl 1987). Is He 3-519 a quiescent *LBV*? A circumstellar shell has also been resolved around the LMC Ofpe/WN9 star Sk -67° 266 (Stahl 1987) and detected spectroscopically for other LMC Ofpe/WN9 stars (Walborn 1982, Stahl and Wolf 1986). Is this just a coincidence or does it strengthen the connection between the Ofpe/WN9 stars first pointed to by HDE 269858 becoming an *LBV* in 1982?

5. The Population of Emission-line Stars in the Magellanic Clouds

By comparison with the Carina region, no *LBVs* are known in Orion or Taurus, also regions of active star formation but with less massive stars present. Perhaps then, one characteristic of an environment in which to find *LBVs* is one with massive stars present in large number. The Magellanic Clouds are rich in massive stars and have a large number of emission-line objects. Some 625 emission-line stars are known in the Large Cloud (Bohannan and Epps 1974) to $M_V = -4.1$ and 231 in the Small Cloud (Bohannan and Doggett 1989) to $M_V = -4.5$. All of these stars have emission-lines much stronger than would be normal for stars of similar luminosity and temperature. The Luminous Blue Variables are very bright, typically brighter than $M_V = -6$. In the LMC, 107 emission-line stars are brighter than $M_V = -6$ compared with 46 in the LMC.

Excluding the Wolf-Rayet stars, the luminous emission-line stars in the Magellanic Cloud can be categorized by the visibility of absorption lines (Bohannan 1986, Doggett and Bohannan 1989):

e Hα and sometimes Hβ in emission with higher Balmer lines possibly weakened by emission. Spectral classification easily assigned by absorption lines.

pe Emission extends to Hβ and possibly to Hγ. Spectral classification assigned with some uncertainty from absorption lines. Different lines may give different sub-type.

ex He I λ4471Å in emission as well as other triplet He I lines. Balmer emission stronger than in *e* or *pe*. Classification usually from absorption component of P Cygni features. (P Cygni would be B1 Iex).

extr No or few absorption features present. Types assigned by presence of certain emission lines :
- O — He II 4686Å.
- B — HeI 4471Å strong, no He II 4686Å.
- A — strong Fe II and [Fe II]. (S Dor would be *A extr*)

There are no known *LBVs* in the SMC, although there are several stars with very strong emission lines. The known *LBVs* in the LMC vary over the OBA *extr* categories. Most curiously, there no stars in the B Iex category in which P Cyg is now observed. HDE 269858 was in the O *extr* category when it was a star of the Ofpe/WN9 class before it erupted and it is now evolved to be the A *extr* category. The 13 LMC stars of OBA *extr* categories not known as *LBVs* may yet become *LBVs* and should be frequently monitored so that we can observe the early phases of high mass loss. The distribution of the OBA *extr* stars in temperature and luminosity (Table 2) is essentially the same as that for the known *LBVs*. Note that some are found at relatively low luminosities. Moreover, one of the first stars to be characterized as Ofpe/WN9, BE 381, is at a luminosity well below the limit of this sample of emission-line stars.

This research was supported in part by grants from the National Science Foundation, most recently AST 88-02937. Support of early stages of this work by a Small Research Grant from the American Astronomical Society is gratefully appreciated.

Table 2:
Distribution of Spectral Types of Luminous Emission-line Stars in the Magellanic Clouds

Type	Totals	Number of Stars with M_V =							
		-9.5	-9.0	-8.5	-8.0	-7.5	-7.0	-6.5	-6.0
SMC:	**31**		**1**	**1**	**2**	**5**	**6**	**9**	**7**
O-B *e*	6				1	2	2	1	
> A *e*	12		1	1	1	1	3	4	1
BA *pe*	3							2	1
BA *ex*	0								
O *extr*	0								
B *extr*	1								1
A *extr*	3					1		1	1
WR	4					1	1		2
unknown	2							1	1
LMC:	**107**	**3**	**2**	**10**	**13**	**11**	**21**	**32**	**14**
O-B e	16			1	1		6	5	3
> A e	6	1		3	1			1	
BA pe	26		2	1	7	4	5	6	1
BA ex	0								
O extr	6						3	2	1
B extr	3				1	1		1	
A *extr*	8			2	1	1	2	2	
WR	20	2		1	1	3	3	6	4
unknown	22			2	1	2	3	9	5

References

Bohannan, B. 1986, in *IAU Symposium 116, Luminous Stars and Associations in Galaxies,* eds. C.W. H. de Loore, A. J. Willis and P. Laskarides (Dordecht: Reidel), p. 227.

Bohannan, B., and Doggett, J. B. 1989, in preparation

Bohannan, B., and Epps, H.W. 1974, *Astr. Ap. Suppl.* **18**, 47.

Bohannan, B. and Walborn, N.R. 1989, submitted to *Ap.J.*

Conti. P.S. 1984, in *Observational Tests of the Stellar Evolution Theory,* eds. A. Maeder and A. Renzini, (Dordecht: Reidel), p. 233.

Doggett, J. B. 1988, M.A. thesis, Uinversity of Colorado.

Doggett, J. B., and Bohannan, B. 1989, in preparation.

Glass, I. S. 1984, *M.N.R.A.S.* **209**, 759.

Feast, M. W., Thackeray, A. D., and Wesselink, A. J. 1960, *M.N.R.A.S.* **121**, 337.

Henize, K. G. 1956, *Ap. J. Suppl.* **2**, 315
Henize, K. G. 1976, *Ap. J. Suppl.* **30**, 491.
Henize, K. G., and Carlson, E. D. 1979, *Vistas in Astronomy* **23**, 213.
Hubble, E., and Sandage, A. 1953, *Ap. J.* **118**, 353.
Humphreys, R. M. 1987, in *Instabilities in Luminous Early Type Stars*, eds. H. J. G. L. M. Lamers and C. W. H. de Loore, (Dordecht: Reidel), p. 3.
Humphreys, R. M., and Davidson, K. 1979 *Ap. J.* **232**, 409.
Lamers, H. J. G. L. M. 1987, in *Instabilities in Luminous Early Type Stars*, eds. H. J. G. L. M. Lamers and C. W. H. de Loore, (Dordecht: Reidel), p. 99.
Maeder, A. 1983, *Astr. Ap.* **120**, 113.
McGregor, P.J., Hillier, D.J., and Hyland, A.R. 1989, submitted to *ApJ*.
Stahl, O. 1987, *Astr. Ap.*, 909.
Stahl, O., Wolf, B., Klare, G., Cassatella, A., Krautter, J., Persi, P., and Ferrari-Toniolo, M. 1983, *Astr. Ap.* **127**, 49.
Stahl, O., and Wolf, B. 1986, *Astr. Ap.* 158, 371.
Stahl, O., and Wolf, B. 1987, *Astr. Ap.* **181**, 243.
Thackeray., A. D. 1974, *M. N. R. A.S.* **168**, 221.
Walborn, N.R. 1977, *ApJ.* **215**, 53.
Walborn, N.R. 1982, *ApJ.* **256**, 452.
Wolf, B., Appenzeller, I., and Stahl, O. 1981, *Astr. Ap.* **99**, 351.
Wolf, B. Stahl, O., Smolinski , J., and Cassatella, A. 1988, *Astr. Ap. Suppl.* **74**, 239.

Bruce Bohannan

DISCUSSION

Friedjung: After Roberta Humphreys' talk I thought I understood what an LBV is. Now I am not so sure. Can you distinguish what physical mechanism occurs just by looking at the spectrum?

Bohannan: No. To be an LBV a star needs something in addition to a spectrum. Remember that R127 was just an Of/WN star until in 1982 it was seen to brighten and undergo remarkable spectral change. The LMC stars I showed here, beyond the Famous LBV's, are only **potential** LBV's.

Davidson: A good definition of LBV's would concentrate on their unique physical characteristic, eruptions. We could simply say that an LBV is a star, in roughly the right part of the H-R diagram, which has large eruptions. Then maybe there are some LBV's that we don't notice because we're living in the wrong century. But so what? -- the same is true of supernova progenitors and we accept that.

Humphreys: This morning I tried to define an LBV. LBV's are characterized primarily by their instability -- their alternating periods of eruption and quiescence. Spectroscopic similarity does not mean that a star is an LBV! There are many luminous emission-line stars in the Galaxy and Magellanic Clouds. They are not all LBV's.

Maeder: It is certainly interesting to try to extend the group of stars sharing the properties of LBV stars, but we need very clear designations for any new groups. Otherwise we shall enter 10 years of confusion. I strongly recommend that we restrict the term LBV to those stars that are known to have experienced strong outbursts. The new stars that you have found could be called LBV **candidates.**

Lamers: I do not like the extension of the LBV class to "stars with types similar to famous LBV's." This criterion may bring a group of rather different stars within the definition. Since we want to concentrate on "the physics of LBV's," at this colloquium, we should be sure that we talk about genuine LBV's -- even if this results in a small number of stars and very incomplete samples.

Bohannan: It certainly was not my intention to imply that the set of LBV's in the LMC should be expanded to include all of the emission-line stars. Rather I am suggesting that those emission-line stars may represent a parent population of the LMC LBV's, stars that may become LBV's in the future or that are now quiescent LBV's.

De Jager: The main characteristic of LBV's is their episodical eruptions. Hence a search for new examples should go along with that criterion. But there is another criterion, not yet mentioned in the present discussion: photometric monitoring of LBV's over a few years has shown that their brightness variations have larger amplitudes than supergiants. Hence a search based on photometric monitoring could help to discover candidate LBV's.

Bohannan: True, there need to be both spectroscopic characteristics and large-scale photometric variations. My list of suggested candidates may help to guide those making the photometric observations.

Heap: A purely spectroscopic definition of LBV's might include some nuclei of planetary nebulae, such as the central star of He 2-131 (O8 feq).

Bohannan: That's true. However, in the LMC one has a handle on the luminosity and this provides a natural bias against highly evolved low-mass objects.

Appenzeller: How did you derive the M_{bol} values plotted in your H-R diagram, and how reliable are these values?

Bohannan: Some that have temperatures similar to B supergiants come from UV continuum-fitting with model-atmosphere flux distributions, similar to what the Heidelberg group of Wolf and Stahl employed. I get slightly higher temperatures because I calculate fluxes from a Kurucz atmosphere rather than using the published grid. I assume that the temperatures of Of/WN stars are like O8 I stars.
Most of the e and pe star temperatures were derived from a *spectral type vs.* T_{eff} calibration. The value for R 84 comes from the detailed analysis by Schmutz *et al.* described at this colloquium.

Conti: M_V is sufficiently faint for some of the LMC stars that revised bolometric corrections cannot bring them above the red supergiant limit. So they are faint, like R 71.

Shore: Adding to Bohannan's point about catalog incompleteness, Walborn has pointed out that S 83/LMC does not exist; it was MWC 112, but when Henize was constructing his catalog it was not in emission. Perhaps Sk -69° 202, the progenitor of SN 1987A, was like this.

M.-C. Lortet asking a question. Near her:
van der Hucht, Hearn, Koenigsberger.

Smolinski, Maeder, Hauck, *et al.*

AGE OF LUMINOUS BLUE VARIABLE CANDIDATES FROM THEIR STELLAR AND NEBULAR ENVIRONMENT.

M.-C. LORTET
Département d'Astronomie extragalactique et de Cosmologie
Observatoire de Paris
F-92195 Meudon Principal Cedex, France

ABSTRACT. The study of the context of LBV stars, that is close neighbourhood (companions) and spatial relationship to young objects such as cepheids, stellar associations, red supergiants, enriched OB supergiants, HII regions, WR stars can provide useful clues on their true luminosity, their age thus eventually initial mass and evolutionary status.

The stars considered here are LBV candidates in a broad sense. The starting list is a compilation of the following families (mostly defined from the Atlas of high-resolution spectra of Stahl et al., 1985) : S Dor-like stars (four in the LMC, none in the SMC), P Cyg type stars (prototype R 81), Of or Of/WN9 or WN9/Of (late-type O), as defined by Walborn (1982) and now extended by Conti and Bohannan (see Walborn in these Proceedings), B[e] stars (Zickgraf et al., 1986 and Zickgraf, these Proceedings), dust shells ; I added Peculiar, B pec and B extr from recent spectroscopic surveys (Conti et al., 1986 ; Garmany et al., 1987) and stars with composite spectra (Cowley and Hutchings, 1978). The sample consists of 52 and 31 stars in the LMC and the SMC respectively, and is detailed in a catalogue (Lortet, 1988) which also gives indications on their environment.

The surveys for recognizing these families are all biased and incomplete, so that no statisticial approach is possible. We will mostly be concerned with the LMC.

1. RELATIONSHIP TO OB STELLAR ASSOCIATIONS, HII REGIONS AND RED SUPERGIANTS

A few general ideas have already emerged from the detailed study of regions of recent, but not active, star formation like Shapley II which is especially rich in LBV candidates (about 1° in diameter, see van den Bergh, 1981). Lortet and Testor (1987, 1988) found the LBV candidates of this region and a few additional areas to be associated with a rather old environment, filamentary, faint, extended nebulae, if any, and with either WN8 or later type WR stars or faint WNE, as well as red supergiants.

They also commented upon the case of R 84 (=HD 269227 = Sk-69 79 =

K. Davidson et al. (eds.), Physics of Luminous Blue Variables, 45–49.

Brey 18), which is a member of an aged association (Schild, 1987) and composite with a brighter M supergiant.

The examination of a large sample of LBV confirms these results. Many other examples of shells or filamentary nebulae related to LBV candidates are found, for instance : N9 (=DEM 31) around Sk–67 17 (=HD 32034, P Cyg type) and near the WR Brey 5 ; the southernmost part of N 30, related to Sk–67 64 (B[e], not far from Brey 17) ; [OIII] and Hα filaments, part of N 74 around Sk–67 266 (Of/WN9)).

LBV candidates as a rule are inside or less than 120pc from stellar associations (see Table 2 in Lortet, 1988). Outstanding exceptions are : a few B extr (may be an older population), two B[e] type stars, Sk–67 23 (=S12, distance 585pc) and Sk–69 46 (=R 66, distance 225pc), and the S Dor-type star R 71 which is discussed below.

From the point view of stellar evolution, the relationship of LBV candidates to red supergiants is crucial. As projected on the sky more than half of the LMC LBV candidates are definitely associated to red supergiants. The proportion increases to about 70% when excluding the B extr stars and those stars the only peculiarity of which is to be suspected of variability, Appendix in Stahl et al., 1984). Spatial association to WN early and WN8 and later is also conspicuous. Examples of such clusterings are :

Sk–69 171 (P Cyg type) and Sk–69 175 (=HD 269687) (Peculiar, newly classified as Of/WN9) near Brey 44a (WN8–9) and at least 7 red supergiants of known radial velocities close to = 275 km s^{-1}, of which two recently measured (Lindgren, private communication). This group had already been noticed by Lortet and Testor (1988).

Sk–67 266 (Of/WN9), three WN4 stars and red supergiants in or near LH 116.

Brey 64 (=BE 381) with four WNE stars and red supergiants in LH 85 + 89.

2. CLOSE NEIGHBOURHOOD

The existence of close companions seems rather frequent for the S Dor-like stars (types S, Of/WN9, P, B[e]). A systematic search and study is important for several reasons :

a) The existence of close companions if ignored lead to an overestimation of the luminosity of the star. Many similar instances exist for hot stars (overestimation by 0.5 to 3 magnitudes).

b) Search for variability requires that the close neighbourhood be first imaged. It may be dangerous to compare measurements made by different authors or with different positions and diameters of the aperture.

Table 1 describes the close neighbourhood of a few LMC LBV candidates. Two identifications and the HD number are given. The last column refers to a remark or a map. SL objects are possible compact clusters, RM1– and RM2– red supergiants and probable red supergiants respectively from Rébeirot et al. (1983). Otherwise the nomenclature is explained in Lortet (1986). Apart from the composite spectra (R 84 is one of those), close association with a red star is found in two cases. It would be important to check the consistency of the radial velocities and also to obtain even a crude photometry of the cluster stars.

Table 1. Multiplicity or peculiarities of LBV candidates in the LMC (classified by subtypes).

Sk	Other	HD	Type	Close neighbourhood	Ref.
Sk–69 94	S Dor	35343	S	In a cluster (diam. ~ 30")	A
Sk–69 142a	HV 5495	269582	S, OF/W	W10–35, 36, 39, inside 6"	B
Sk–69 220	R 127	269858f	S, OF/W	R 127B within 3"W, R 128 wihtin 18"	C
Sk–69 79	R 84	269227	W/OF + M	Composite with RM2–54, unresolved	–
Sk–69 249C	Brey 91	269927C	W/OF	Sk–69 249A and B, OB supergiants	D
Sk–69 92	R 85	269321	P, V	In a group of three stars inside 20", one is RM2–60	E
Sk–69 171	R 110	269662	P	SL 530: few stars, inside 20"	–
(Sk–69 147a)	S 111	(HD 269599)	B[e]	Cluster 10" in diameter	F
–	R 108	–	COMP	Composite with RM2–88	–
Sk–70 29	R 67	32763	COMP	RM2–29 at 5"	–
Sk–68 162	-	–	B extr	RM1–811 13" SE	–
Sk–71 29	HV 5827	–	B extr	N 202 (53"x59")	–
Sk–67 181	–	269736	V	Cluster SL 564, diameter 20"	–

A, Leitherer et al., 1985 ; B, in NGC 1983, Westerlund, 1961, Fig. 15 and Table 10 ; C, Stahl, 1985 ; D, Lortet and Testor, 1988, Fig.1 ; E, Brunet et al., 1975, p.136. F, the HD and Sk numbers refer to part of a cluster which also contains a normal OB star, R 105 ; Appenzeller et al., 1984.

3. THE ENVIRONMENT OF R 71 (= HD 269006 = Sk–71 3)

R 71, one of the genuine S Dor–type stars in the LMC though subluminous, is known to be pretty isolated. Close examination confirms that no young object is physically related to it. A rather old cepheid (HV 12433, P = 7.5 day) is located at about 33pc. Red supergiants and other old cepheids are scattered farther than 10′ away. The youngest objects are at about 16–20′ in the south of the stellar association LH 28 studied by Lucke (1972) : an anonymous Hα knot, an O9 II star (Conti et al., 1986), a cepheid HV 2343, of period 42.172d. We notice that the B extr star Sk–71 12 is at 11–15′ from these objects.

4. THE SMALL MAGELLANIC CLOUD

About the SMC, little can be said at present. Projection effects are severe. Excluding the 21 B extr stars, few interesting LBV candidates are known : no genuine S Dor–type star, only two P Cyg stars, three B[e], four Pec and one star suspected of variability (the candle AZVI 415 = R 40). Those stars are poorly related to HII regions and stellar associations (Hodge, 1985) except for the three B[e] stars and AZVI 6 (B pec) which is in association 8 and in the nebula DEM 14. No connection whatever between the 10 S Dor–like and the 8 known WR stars is found. This may be largely a question of incompletion of different kinds of surveys.

5. CONCLUSIONS

The close examination of the context of LBV candidates appears as both a guide and an independent complement to finer studies. The prediction of rather low mass from the moderately old age of the parent association (R 84, Schild, 1987) or environment has been confirmed in a few cases (e.g., R 81, Stahl et al., 1987). It is also in beautiful agreement with the enhanced atmospheric abundances described first for P Cyg (Lamers, 1986) and since then for several stars which thus appear as post-red supergiants.

Indeed, all the LBV candidates and WR stars in Shapley II (especially those quoted in Sect. 1) are potential post-red supergiants. The general questions are now : which are the distinctive observational and physical characteristics of those LBV's and WR stars which are post-red supergiants ? And more generally, into which WR subtypes do LBV's evolve ?

REFERENCES

Appenzeller, I., Klare, G., Stahl, O., Wolf, B., Zickgraf, F.-J. :1984, The ESO Messenger 38, 28.

Brunet, J.P., Imbert, M., Martin, N., Mianes, P., Prévot, L., Rébeirot, E., Rousseau, J. : 1975, Astron. Astrophys. Suppl. 21, 109.

Conti, P.S., Garmany, C.D., Massey, P. : 1986, A.J. 92, 48.

Cowley, A.P., Hutchings, J.B. : 1978, P.A.S.P. 90, 636.

Garmany, C.D., Conti, P.S., Massey, P. : 1987, A.J. 93, 1070.

Hodge, P. : 1985, P.A.S.P. 97, 530.

Lamers, H.J.G.L.M. : 1986, IAU Symp. 116, 157.

Leitherer, C., Appenzeller, I., Klare, G., Lamers, H.J.G.L.M. : 1985, Astron. Astrophys. 153, 168.

Lortet, M.-C. 1986, Astron. Astrophys. Suppl. 64, 303.

Lortet, M.-C. : 1988, Bull. Inf. CDS 35, in press.

Lortet, M.-C., Spite, F. : 1986, Astron. Astrophys Suppl. 64, 329.

Lortet, M.-C., Testor, G. : 1984, Astron. Astrophys. 139, 330.

Lortet, M.-C., Testor, G. : 1987, Proc. Workshop "Starbursts and Galaxy Evolution", eds. T.X. Thuan, T. Montmerle and J. Tran Thanh Van, Ed. Frontières, Paris, 79.

Lortet, M.-C., Testor, G. : 1988, Astron. Astrophys. 194, 11.

Lucke, P.B. : 1972, Ph.D. Thesis, University of Washington.

Rébeirot, E., Martin, N., Mianes, P., Prévot, L., Robin, A., Rousseau, J., Peyrin, Y. : 1983, Astron. Astrophys. Suppl. 51, 277.

Schild, H. : 1987, Astron. Astrophys. 173, 405.

Shore, S.N., Sanduleak, N. : 1984, Ap.J. Suppl. 55, 1.

Stahl, O. 1985, The ESO Messenger 39, 13.

Stahl, O., Wolf, B., Leitherer, C., Zickgraf, F-.J., Krautter, J., de Groot, M. : 1984, Astron. Astrophys. 140, 459.

Stahl, O., Wolf, B., de Groot, M., Leitherer, C. : 1985, Astron. Astrophys. Suppl. 61, 237.

Stahl, O., Wolf, B., Zickgraf, F.-J. : 1987, Astron.Astrophys. 184, 193.

van den Bergh, S. : 1981, Astron. Astrophys. Suppl. 46, 79.

Walborn, N. : 1982, Ap.J. 256, 452.

Westerlund, B. : 1961, Uppsala Astron. Obs. Ann. 5, No. 1.

Zickgraf, F.-J., Wolf, B., Stahl, O., Leitherer, C., Appenzeller, I. : 1986, Astron. Astrophys. 163, 119.

DISCUSSION

Sreenivasan: (1) Are you suggesting that S Dor variables are in fact only 15--20 $M_\odot$ objects? (2) How old do you mean when you say they are older than previously believed?

Lortet: (1) I did not quote figures but indeed **some** may have masses as low as that. (2) "Old" means that some are found near red supergiants and/or in associations containing only B stars, *e.g.* R 84 (=Brey 18) in LH 39, studied by Schild. My guess is that they may be as old as 10 million years.

De Jager: In the discussion following Maeder's talk this morning, I suggested that some LBV's have smaller masses than supergiants at the same location in the H-R diagram, and that they are on the blueward evolutionary track. This seems to fit with your result that LBV's are older than has been assumed so far.

Leitherer: Your result that S Dor variables have lower masses than generally assumed agrees nicely with the mass that we and the Munich group have independently derived from NLTE analyses of R 71, $M \approx 10$--$15\,M_\odot$. S Doradus itself may also have a mass less than about $40\,M_\odot$.

Lortet: These fine results show that the study of the environment may be very effective.

Appenzeller: A mass of 15--20 $M_\odot$ appears reasonable for R 71, which is one of the least luminous LBV's. However, the most luminous LBV's, such as R 127, must be quite massive ($\geq 40\,M_\odot$). Otherwise they would be significantly above their Eddington luminosities.

Lortet: I agree that some LBV's may be more massive than $40\,M_\odot$; R 127 is the hottest of them anyway.

Walborn: Do you find any clear associations between LBV's (including possible quiescent LBV's) and WR stars? If the former turn into the latter on a short time-scale, then such associations are to be expected.

Lortet: This is indeed a very important point. Also, which WR subtypes do LBV's become, if any? The best place for suggesting an answer is in the LMC and especially in Shapley II. There are clear associations between LBV's and red supergiants on one hand, and LBV's and WR stars on the other. My guess is that LBV's may be related to WN 8,9,10 or Of/WN9 stars, or to early-WN's, but not to WN 6 or WN 7.

Gallagher: If some S Dor stars originate from binaries, then delayed evolution may occur. Mass-exchange time scales for a high-mass binary can be longer than the nuclear evolution time-scale of a single star with the same total mass.

Lortet: I agree.

Sally Heap

Peter Conti
and
Francesco Paresce

An Ultraviolet View of the Luminous Blue Variables

Steven N. Shore
Astrophysics Research Center, Dept. of Physics
New Mexico Institute of Mining and Technology
and DEMIRM, Observatoire de Meudon

1. Introduction

It is, perhaps, tautological to say that the luminous blue variables (*LBVs*), being luminous and blue, should be ideal targets for space ultraviolet observation. In fact, many have been observed on occasion and these will be discussed in this article. Mitigating circumstances, however, conspire to make this study difficult. The LBVs are rare and tend to be apparently faint, and since they are massive and formed in the plane in our Galaxy and in active star-forming regions in other galaxies, they tend to be heavily reddened. The stars were not really studied at all in the ultraviolet until the launch of the *International Ultraviolet Explorer Satellite* (*IUE*), with the notable exceptions of photometric observations of several of the galactic stars with ANS and TD-1 and *Copernicus* observations of P Cygni (Hutchings *et al.* 1987). *IUE*, a small aperture (45 cm) spectrographic satellite, covering the spectral range from 1200 to 3300 A, allows for an adequate characterization of the UV properties of these stars.

2. Why Go to the Ultraviolet?

Optical observations show that many of the LBVs are embedded in emission regions and that the optical emission lines from the extended envelopes render it difficult or impossible to say much about the behavior of the underlying star. For mass loss indicators, the optical depths in the absorption components of optical P Cygni lines are not sufficient to properly sample the stellar wind velocity profile; many of the stellar wind lines are formed from excited states. In the ultraviolet, we have a better view of the opacity of the envelope and a clearer probe of the ionization structure. Many of the strong UV lines are resonance transitions, which give a better handle on the radiative balance of the wind. Finally, and perhaps most important, the UV permits the direct measurement of the bulk of the stellar flux so that one can determine stellar bolometric luminosities without reliance on optical properties of normal supergiant atmospheres.

3. Low Dispersion Spectra and Ultraviolet Spectral Types

We do not see the naked star in the ultraviolet anymore than we do in the optical, but more information is available from the UV about the behavior of the envelope because we see a pseudo-photosphere which looks remarkably like a normal supergiant. It is better to say that the UV gives us a picture of what the star would like to believe itself to be,

K. Davidson et al. (eds.), Physics of Luminous Blue Variables, 51–58.

and what its environment knows it as. For instance, it is possible to use the ultraviolet to connect the infrared emission to the bolometric luminosity of the star through IR line fluorescence and dust re-emission, which are driven by the absorption of UV photons (see, for example, McGregor *et al.* 1987). Because this UV continuum is also the spectrum responsible for illuminating any surrounding nebula, direct UV observation is important for chemical analysis of the ejecta.

Ultraviolet Spectral Types (UVST): In the absence of good quantitative stellar atmosphere models for these stars, one can form an ultraviolet spectral type using Si III λ1300, Si IV λ1400, C IV λ1550 and Al III λ1860 (Henize *et al.* 1981, Shore and Sanduleak 1984) for supergiants in the range O8 to B9; additional information about the shell structures is provided by the strong Fe II absorption systems near 1500, 1600, and 2200Å. High resolution observations of the Fe II lines between 2500 and 2600Å sometimes show P Cygni profiles, which probe the outer portions of the envelope and provide information about expansion velocities and wind strengths (*e.g.* Stahl and Wolf 1986). An advantage of the UVST method over model fitting is that it is independent of the reddening assumptions. For detailed application of this method to the LMC/SMC stars see Shore and Sanduleak (1984); recent work has extended to Galactic stars (Shore *et al.* 1988).

General Results: The Si IV λ1400 doublet is typically resolved even at low resolution for most of these stars. R 127 is an unusual exception to this rule, suggesting that the ejection event may be initially at higher velocity than normally observed for the shells in these stars. Ultraviolet observations show that the envelopes often remain stable on long timescales, and that the high mass loss rates are perhaps present for decades. The structure of winds in these stars does not appear to greatly differ between the Galaxy and the LMC. Put more precisely, it appears that the metallicity difference between these galaxies is insufficient to produce any observable difference in the behavior of the stellar mass outflows.

Those stars which optically show the strongest Fe II and [FeII] systems, like R 50 = S 65/SMC, S 22/LMC, S Dor, and He3-407, typically show the most strongly absorbed UV systems as well. Strong optical stellar wind profiles are normally associated with the strongest resonance line absorption in the UV (S 12/LMC, S 61/LMC). Many of the stars which show S Dor-like profiles in the optical, with broad emission wings on the hydrogen lines and narrow, low velocity absorption, show B2 - B4 spectra in the UV; this may be an indication of multiple shell ejection events and could be a useful indicator of those stars which are likely to undergo shell outbursts (like R 127 = S 128/LMC).

It is generally the case that the LBVs show only an optically thick spectrum in the UV. Only a few LMC stars show emission lines: S9 (NIII], Si III]:), S 12 (N IV), S 30 (He II, N III], Si III]:), MWC 112 (N III]:), S 131 (Si IV, C IV, Si III], N III]). Only S 134 shows both narrow, nebula-like emission and P Cyg profiles on the resonance lines. The strongest emission line objects are S 18/ SMC and LMC Anon. For the Galactic stars, only He3-40 shows Mg II emission. Strong Fe II absorption systems are the rule for these stars. The stars in the LMC sample (Shore and Sanduleak 1984) had been chosen from the Henize list of Hα emission line stars, while those observed by Stahl *et al.* (1985) have been selected optically from the same list. Not all of these stars have been known as LBVs, but several have proven to be after the fact.

Disks vs. Cocoons: The Hamburg group has suggested that the B[e] stars are hybrid spectra, and has explained the low excitation, narrow lines as arising from a circumstellar disk, perhaps a low velocity equatorial wind, on the basis of optical spectra. Gallagher (*this conference*) has argued that many LBVs are binaries with dissipative accretion disks. The ultraviolet spectral distributions, at low dispersion, do not match the expectations for disk systems. There are thus two different kinds of disks, and perhaps the UV observations can narrow the choices.

Turbulent disks have a characteristic spectral shape. Assuming that the disk is optically thick, but geometrically thin, and that the energy is released primarily through viscosity acting locally throughout the disk, then the temperature in the disk varies as $T(r) \sim r^{-3/4}$. Assuming that the disk radiates like a Wien radiator, then $F_\nu \sim \nu^{1/3}$ with the upper portions of the disk presenting an effective chromosphere to this hot optically thick surface. The luminosity depends on the stellar mass and the mass accretion rate, and the M/R value for the accreting star. Spectra much like those of the LBVs have been observed in some of the longer period Algol systems, especially 22 Vul and HD 207739 (McCluskey and Sahade 1987). The equatorial winds (hybrid stars) will not show the same dependence of temperature on radius, but it is still to be demonstrated that they reproduce the UV spectra.

Predictions are needed of the UV and optical appearance of so luminous a disk, accreting at the Eddington limit, if this is to be the sole explanation for the UV behavior. For the LMC stars which have been well studied, and for which the reddening is low, like S 111/LMC, the continuum matches well with normal stellar atmospheres. If there are disks, they may be a factor in the emission lines but do not play much of a role in the formation of the ultravolet continuum. If disks are involved with the LBVs, UV observations suggest that they must be excretion, rather than accretion, type.

Only a few systems, like S 18/ SMC (Shore *et al.* 1987, Zickgraf *et al.* 1988), R 99, or S 131/ LMC (Stahl *et al.* 1987) show strong UV emission lines. LMC Anon (Michalitsianos *et al.* 1988, *preprint*) looks similar to η Car and S 18/SMC in these characteristics and is also variable in optical He II λ4686. S 18/SMC appears to require some accretion from a companion to power the far-UV radiation source (Shore *et al.* 1987; Zickgraf *et al.* 1988, *preprint*).

4. Some Galactic LBVs

A good sample for Galactic stars is the Henize sample of P Cygni and B[e] stars studied optically in some detail by Carlson and Henize (1979) and in the IR by McGregor *et al.* (1987). They consist of a sample of optically related stars, which generally show strong emission lines of H and He I, sometimes N and C ions, and Fe I and [Fe II] as well. They have been observed to show IR excesses in a few cases, and at least one of them, He3-1482 = HD 316285, is a radio source, like P Cyg. Since they are all fainter than $V = 9^m$, they appear to be at large distance and to have very large IS reddening.

Several of these stars have been observed to show extended shells, especially He3-517, but for this star there are no UV spectra (Stahl, *this conference*). A few are hot IRAS point sources. η Car, has been observed on several occasions with IUE. It shows strong nitrogen enhancement in the UV spectra of the ejecta (Davidson *et al.* 1982), in agreement with the optical spectra. Another well studied star is AG Car, which also shows an extended asymmetrically structured shell. High dispersion spectra are available for this star, which show that it has a strong stellar wind (Johnson 1982). Other stars which have been studied in detail in the ultravolet are MWC 300 (Wolf and Stahl 1985), P Cygni (Lamers *et al.* 1985), and GG Car (Brandi *et al.* 1987) (which may be a main sequence star although it has many of the LBV properties and the luminosity is poorly known).

R 126 = S 127/LMC shows the strongest resonance lines of the B[e] stars in the LMC. This star provides a useful example of the UVST method. Its properties overlap part of the range of R 127, but the absorption lines are consistently stronger; it is not notable as a photometric variable. The Galactic star, He3-395 = HD 89249 shows nearly identical line strengths (Si III, Si IV, C IV, and Al III are within about 15 percent). R 126 has been called a rotationally unstable hybrid star (Zickgraf *et al.* 1986), but it is surprising to find that it has a *sosie* in the Galaxy. The UVST for He3-395 is B0.5±0.5, with an effective temperature of about 27000 K, and a reddening of $E(B-V) = 0.8$; optical photometry and spectra give 25000 K and $E(B-V) = 0.8$. Assuming average reddening, this gives a distance of about 2 kpc and a bolometric luminosity of about $L \approx 10^5 L_\odot$, fainter than R 126. Few of the B[e] stars have been observed to show strong UV variability, with HR Car and R 127 being notable exceptions. Several of the shells have been quite stable, notably S 22/LMC and S 65/SMC.

Some additional results for Galactic stars are: He3-365 ($logT_{eff} = 4.4, A_V \approx 2.9, logL/L_\odot \approx 4.9$), He3-1138 ($logT_{eff} = 4.2, A_V = 5.6, M_{bol} \approx -9.6$), He3-1300 ($logT_{eff} = 4.4, A_V \approx 5.0, M_{bol} \approx -8.3$), He3-1330 ($logT_{eff} = 4.6 : (B1.5), A_V \approx 4.0, M_{bol} \approx -8.8$). For most of the other stars in the Carlson - Henize sample there are no good photometric observations available.

5. The Large Magellanic Cloud

The disadvantages introduced by the distance of the LMC, both of faintness and of crowding, are largely offset by its distance above the galactic plane and the certainty in the distances to the associated LBVs. In the LMC, we see all of the stars in their environmental context and with little foreground obscuration. This is very important, because for these stars we can address the questions of dust formation and mass expulsion that would not be possible for their Galactic counterparts.

By far the best studied LBV in the LMC subsequent to the launch of IUE is R 127 = S 128 = HDE 269858*f*. First observed as an O Iapef star by Walborn, it was discovered to be a B2 Ia by IUE observations in 1982 (Shore and Sanduleak 1984, Stahl and Wolf 1983, Stahl and Wolf 1986); it may be the first case of an LBV being found in another wavelength region than optical. Thanks to the longevity of IUE, it is possible to trace most of the history of this star from the time of its first detection as a B star through its present status as one of the most optically luminous stars in the LMC and an A2Ia$^+$ spectrum similar to S Dor. During the years 1982 - present its bolometric output has remained essentially constant. When first observed, it showed strong Si IV and Al III lines, with little C IV absorption and no emission. The Fe II lines subsequently deepened and obliterated the hotter photospheric features without the appearance of strong emission lines. With the increase in brightness in the long wavelength end of the IUE range, high dispersion spectra revealed the P Cyg profiles on the Fe II (42) multiplet.

The prototypical LBV, S Dor, is a bit more difficult to study in the UV because of the presence of a hot companion, a B0 star which is several magnitudes fainter in the optical but about equal to S Dor at 1600 A (Wolf *et al.* 1980). This complicates the interpretation of the IUE spectra. Low resolution studies have deconvolved the spatial blend of these two stars showing that S Dor is aproximately a B4 spectrum, with deep absorption bands of Fe II (Leiterer *et al.* 1985).

The best studied stars are what are now called the B[e] stars (Stahl *et al.* 1985, Zickgraf, *et al.* 1986), also called the "Zoo" sample by Shore and Sanduleak (1984). Many of these stars are known to show only microvariability, but since R 127 has revealed its LBV nature in recent years, there is every reason to suspect that many new LBV candidates are lurking in this class of stars. Several of the B[e] stars have been studied at multiple wavelengths. S 22/LMC (Besammar *et al.* 1983) shows extremely strong Fe II absorption in the UV and one of the strongest Fe II and [Fe II] optical spectra. They argue that the continuum is well fitted by a power law and could arise from an accretion disk. S 134/LMC was discussed by Shore and Sanduleak (1983), who concluded that the star has a dusty envelope, with enhanced 2200Å absorption, and a high mass loss rate; Stahl and colleagues argue that this is a hybrid spectrum. This star's UV spectrum is unique, displaying a strong P Cyg profile on N V and on C IV with possible emission at Si IV. The probable mass loss rate for this star is about $10^{-5} M_{\odot} yr^{-1}$.

Wolf, *et al.* (1987) have performed studies of R 84 and S 61. R 84 = S 91 is a star with a massive red companion; Shore and Sanduleak (1984) obtained a spectral type of B1 or earlier for the blue component. The ultraviolet absorption lines are strong with no obvious P Cyg structure; the Si IV lines show some structure, the lines are separated and the wind slow, in accord with the general behavior of the absorption systems in the galactic and LMC supergiants. C IV shows no obvious P Cyg structure. Possible structure is observed in the Al III 1860 doublet, which may be stronger than the galactic comparison star HD 188209. S 61 displays strong P Cyg absorption on many of the optical profiles but shows no emission lines in the UV. The Al III lines show strong P Cyg structure, even at low S/N. There is a hint of an emission component on the Si IV doublet, which is well resolved. In general, the emission features in the UV are weaker for these two stars than in their likely galactic comparison stars. All of the UV absorption lines are blue - shifted by about 200 km s^{-1}, suggestive of a pseudo - photosphere.

Stahl and Wolf (1987) have studied HD 37836 = S 124/ LMC = R 123. which they have also shown to be surrounded by a circumstellar envelope of high density. This paper also includes a study of S 131/ LMC and HDE 269445. High dispersion data for S 124/ LMC shows strong absorption at C IV and Si IV, showing terminal velocities of about 2500 km s^{-1}; this is very similar to what has been seen in HD 38489 and quite high for most of the S Dor-type variables. For S 131, for which excess dust absorption is seen at 2200Å, the reddening may not be correct; this may have caused an underestimation of the temperature.

The eclipsing system R 81 has been studied by Stahl *et al.* (1987), which is a prime target for UV observations during eclipse ingress and egress. The techniques employed for the eclipsing WR systems would be well applied here.

6. Dust and Shells

One curious result of the LMC survey is that many of the stars show a strong absorption feature at λ2200Å which is not present in neighboring stars in their parent associations, and also not present in the field. In general, *those stars which show strong emission lines in the ultraviolet are also the stars for which the dust feature is enhanced and for which the reddening curve looks "galactic"*. For a number of LBVs, IRAS observations, and groundbased follow-up, have revealed extended nebulae. AG Car and He3-519 (Stahl 1986, Stahl and Wolf 1986, McGregor *et al.* 1988), and S 134/LMC, S 9/LMC and S 12/LMC (Stahl *et al.* 1984). None of these stars have turned up as IRAS sources, suggesting that the dust observed in the UV may not occupy a large volume or may not have sufficient mass to produce a great deal of emission. Several stars for which the dust is not a strong absorber at 2200Å show up as extended shell sources; for these stars, the dust may be more "Magellanic" in nature (Hutchings *et al.* 1987).

Near IR signatures of dust have been found in several B[e] stars (Stahl *et al.* 1984). A number of the stars show evidence for dust shells also show CO, and possibly TiO, emission (see Zickgraf *et al.* 1988, *preprint*). When the dust is formed, and how it is related to the hot wind phase we now see, are open questions.

7. Hubble - Sandage Variables

Humphreys, *et al.* (1984) have obtained IUE and groundbased observations of the Hubble - Sandage variables in M31 and M33. The stars look like the cooler members of the Zoo stars, especially S22/ LMC and S71/ LMC. They are especially notable for the lack of strong absorption lines, no emisison lines, and flat continua in the merged spectra. The UV observations show that these stars are systematically selected for in optical studies when they are heavily blanketed in the UV and thus brightest in the visual. If the photometric variations merely reflect the changes in the bolometric correction, one would expect that the ultraviolet spectrum should be later than the stars selected on the basis of their strong optical emission lines.

8. Some Implications of UV Observations

The presence of LBV-type stars poses in understanding the composite spectra of star-forming galaxies. For instance, in the blue compact dwarf systems, population models have not included the effects of a few massive, bright, highly evolved supergiants on the derived initial mass function (Kunth *et al.* 1985). One B or A-type massive supergiant can strongly skew the IMF toward the middle mass range, and could be responsible for some of the anomalous mass functions that have been suggested for the blue compact dwarf galaxies. For instance, the spectrum of Haro 3 is nearly identical with the 1982 spectrum of R 127 alone, and the inclusion of several LBVs in the spectra of Haro 2 and Mk 59 produce spectra which match very well. One need only recall the experience with R136*a* in 30 Dor to see how difficult the problem of determining the mass function can be when LBV or B[e] stars are involved.

9. Conclusions

Ultraviolet observations have made it possible to better understand the placement of the LBVs and B[e] stars on the HR diagram and to detail the characteristics of their photospheres and winds. They should be even more important once Space Telescope is launched, and we are able to obtain good high resolution spectra of mass loss diagnostic features. This is a field still in its infancy. Continuing ultraviolet monitoring of the Galactic and Magellanic Cloud stars is urgently needed to provide a database for future work on these beasts. Also, model atmospheres are needed, as well as quantitative predictions of high resolution line profiles in order to take advantage of the expanding horizon of UV observational possibilities.

ACKNOWLEDGEMENTS

I wish to thank K. Davidson, R. Humphreys, H. Lamers, E. Fitzpatrick, J. Gallagher, D. Hunter, N. Walborn, O. Stahl, M. Shara, Y. Kondo, B. Bohannan, C. Garmany, P. Conti, F. Viallefond, and D. Kunth for enjoyable discussions, and to apologize to those whose work, for want of space, has not been cited (see also Mead *et al.* 1986). I especially thank my long-time collaborators N. Sanduleak, D. N. Brown, and G. Sonneborn. Some of this work, especially on population synthesis, has been in collaboration with P. Dyer. This work has been supported over the years by NASA and the ENS (Paris).

REFERENCES

Bensammar, S., Friedjung, M., Muratorio, G., & Viotti, R. 1983, *Astron.Astrophys.* **126**, 427.

Brandi, E., Gosset, E., & Swings, J.-P. 1987, *Astron.Astrophys.* **175**, 151.

Carlson, W., & Henize, K. 1979, *Vistas in Astronomy* **23**, 213.

Davidson, K., *et al.* 1982, *Astrophys.J.Letters* **254**, L47.

Davidson, K. 1987, *Astrophys.J.* **317**, 760.

Henize, K., Wray, J.D., & Parsons, S.B. 1981, *Astron.J.* **86**, 1658.

Humphreys, R.M., *et al.* 1984, *Astrophys.J.* **278**, 124.

Hutchings, J.B., Lequeux, J., & Wolf, B. 1987, in *Exploring the Universe with the IUE Satellite*, ed. Y. Kondo (Dordrecht: Reidel), p. 605.

Johnson, H. 1962, *Astrophys.J.Suppl.* **50**, 551.

Kunth, D., Thuan, T.X., Tran Thanh Van, T. (eds.) 1985, *Star-Forming Dwarf Galaxies and Related Objects* (Paris: Editions Frontieres).

Lamers, H.J.G.L.M. *et al.* 1985, *Astron.Astrophys.* **149**, 29.

Leitherer, C. *et al.* 1985, *Astron.Astrophys.* **153**, 16.

McGregor, P.J., Hyland, A.R., & Hillier, D.J. 1988, *Astrophys.J.* **324**, 1071.

McGregor, P.J., Hillier, D.J., & Hyland, A.R. 1988, *Astrophys.J.*, **334**, 639.

McCluskey, G., & Sahade, J. 1987, in *Exploring the Universe with the IUE Satellite*, ed. Y. Kondo (Dordrecht: Reidel), p. 427.

Mead, J.M., Brotzman, L.E., & Kondo, Y. 1986, *IUE Newsletter*, Nr. 30.

Shore, S.N. & Sanduleak, N. 1983, *Astrophys.J.* **273**, 177.

Shore, S.N. & Sanduleak, N. 1984, *Astrophys.J.Suppl.* **55**, 1.

Shore, S.N., Sanduleak, N., & Allen, D.A. 1987, *Astron.Astrophys.* **176**, 59.

Shore, S.N. *et al.* 1988, in *A Decade of UV Astronomy with IUE*, ed. E. Rohlfs (ESA SP-281), p. 417.

Stahl, O. & Leitherer, C. 1987, *Astron.Astrophys.* **177**, 105.

Stahl, O., Leitherer, C., Wolf, B., & Zickgraf, F.-J. 1984, *Astron.Astrophys.* **131**, 307.

Stahl, O. *et al.* 1985, *Astron.Astrophys.Suppl.* **61**, 237.

Stahl, O. & Wolf, B. 1986, *Astron.Astrophys.* **154**, 243.

Stahl, O. & Wolf, B. 1986, *Astron.Astrophys.* **158**, 371.

Stahl, O. & Wolf, B. 1987, *Astron.Astrophys.* **181**, 293.

Stahl, O., Wolf, B., & Zickgraf, F.-J. 1987, *Astron.Astrophys.* **184**, 193.

Walborn, N.R. 1982, *Astrophys.J.* **256**, 452.

Wolf, B., Appenzeller, I., & Cassatella, A. 1980, *Astron.Astrophys.* **88**, 15.

Wolf, B. 1987, in *IAU Symposium 122*, eds. I. Appenzeller & C. Jordan, p. 409.

Wolf, B. & Stahl, O. 1985, *Astron.Astrophys.* **148**, 412.

Wolf, B., Stahl, O., & Seifert, W. 1987, *Astron.Astrophys.* **186**, 561.

Zickgraf, F.-J., Wolf, B., Stahl, O.Leitherer, C., & Klare, G. 1985, *Astron.Astrophys.* **143**, 421.

Zickgraf, F.-J., Wolf, B., Stahl, O., Leitherer, C., & Appenzeller, I. 1986, *Astron. Astrophys.* **163**, 119.

DISCUSSION

Hillier: A comment on HD 326823. This object appears to be different from other stars in Henize's list. It is very hydrogen-deficient and its red and infra-red spectra are very different from the other objects that you mentioned. Also, high-resolution red spectra indicate that the profiles are split and asymmetric, indicating that the envelope is not spherical.

Zickgraf: For R126 a high-velocity component in the wind, with maximum velocity ~ 1800 km/s, is clearly indicated. Line widths in the optical region, on the other hand, show a second wind component with low velocity (several times 10 km/s).

Shore: When I say slow wind, I mean both the optical (~ 200 km/s) and UV (~ 1000 -- 1500 km/s). This is slow compared to S134/LMC (= HD 38489), which is a similar star in physical parameters but which has a wind speed of about 2500 km/s.

Schulte-Ladbeck: Polarimetric monitoring may be able to distinguish observationally between the "pipe" and "torus" models. In the pipe model the puffs that come off asymmetrically should cause variable polarization, while in the torus model the dust ring might not show time-dependent polarization. Polarization would depend on the inclination in the latter case, providing another test for a torus model, *i.e.*, large polarization for edge-on examples and small polarization for pole-on objects.

Shore: Lovely point, and I agree. The torus, however, may be unstable. Models for Newtonian, radiation-dominated tori around active galactic nuclei are generally found to be unstable and tend to form rings and jets; so the polarization results may be more ambiguous than one would have hoped. It is true, however, that polarization gives the most immediate accessible information about the geometry.

Steve Shore

PHOTOMETRIC VARIABILITY OF LUMINOUS BLUE VARIABLES IN QUIESCENT STATE

C. STERKEN (*)
Astrophysical Institute
University of Brussels (VUB)
Pleinlaan 2
1050 Brussels, Belgium

ABSTRACT. This paper reviews the major new facts on microvariability of LBV's of the last few years, and discusses some observational requirements for the future.

1. INTRODUCTION

LBV's are photometrically variable with a large range of amplitudes (hundreths of a magnitude to several magnitudes) and with a vast range of timescales (hours to several decades). The amplitudes of the variations seem to increase with the timescales at which they occur. Lamers (1987) reviewed the three types of variations, viz. the large variations (which are associated with the eruptions, and which are seen in some, but not in all LBV's, and which occur on timescales of centuries), the moderate variations (seen on timescales of decades, and which occur at irregular intervals), and the small-scale photometric variations, also called microvariations, which have been found in all LBV's which have been observed sufficiently accurately, and which are also known to be present in the case of normal supergiants.

2. THE CHARACTER OF THE SMALL-SCALE PHOTOMETRIC VARIATIONS

The variability of Ia supergiants was discovered by Maeder and Rufener (1972). No lightcurves were available at that time, but the semi-periodical variations seen in the spectroscopically studied supergiants seemed to support the picture that there is a linear relation between the characteristic times of the spectroscopic variations and the luminosities of the stars, in the sense that the brightest stars have also the longest semi-periods. These relations depend on the spectral types of the stars.

During the following years, several investigators systematically monitored supergiants (for a review, see de Jager 1980, chapter 8.2). All well-studied supergiants show light fluctuations on timescales

---

(*) Senior Research Associate, N.F.W.O. Belgium

K. Davidson et al. (eds.), Physics of Luminous Blue Variables, 59–66.

of days to years with amplitudes of hundreths of a magnitude to several tenths. The variations are strongest in hypergiants.

These variations are not strictly periodic, and the characteristic period is not always unambiguously visible. Van Gent and Lamers (1985), for instance, reanalysed available data on photometric variations of P Cygni, and concluded that the range of timescales goes from 25 days to 60 days. These timescales are not constant in length.

The variations reported in the 1970's have one common aspect: with a few exceptions, the observations cover in time only a couple of months, and they are all distributed over one or two observing seasons. A more complete picture however emerged from a few cases of long time-baseline photometry of LBV's and some less luminous supergiants. These investigations were mainly the work of van Genderen and his collaborators at the University of Leiden, and by Wolf and his team at the Landessternwarte in Heidelberg.

Van Genderen, during the early seventies, started a systematic high-precision VBLUW photometric study of several very luminous LMC supergiants. Due to this intense observing, long series of homogeneous photometric data are now available for analysis.

In the case of R71 the microvariations are characterised by regular and smooth ups and downs with an amplitude of about 0.1 mag, and with a semi-period of about 23 days for the 1983-1985 period, and of about 14 days for the 1986-1987 period (van Genderen et al. 1985, 1988a). This clearly illustrates that this star, in quiescent state, changes its period of light variation. It would be interesting to know if these cycle length variations have any relation with an eventual outburst which might build up during the next five to seven years. The colours vary too, but the amplitudes of variation are a few orders of magnitude smaller than the light amplitudes. The colour curves are usually in phase with the lightcures, and the light amplitudes tend to increase with decreasing wavelength.

A star similar to R71 for what concerns its variability, is AG Car. Van Genderen et al.(1988a) give an extensive discussion of its variability. The time interval between successive eruptions is of the order of 5 years, and the brightness rises seem to be characterised by a very steep rising branch (about 0.01 mag/day in V), and a descending branch which gradient is about 0.002 mag/day (this is still two times as large as the great descend in light seen in η Carinae between 1856 and 1868!). The microvariations in quiescent state have a characteristic time of about 14 days, and there seem to be large cycle-to-cycle shifts in the mean light level present. This was not the case for R71. Whether this is a signature of activity which is related to the higher gradients of rise and descend as is seen in R71 is not clear. The colour variations of AG Car on the other side are very alike those seen in R71.

Van Genderen and Thé (1987) report new photometry of η Carinae for the years 1983-1986, and show that a new light maximum was present in 1986. This maximum has a similar height as the maximum seen in 1981-1982. The problem of these data is that the time resolution is only sufficient to reveal the occurence of such maxima, but gives little information about the microvariations.

S Doradus received a lot of attention by the Leiden and Heidelberg observers (van Genderen 1979, 1982; Stahl and Wolf 1982, Leitherer et al. 1985). In 1983 Leitherer et al. reported that S Dor reached a maximum brightness of V=9.2, the highest value ever measured. This maximum was followed by a typical fast decline. In 1985 there is evidence for a second maximum, but it is not clear whether this is a maximum, or just a wiggle as was seen in 1965 and 1968. If this is a regular microvariability, it may well have a characteristic time of about 100 days.

Another LBV in the LMC, R127, is now under continuous monitoring since Stahl et al. (1983) discovered that this star had become an S Dor star in 1982. No data are available in quiescent state, but observations obtained in 1986 showed it in outburst (0.001 mag/day).

The highlight in LBV monitoring undoubtedly is the discovery that R81 is an eclipsing binary (Stahl et al. 1987). From data collected by 20 different observers during more than 600 allotted nights of telescope time, they saw that dips in the lightcurve with a depth of 0.4 mag occurred at regular intervals. Appenzeller (1972) had observed a similar brightness decrease in 1972, and his observations were fully compatible with the data used by Stahl et al. The resulting period is 74.59 ± 0.01 days. There is no secondary minimum visible, but there is a pre-eclipse dip at phase 0.8. The lightcurve indicates a contact system with complicated mass motions. Strangely, the irregular intrinsic variations of R81 itself are unusually small for a supergiant.

Van Genderen et al. (1988b) discuss two other interesting cases of microvariations which may have a causal connection with the geometric configuration: HD57060 and HD167971. HD57060 = UWCMa is a binary consisting of an O8 supergiant star and an O or B type main sequence star in synchronous revolution with a period of 4.39 days. According to the authors the system shows abnormally strong microvariations which may be connected with the strong distortion of the star filling its Roche lobe, and the intensity of the mechanism causing the microvariations. HD167971 is an interacting binary consisting of two main-sequence O stars, with extra light from a third O8 supergiant which may or may not be a member of the system. This system was studied in detail by Leitherer et al. (1987) who determined the period P=3.32 days. This system shows (according to van Genderen et al. 1988b) normal microvariations.

3. THE AMPLITUDES AND THE PERIODS OF THE MICROVARIATIONS

Van Genderen (1988) discusses the maximum light amplitude (MLA) of the microvariations of nearly 100 massive stars with spectral types O3-F8. The data are taken from visual lightcurves from the literature and from his own unpublished material. All involved stars are more massive than 15 Mo. He did not use the six massive B[e] stars in the LMC (Zickgraf et al. 1986) because they are fast rotators with peculiar stellar wind properties. Also η Carinae was ommitted for various reasons, as well as the distorted massive stars in close binary systems. Van Genderen sees a clear separation between normal α Cygni variables and the S Dor stars and those suspected as such (with, as exceptions, HR Car and R81): the light amplitudes are systematically much larger for the latter than for the former. There are also two types of oscillation: variations of S Dor stars are smooth, while those for normal α Cyg variables often are more bumpy. From this he concludes that there is a marked difference between the outer layers of S Dor (QS) variables and the other blue super-and hypergiants.

If we plot the positions of the well-observed LBV's in the Mv/Sp diagram of Burki (1978) we see that the positions agree well with the Sp/Mv/P relation given by him. In other words: α Cyg variables and S Dor variables are not separated in the Sp/Mv plane: the periods agree with the lines for calculated periods given by Burki.

How then the amplitudes deviate, and not the periods? One explanation simply can be that the order of magnitude of the semi-period, even in the case of the S Dor variables, can be accurately determined, while the determination of the amplitude is more dependent on the frequency at which observations were taken. Take for example R81, with a period of 2.5 months, which is almost a full observing season. One week of interruption around the time of minimum (due to bad weather, or due to the observing schedule) would be sufficient for an eclipse to go undetected, and to suggest that the star has microvariations with an amplitude of not more than 0.1 mag. This would move the star outside the region indicated in figure 1 of van Genderen (1988). Stars with higher amplitudes could be undetected binaries. Especially interacting binaries, like HD167971 and HD57060 with periods of a couple of days may, when observed at too low frequency, give the impression of microvariability. But the reverse is also true: the gradients of decline during the eclipse of R81 are alike those which are measured at the occasion of ingress or egress from an eruptive phase of an LBV, so a poorly defined lightcurve with suppressed microvariability might suggest an eruptive phase instead of an eclipse. After all, the orbital period of R81 is of the same order of length as the semi-periods present in the light variations of some of the not so well observed Ia supergiants.

4. THE CAUSES OF THE MICROVARIABILITY

The modulation of the lightcurves, and the radial velocity variations, are often interpreted as evidence for oscillations in the photosphere (Leitherer et al. 1985, Wolf 1986, Lamers 1987, van Genderen 1988). These speculations are supported by the presence of a P/L/C relation and the rough constancy of the pulsation constant Q. Van Genderen (1988) believes that the microvariations are caused by an oscillation mechanism which is only confined to the surface layers of super-and hypergiants. But Harmanec (1987) argues whether the existence of a PLC relation necessarily implies the presence of some kind of pulsation. He tested the possibilities that the variations may be caused by rotation of a spotted star, or by orbital motion, and he concludes that both alternative interpretations give rise to a similar PLC relation, but that the spotted star hypothesis is untenable.

The binary hypothesis seems to be as plausible as the possibility of pulsation: Harmanec points out the large uncertainties in deriving the PLC relation, such as the estimate of the semi-period, the mass, effective temperature and radius.

Direct observational support of the binary thesis are the small radial velocity variations (up to 20 km/s), the shallow eclipses, the small orbital light variations due to ellipticity effects, and the cycle-to-cycle variations of light- and velocity curves caused by mass transfer variations.

But there are evenly simple observational arguments which would, in spite of the large scatter in the PLC relation, support the pulsational viewpoint. For example the interesting point of a causal connection between the strength of the microvariations (oscillations) and the binary nature of the star. Van Genderen et al.(1988b) observed that the primary of HD57060 shows abnormally strong microvariations, which are either due to the supergiant, or to the comparison, or to both. This is in apparent contradiction with the case of R81, where the microvariations are abnormally small. It is tempting to look at the interrelation between pulsation and binary nature in the much less luminous pulsating B stars. Fitch (1967) showed that in the ß Cephei star σ Sco the long-period modulation observed in the light variations appears to originate in disturbances of the normal pulsation mode by tidal deformation produced in the outer layers by a faint companion. Waelkens and Rufener (1983) investigated the photometric variability of close binary stars, and they concluded that the presence of a close companion prevents a star in the ß Cephei instability region from pulsating with a large amplitude. This result would support the effect seen in R81, but it contradicts van Genderen's conclusions for HD57060.

Extrapolating the positions of Burki's lines of equal period towards periods of a couple of hours approximately brings us down to the main sequence, where we know that there are several groups of stars which exhibit pulsational variations, such as ß Cephei stars, 53 Per stars, etc. This region corresponds to the locus of very small amplitude variability in van Genderen's MLA/Teff diagram. If pulsation is the cause of the microvariability in LBV's and supergiants, it would suggest that in the PLC or P/M/Teff space there exists a continuous smooth surface which links the characteristic parameters, and which goes down right to the main sequence with shorter and shorter periods of pulsation. Indeed, some of the observed phenomena in pulsating B stars are remarkably similar to some of the properties of microvariations seen in super- and hypergiants, such as

- progressions in radial velocity
- increasing light amplitude with decreasing wavelength
- modulated lightcurves
- in some specific cases the light amplitudes stand out amidst other stars of the same class of variability (e.g. BW Vul)
- there are changes of the period (which, in some cases, are sudden)
- the origin of the pulsations is in the atmospheres
- the amplitudes are variable (Spica, 16 Lac)
- the pulsation mechanisms remain unknown

5. WHICH OBSERVATIONS ARE CRUCIAL?

Almost every paper on LBV's ends with an explicit plea for more observations of a specific kind. But it is obvious that the breakthrough in this research could only occur after an enormous amount of painstaking systematic observations over long periods of time had been collected. It is now clear that this kind of approach must be continued and even extended on a larger scale. Two fundamental subjects must be tackled:

i. A search for new LBVs

One should compile a list of LBV candidates from an inspection of observational HR diagrams and from a search of archive plates and catalogues for photometric anomalies. These candidates should be monitored photometrically in one stable photometric system for several years. The time resolution for the monitoring should not be lower than 2 to 3 days. This monitoring should be supplemented by spectrographic (ground-based and space) backup in order to permit the discrimination between eruptions, pulsational variations, and changes caused by orbital motion in a binary system. One should focus on short-period ellipsoidal and eclipsing binaries, solve the light curves and study the residual microvariations. Special attention should be given to the detection of multiple periods, and to period changes.

ii. Determination of the stellar parameters

Masses must be derived from eclipsing binary studies, and also luminosities and distances. Accurate knowledge of distances is imperative, because this yields clear information on the existence of period-luminosity relations, and the usefulness of these stars as distance indicators. Therefore, systematic research on extragalactic LBVs is the best choice. It is clear that the best place to work is the LMC (and eventually the SMC). The LMC as a laboratory is optimal because of its advantageous declination which allows most favourable seasonal coverage using high-technology instruments which are available at different locations in the southern hemisphere.

The search phase of this project could be accomplished in less than three years, while the monitoring part would need at least ten years of intensive observations.

REFERENCES

Appenzeller, I.: 1972, Publ. Astron. Soc. Japan 24, 483
Burki, G.: 1978, Astron. Astrophys. 65, 357
de Jager, C.: 1980, The Brightest Stars (Reidel Publishing Co)
Fitch, W.S.: 1967, Astrophys. J. 148, 481
Harmanec, P.: 1987, Bull. Astron. Inst. Czechosl. 38, 52
Lamers, H.J.G.L.M.: 1987, in Instabilities in Luminous Early Type Stars, eds. H.J.G.L.M. Lamers, C.W.H. de Loore (Reidel Publ.Co), 99
Leitherer, C. Appenzeller, I., Klare, G., Lamers, H.J.G.L.M., Stahl, O., Waters L.B.F.M., Wolf, B.: 1985, Astron. Astrophys. 153, 168
Leitherer, C., Forbes, D., Gilmore, A.C., Hearnshaw, J., Klare, G., Krautter, J., Mandel, H., Stahl, O., Strupat, W., Wolf, B., Zickgraf, F.-J., Zirbel, E.: 1987, Astron. Astrophys. 185, 127
Maeder, A., Rufener, F.: 1972, Astron. Astrophys. 20, 437
Stahl, O. Wolf, B.: 1982, Astron. Astrophys. 110, 272
Stahl, O., Wolf, B., Klare, G.,Cassatella, A., Krautter, J., Persi, P., Ferrari-Toniolo, M.: 1983, Astron. Astrophys. 127, 49
Stahl, O., Wolf B., Zickgraf, F.-J.: 1987, Astron. Astrophys. 184, 193
van Genderen, A.M.: 1979, Astron. Astrophys. Suppl. Ser. 38, 381
van Genderen, A.M.: 1982, Astron. Astrophys. 112, 61
van Genderen, A.M.: 1988, Astron. Astrophys. In Press

van Genderen, A.M., Steemers, W.J.G., Feldbrugge, P.T.M., Groot, M.,Damen, E., van den Boogaart, A.K.: 1985, Astron. Astrophys. 153, 163
van Genderen, A.M., Thé, P.S.: 1987, in Instabilities in Luminous Early Type Stars, eds. H.J.G.L.M. Lamers and C.W.H. de Loore (Reidel Publishing Co) p 253
van Genderen, A.M., Thé, P.S., Augusteijn, Th., Engelsman, E.C., van der Grift, E.W., Prein, J.J., Remijn, R.L., Steeman, F.W.M., van Weeren, N.: 1988a, Astron. Astrophys. Suppl. Ser., in press
van Genderen, A.M., van Amerongen, S., van der Bij, M.D.P., Damen, E., van Driel, W., Greve, A., van Heerde, G.M., Latour, H.J., Ng, Y.K., Oppe, J., Wiertz, M.J.J.: 1988b, Astron. Astrophys. Suppl. Ser., in press
van Gent, R.H., Lamers, H.J.G.L.M.: 1986, Astron. Astrophys. 158, 335
Waelkens, C., Rufener, F.: 1983, Astron. Astrophys. 121, 45
Wolf, B.: 1986, in Luminous stars and associations in galaxies, eds. C.W.H. de Loore, A.J. Willis and P. Laskarides p 151
Zickgraf, F.-J., Wolf, B., Stahl, O., Leitherer, C., Appenzeller, I.: 1986, Astron. Astrophys. 163, 119

DISCUSSION

Moffat: One could take your comparison between variations in supergiants and in β Cephei stars further by including polarization. This leads to problems if one claims non-radial pulsations (NRP) in both. Some β Cep stars show $\Delta m \approx 0^m_.1$ and $\Delta P \approx 0.01\%$ (Watson 1983, *Astrophs. Sp. Sci.* **92**, 293). Some supergiants show $\Delta m \approx 0^m_.1$ but $\Delta P \approx 0.5\%$ (!) for stars like P Cyg showing micro-variations (see work by Hayes; Lupie & Nordsieck). Even if a different density-temperature combination prevails in each kind of star, one might expect $\Delta m/\Delta P \approx$ constant with NRP. Thus I have difficulty assigning NRP to blue supergiants/LBV's.

Sterken: There are no polarization measurements of LBV's on the micro-variation timescale. Concerning β Cep stars, theoretical considerations make one expect polarization variations but the observational evidence so far is insufficient.

Moffat: Hayes has studied short-timescale polarimetric variations of P Cyg (1985, *Astrophys. J.* **289**, 726).

Friedjung: As far as I am aware the cause of the instability of β Cep stars is not known. Therefore aren't you putting together two badly-known phenomena?

Sterken: I realize that the two groups of stars are very different and that the observed phenomena are poorly known. My talk illustrated the observational similarities, which support the interpretation of pulsation in LBV atmospheres. Whether the excitation mechanism is the same for both groups is another problem, which can be solved only after we know more about the excitation mechanism in β Cep stars.

Maeder: Supergiants having a characteristic pulsation time longer than average in view of the standard *P-L-C* relation should also exhibit significant nitrogen enhancements, provided that the pulsation mode is the same for stars on the blueward and redward tracks.

De Groot: With respect to Maeder's remark, remember that the time-scale of variations that we see in P Cyg is at least twice as long as expected for a normal blue supergiant. On other grounds we already concluded that P Cyg is moving to the left in the H-R diagram (*Astr. Astrophys.* **128**, 299). So here is a clear example.

Gallagher: Could there be a difference in micro-variability characteristics between two kinds of binaries, those where the extended envelope is contained within the system (as in R 81, where eclipses occur) *vs.* those where both components are buried within an extended envelope (common-envelope systems)? How would this affect the interpretation of micro-variability?

Sterken: Micro-variability might be influenced by the binary configuration even if it is caused by pulsation. I do not know what would be seen in the case of a common envelope.

Lamers: Could the smaller micro-variations in binaries be explained by the fact that two stars are in the photometer aperture? If only one star in a pair is variable, the relative amplitude will then be smaller than for the varying star by itself.

Sterken: In R 81 the secondary component does not contribute much to the light (the secondary minimum is not visible). So there is no doubt in that case.

Walborn: UW CMa (= HD 57060) is a peculiar interacting binary with gas streams which may be the source of the variations, rather than the stars themselves.

Sterken: Mass exchange may cause additional variability, as Van Genderen has pointed out.

MODEL ATMOSPHERES AND QUANTITATIVE SPECTROSCOPY OF LUMINOUS BLUE STARS

R.P. KUDRITZKI, A. GABLER,
R. GABLER, H.G. GROTH,
A.W.A. PAULDRACH, J. PULS
Universitäts-Sternwarte München

ABSTRACT. The basic physics of model atmospheres for Luminous Blue Stars (LBS) are discussed. The discussion comprises the model description of both subsonic photospheres and supersonic winds and the application of model calculations for the purpose of quantitative spectroscopy.

1. THE BASIC PHYSICS OF PHOTOSPHERES

1.1. The definition of "photospheres" and "winds"

Since the term "photosphere" has been used several times during this meeting with a different meaning, we want to give our definition at the beginning of this paper. Photospheres are the subsonic layers (wind outflow velocity $v(r) \ll v_{sound}$), which in the case of OB-stars normally emit the optical and UV continuum, the weak lines and the wings of the hydrogen and helium lines. The layers of supersonic outflow we call "winds".

In extreme cases - such as LBVs in outburst - it might occur that photospheres do not exist and that the whole continuum is also formed in the wind. However, in all normal situations of OB-stars including the quiescent phase of LBVs this definition is very useful. It allows us to approximate the density structure of the photospheres of LBS by a simple hydrostatic description which is sufficiently accurate (see Kudritzki, 1988, hereafter Ku 88). In the following we discuss the basic physics of these photospheres.

1.2. Departures from LTE

LBS have an intense radiation field and rather low photospheric density. This means that collisions between electrons and ions are rare compared with radiative transitions. As a consequence NLTE effects dominate throughout the entire atmosphere. Although this has been well known for almost 20 years, it is worthwhile to give a significant exam-

K. Davidson et al. (eds.), Physics of Luminous Blue Variables, 67–82.

ple. Fig. 1 shows the H_γ line profile in LTE and NLTE calculated for T_{eff} = 45000 K and log g = 3.6, parameters typical for an O4f-star. Similar striking differences are found for the HeI and HeII lines. It is thus obvious that LTE completely fails to describe the observed spectra (for a recent discussion, see Ku 88).

Since NLTE effects are expected to become smaller towards higher log g and lower T_{eff}, it is important to investigate where they become negligible with respect to the model atmosphere structure. This is summarized in Fig. 2. The borderline in these figures is the curve where the difference between NLTE and LTE H_γ equivalent width is equal to 15%. In the NLTE domain the differences are larger, whereas they become smaller in the LTE domain. We see that for stars with $M_{ZAMS} \geq 25\ M_\odot$ a NLTE treatment of their photospheres is inevitable. We note, however, that even in the LTE domain, careful NLTE calculations for complex metal ions are needed if one intends to determine accurate abundances from the metal lines (for examples see Ku 88).

1.3. Helium abundance

Almost as important as NLTE effects is the influence of the helium abundance. In particular, for LBS close to the Eddington-limit the observable quantities crucial for diagnostics like hydrogen lines, Balmer jump, (u-b) or (U-B) depend strongly on the helium abundance (see Kudritzki, 1973). In Fig. 3 we show a typical example for the case of A0 Ia supergiants (T_{eff} = 10^4 K, log g = 1): Hydrogen lines and Balmer jump become stronger with increasing helium abundance (i.e. decreasing hydrogen abundance)! Obviously a change from N_{He}/N_H = 0.1 to 1.0 is equivalent to an increase Δ log g≈0.5 for models with normal helim content. If we compare this apparent log g shift at T_{eff} = 10^4 K with the evolutionary tracks it is equivalent to a shift from the 60 $M_\odot$ track down to 25 $M_\odot$ or equivalent to Δ log $L/L_\odot \approx 0.8$. It is clear that this is of enormous relevance for the use of blue supergiants as distance indicators!

The reason for this effect is the influence of N_{He}/N_H on effective gravity and density divided by opacity. A simple estimate for the pressure at optical depth unity yields (for details see Kudritzki, Groth, Humphreys, 1988, Astron. Astrophys., subm.):

$$P(\tau=1) \approx \frac{m_H}{\sigma_E}\ g\ (1-\Gamma_o + 4\ N_{He}/N_H),$$

where $\Gamma_o = \dfrac{7.52}{g} \left(\dfrac{T_{eff}}{10^4}\right)^4$ (g in cm s^{-2}, T_{eff} in K).

Obviously, the helium abundance is as important as gravity for the atmospheric pressure structure for models close to the Eddington limit .

This effect might be the solution of a longstanding problem (Fehrenbach and Duflot, 1972, Humphreys, 1983, Dubois, 1986, Azzopardi, 1981). There exists a large group of B8 to A5 supergiants in the LMC as well as in the SMC with magnitudes that clearly point to luminosity class Ia. However, H_γ and H_δ are much too strong and U-B much too red (0.2 to 0.3 mag) and point to luminosity class II. This leads to a situation in which the B,A supergiants although brightest in M_v are obviously not useful as distance indicators. However, evolutionary tracks show that enhanced helium abundance in the photosphere of supergiants is to be expected (see also the following section), so that our effect can easily explain this observed discrepancy in principle. Of course, this has to be proven by detailed quantitative spectroscopy of A supergiants. Such a project is presently under way (Kudritzki, Groth, Humphreys). If successful, it will provide the means to use these optically brightest objects as distance indicators once more.

2. QUANTITATIVE SPECTROSCOPY OF LBS

2.1. Analysis method

The principle of the analysis is to use the profile fit of the hydrogen lines H_γ, H_δ for the determination of gravity and the profiles of HeI and HeII lines for the simultaneous determination of T_{eff} and helium abundance (for details see Ku 88 and references therein). Alternatively, for B-supergiants the use of other ionization equilibria (Si IV/III/II) or the energy distribution is necessary for T_{eff}, if the objects are too cool to show HeII lines. Note, however that here also log g has to be determined simultaneously, since the gravity influences the shape of the energy distribution significantly (Kudritzki et al., 1983, Fitzpatrick, 1987). Note also that the energy distribution method fails for O-stars (Hummer et al., 1988).

An important remark is necessary for all those objects, that have not yet been analysed quantitatively. Normally, an effective temperature is assigned in these cases according to their spectral type, which, following Conti and Alschuler (1971) and Conti and Frost (1977), is defined by log W': = log W_λ(HeI 4471) - log W_λ (HeII 4542). The dependency of log W' on both T_{eff} and log g is usually neglected. Fig. 4 shows that this simultaneous T_{eff} and log g dependence is important, in particular close to the Eddington-limit. This means that without a simultaneous gravity determination effective temperatures are extremely uncertain.

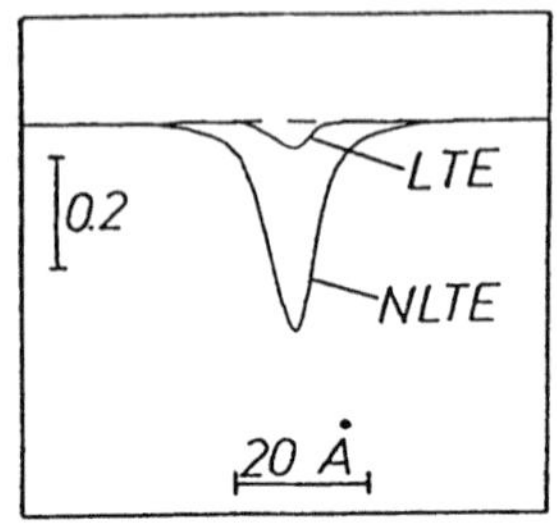

Fig. 1: Photospheric profile of H_γ calculated in LTE and NLTE for a typical O4f-star.

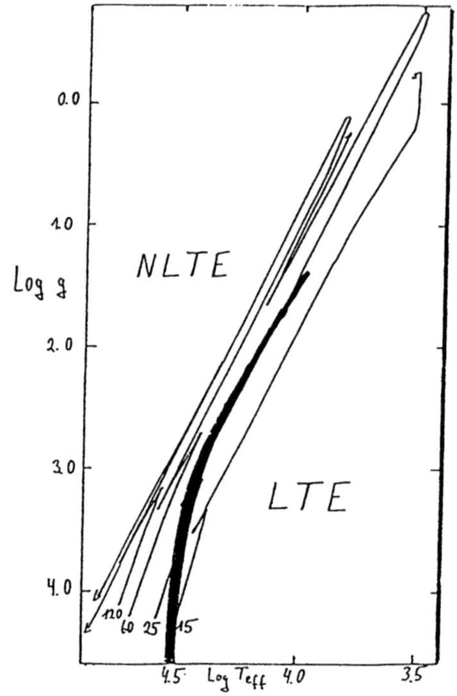

Fig. 2b: NLTE and LTE domain in the (log g, log T_{eff})-plane.

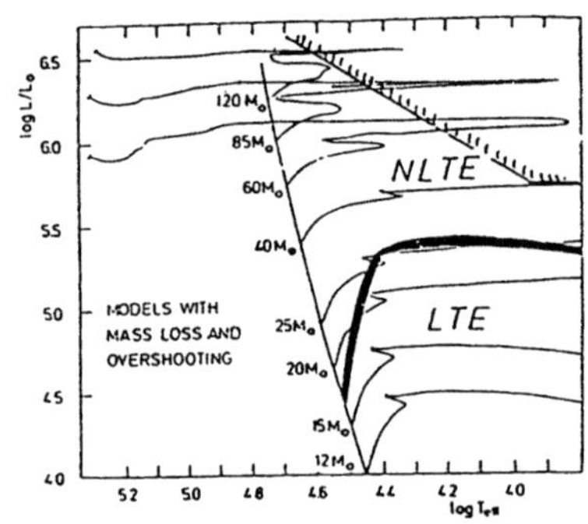

Fig. 2a: NLTE and LTE domain in the HR-diagram compared with tracks by MM (Maeder and Meynet, 1987).

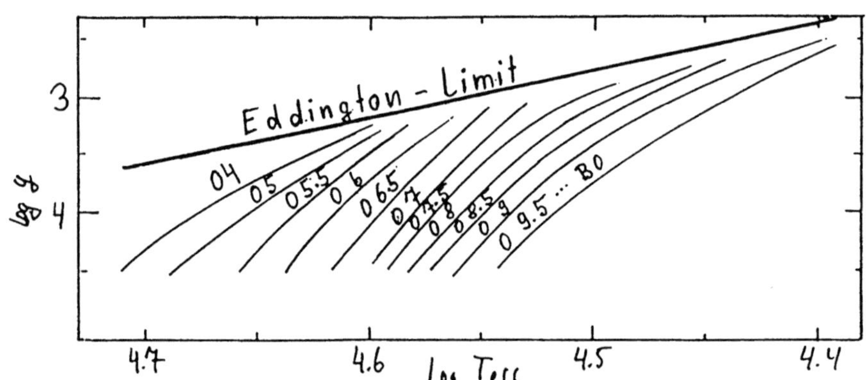

Fig. 4: The domain of the O-star spectral types in the (log g, log T_{eff})-diagram. The spectral types are defined by the value of log W' = log W_λ (HeI 4471)-log W_λ (HeII 4542) (see Conti and Frost, 1977). Note that the spectral type depends on both T_{eff} <u>and</u> log g.

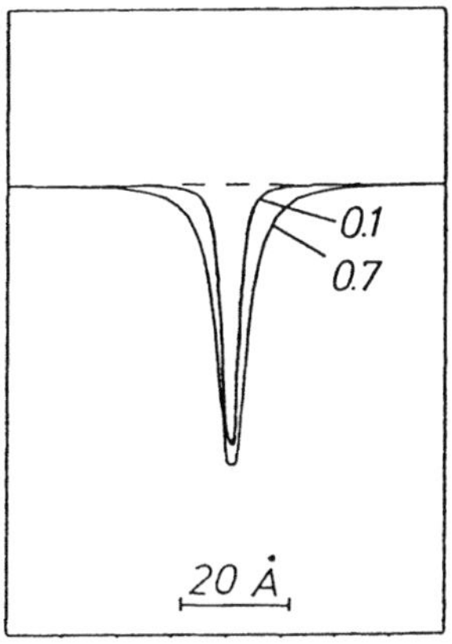

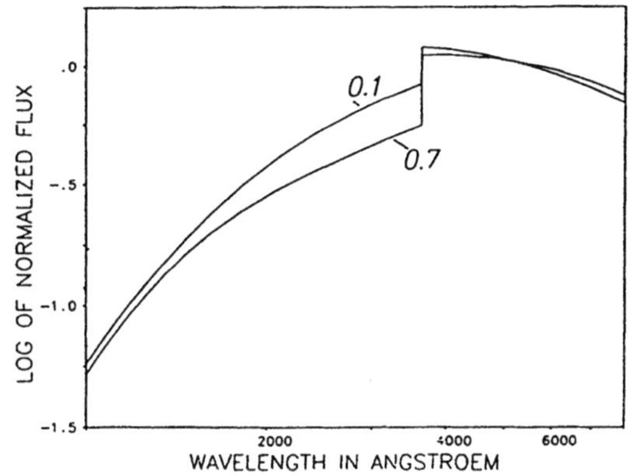

Fig. 3a: H_γ profiles (left) and energy distribution (right) of a typical AO Ia supergiant photospheric model with two different values of N_{He}/N_H.

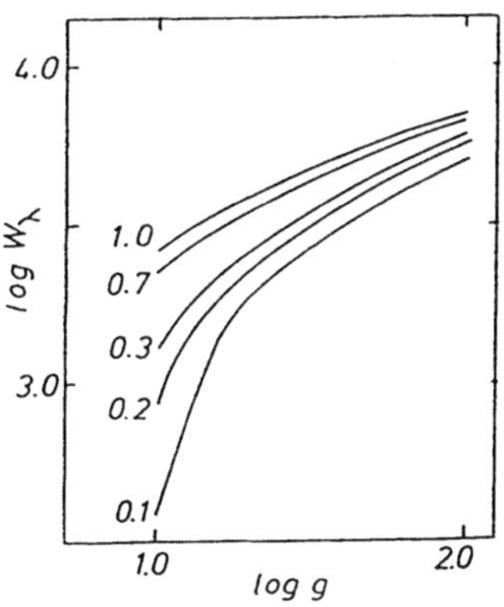

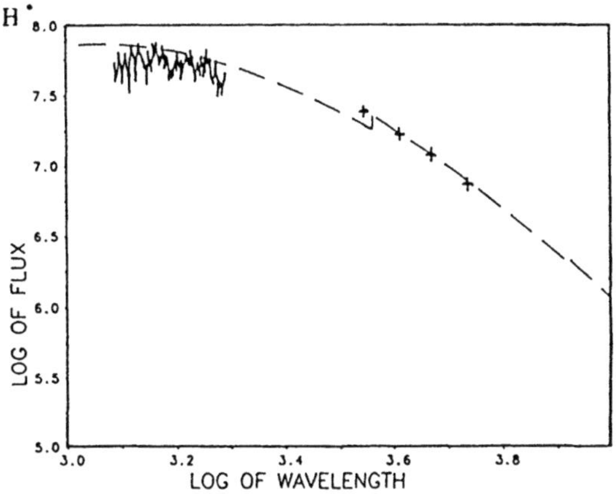

Fig. 3b (upper left): log of H_γ equivalent width for models at T_{eff} = 10^4K and different log g and N_{He}/N_H. (3a and 3b from Kudritzki, Groth, Humphreys, in prep.).

Fig. 5a(upper right): The observed energy distribution of R 71 compared with an unblanketed NLTE model of T_{eff} = 14000 K, log g = 1.6, y = 0.09.

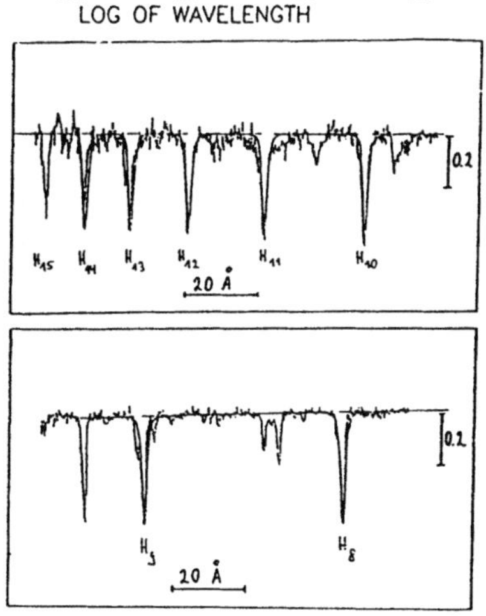

Fig. 5b (right): Fit of the Balmer lines of R 71 with the final hydrostatic NLTE model.

An additional important effect is the "wind-blanketing" (see Abbott and Hummer, 1985). As a consequence, the spectral types show an additional dependence on the stellar mass-loss rate. For the sake of brevity, we omit a fuller discussion at this point.

2.2. O-stars in Galaxy, LMC, SMC

The application of the method of detailed photospheric NLTE analysis on O-stars has been reviewed very recently by Ku 88. We therefore summarize only the most important issues: Very massive stars with $M \geq 100\ M_{\odot}$ exist not only in the Galaxy (Kudritzki, 1980, Simon et al., 1983) but also in the LMC (Gehren et al., 1986) and even in the SMC (Kudritzki et al., 1988a). For supergiants like ζ Puppis (O4 f), α Cam (O9.5 Ia) in the Galaxy or Sk 80 (O7 Iaf) in the SMC photospheric helium enrichment is very common (Kudritzki et al., 1983, Bohannan et al., 1986, Voels et al. 1988, Kudritzki et al., 1988b). This means that the effects of mixing are stronger than normally predicted by evolutionary calculations.

2.3. B-supergiants in Galaxy, LMC and SMC

As part of an extensive spectroscopic project on the formation and evolution of massive stars in parent galaxies of different metallicity, B-supergiants in Galaxy, LMC and SMC have been studied (R.P.Kudritzki, H.G. Groth, K. Butler, D. Husfeld, S. Becker, F. Eber, E. Fitzpatrick). The results obtained so far are summarized in Table 1:

Table 1a: the objects

star	spectral type	V	B-V	galaxy
Sk 21-65°	B0 Ia	12.02	-0.14	LMC
Sk 41-68°	B0.5Ia	12.01	-0.14	LMC
HDE 268685	B1 Ia	11.52	-0.10	LMC
Sk 213-69°	B1 Ia	11.97	-0.10	LMC
HDE 269504	B2 Ia	11.95	-0.09	LMC
Sk 159	B0.5Ia	11.88	-0.13	SMC
Sk 119	B2 Ia	12.19	-0.09	SMC
ε Ori	B0.Ia	1.70	-0.19	GALAXY
κ Ori	B0.5Ia	2.04	-0.18	GALAXY

Table 1b: stellar parameters

	T_{eff} K	log g cgs	y	$R/R_\odot$	log $L/L_\odot$	$M_{ZAMS}/M_\odot$	$M_{grav}/M_\odot$
Sk 21-65°	27000	2.70	0.35	30	5.63	27	16
Sk 41-68°	25000	2.60	0.23	32	5.55	22	15
HDE268685	22000	2.35	0.23	45	5.63	27	16
Sk213-69°	22000	2.40	0.15	48	5.68	30	21
HDE269504	20000	2.25	0.23	40	5.36	14	10
Sk 159	25000	2.55	0.35	47	5.89	47	29
Sk 119	20000	2.30	0.15	51	5.57	24	19
ε Ori	26000	2.75	0.20	29	5.54	22	17
κ Ori	25000	2.70	0.20	28	5.44	18	14

Note that $y = N_{He}/C\ (N_H + N_{He})$.

M_{grav} is obtained from the gravity, the radius from distance and observed dereddened magnitude (for further details see Kudritzki et al., 1987). The typical uncertainties are $\Delta T_{eff} = \pm 1000$ K, $\Delta \log g = \pm 0.1$, $\Delta y = \pm 0.07$, $\Delta R/R_\odot = \pm 10\%$, $\Delta \log L/L_\odot = \pm 0.1$, $\Delta M_{grav}/M_\odot = \pm 30\%$.

The most striking result is that <u>all</u> objects are helium enriched with a large scatter in y. This again confirms observationally that mixing must be more effective than commonly thought. (For metal abundances see Kudritzki et al., 1987).

2.4. Quantitative spectroscopy of the "subluminous" LBV R 71 at minimum

Up to now LBVs have never been the subject of detailed quantitative photospheric NLTE spectroscopy. The major reason for this is that it is hard to find appropriate "photospheric" lines that are undisturbed by the emission or absorption of the surrounding wind. However, a good candidate is R 71, which has been studied by Wolf et al. (1981) during minimum on the basis of IIaO Coudé plates and IUE spectra. Their estimates for the properties of R 71 in minimum stage were: $\log L/L_\odot = 5.3$, $R/R_\odot = 81$, $T_{eff} = 13600$ K, $v_\infty = 130$ km s^{-1}, $\dot{M} = 3 \times 10^{-7}\ M_\odot$/yr. With this luminosity the object is about 0.6 dex below the Humphreys-Davidson limit.

Stahl, Wolf and Zickgraf have recently secured several ESO 3.6m CASPEC CCD spectra of high resolution and high signal to noise, which extend almost up to the Balmer edge and which were taken during different minimum stages. Since the higher Balmer lines in this phase are obviously undisturbed by wind effects, these spectra make it possible to determine the gravity of R 71 directly and thus to impose an important additional constraint on its evolutionary status. Our analysis technique was therefore applied (H.G. Groth, R.P. Kudritzki) to R 71 in the usual way:

The effective temperature is found from the dereddened (E_{B-V} = 0.1 to 0.15) energy distribution using photospheric NLTE models (Fig. 5a). We obtain T_{eff} = 14000 K (Note that with a blanketed LTE model we would get 13500 K). Log g is determined from all the Balmer lines from H_{15} down to H_β. Simultaneously the helium abundance is obtained from the weakest HeI absorption lines. The final result (see Fig. 5b) is log g = 1.5, y = 0.30. Thus R 71 is a further example of an evolved massive star with a helium enriched atmosphere! Placing the object in both the HR and the (log g, log T_{eff})-diagram we get the same answer: R 71 is most probably an object with a ZAMS mass of about 25 $M_\odot$ on its way back from the RGB. Its present mass of 8 to 10 $M_\odot$ (log g 1.5 to 1.6) confirms this conclusion.

We will continue with this project and determine metal abundances in future work.

3. WINDS OF LUMINOUS BLUE STARS

3.1. Observed wind properties and the recent progress of the theory

As has been proved convincingly by Lucy and Solomon (1970) and Abbott (1979), the winds of stars more massive than 15 $M_\odot$ are basically radiation driven (for a recent discussion see Ku 88). Castor, Abbott and Klein (1975, "CAK") were the first to formulate the theory of line driven winds in a self-consistent way. In its later version (Abbott, 1982) the theory used a realistic line list of 250000 lines of H to Zn in ionization stages I to VI. The predictions of the theory were:

$$v_\infty = (\alpha/(1-\alpha))^{1/2} \; v_{esc}$$

$$\dot{M} \sim L^{1/\alpha} \; (M(1-L/L_E))^{1-1/\alpha}$$

The parameter α comes out as a result of the theory and lies between 0.5 and 0.7. L_E is the Eddington luminosity. As is well known (see Ku 88), these predicted proportionalities are roughly reproduced by the observations. However, the old CAK theory fails with respect to the proportionality constants. Predicted mass-loss rates are too high by a factor of three and theoretical terminal velocities too low by a large factor. Since v_∞ can be measured with high precision, the latter discrepancy cast enormous doubts on the theory.

However, Pauldrach et al. (1986) were able to prove that this failure was caused by the "radial streaming approximation" made by CAK. Relaxing this approximation they showed that radiation driven wind theory can reproduce the observed $\dot{M}$ and v_∞. In a next step, Kudritzki et al. (1987) applied the theory to low metallicity O-stars as are to be found in the SMC and LMC and reproduced the lower v_∞ observed for these objects by Garmany and Conti (1985).

The most important improvement concerns the calculation of the occupation numbers in the stellar wind plasma. For correct radiation hydrodynamics and spectral diagnostics accurate calculations for all ions driving the wind are inevitable. For a long time occupation numbers in the wind were computed in a very approximate way. Recently the first realistic calculations became available. Pauldrach (1987) adopted a cool wind ($T_W \approx T_{eff}$) and treated the full multi-level NLTE problem of all relevant elements and ions including electron collisions self-consistently with the radiation driven wind hydrodynamics. He includes 26 elements, 133 ionization stages, 4000 levels and 10000 radiative bound-bound transitions in the rate quations and the correct continuous radiation field for the bound-free rates calculated from the spherical transfer equation. Puls (1987) extended these improvements significantly by including "multi-line effects". These arise from the velocity induced line overlap, which causes radiative coupling between different ions and possible multiple momentum transfer from photons to the wind plasma. The first application of this improved theory to the case of the O4f-star ζ Puppis shows excellent agreement with respect to the dynamical quantities $\dot{M}$ and v_∞. In addition - and this is the really important result - an enormous shift towards higher ionization stages is obtained due to the detailed NLTE treatment. Thus, it was possible to reproduce for the first time the observed high ionization features of OVI, NV, CIV etc. by cool wind models. A typical example is given in Fig. 6, which shows how perfectly the observed NV profile in the UV-spectrum of ζ Puppis can be reproduced by the self-consistent radiation driven NLTE wind model atmospheres. Note that this is not a profile fit, where $\dot{M}$ and v(r) have been adjusted as parameters. It is the result of a self-consistent calculation, that depends only on the choice of the stellar parameters log $L/L_\odot$, T_{eff} and log g.

We have now applied calculations of this type along evolutionary tracks for massive stars to investigate whether our wind models also reproduce the change of spectral morphology during stellar evolution as observed by Walborn and collaborators (Walborn and Panek, 1984a,b, 1985; Walborn and Nichols-Bohlin, 1987; Walborn et al., 1985). Fig. 7 demonstrates the extreme behaviour of the SiIV resonance lines, which exhibit a pronounced luminosity effect both in theory and observation. The physical reason for this effect, which will prove to be of enormous potential for the lumi-

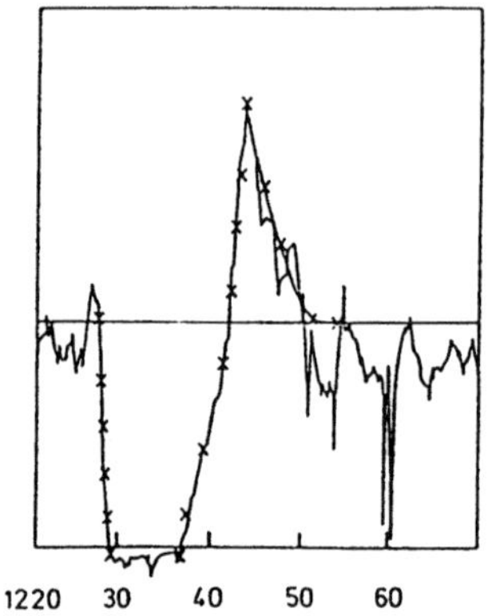

Fig. 6: The IUE high resolution profile of NV λ 1240 of ζ Puppis compared with the result from a self-consistent radiation driven wind model (crosses) (from Puls,1987).

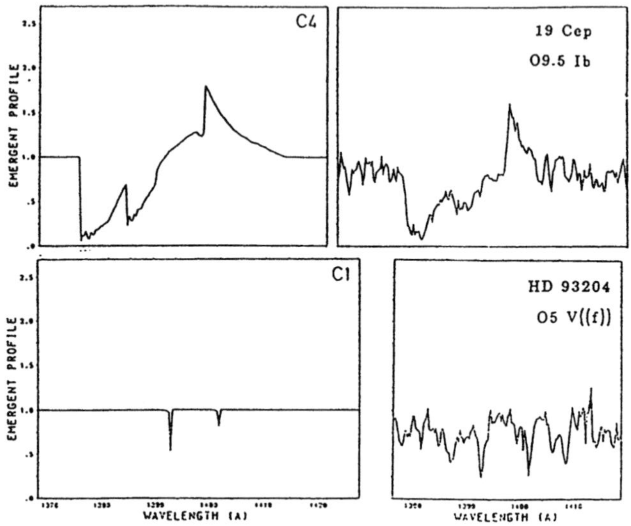

Fig. 7: The strong observed luminosity effect of SiIV as reproduced by the wind models. The observations are taken from Walborn et al. (1985). (For discussion see text).

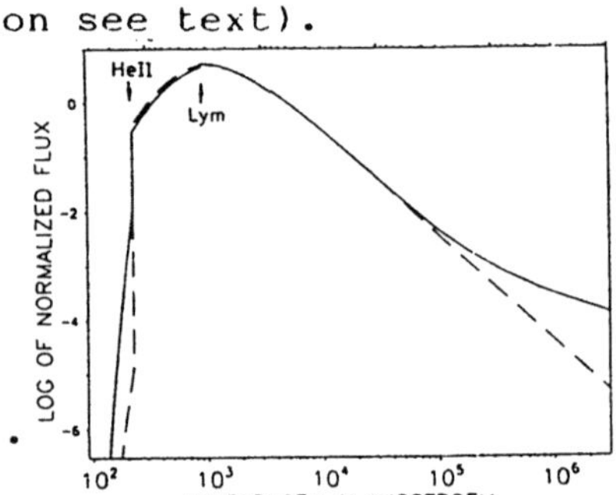

Fig. 8a: Energy distribution of model A (fully drawn) compared with a hydrostatic model (dashed).

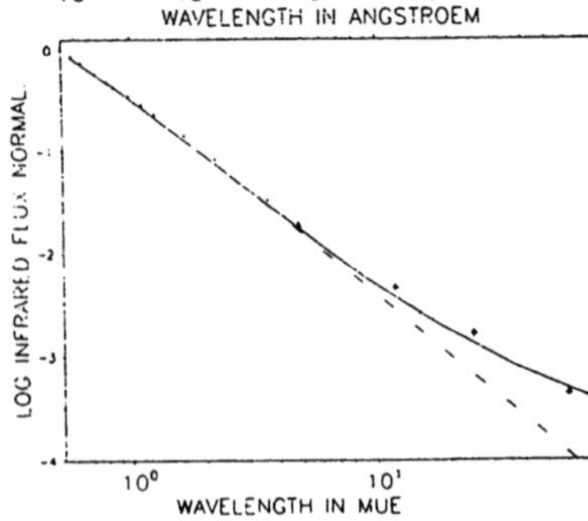

Fig. 8b: Enlarged optical and IR part of Fig. 7a compared with the observations for the O4f star ζ Puppis.

nosity classification of extragalactic O-stars using HST, is that in the winds of O-stars most of the silicon is SiV. Supergiants lie close to the Eddington limit and thus have dense winds, which lead in turn to strong recombination towards SiIV. Consequently, the SiIV wind features show up in the supergiants. The behaviour of NV, NIV, CIV can also be reproduced by the models, but this is not shown here for sake of brevity. We thus conclude that the status of the theory is very promising with respect to normal OB-stars. The question, however, is how does it work in the case of LBVs.

3.2. Wind models for LBVs

We have applied our improved radiation driven wind code to two typical LBV-stars in quiescence, R 71 and P-Cygni. We start the discussion with R 71:

As stellar parameters for R 71 in quiescence we have chosen (see section 2.4.) $\log L/L_\odot = 5.3$, $\log g = 1.6$, $T_{eff} = 14000$ K. The radiation driven wind hydrodynamics (including the self-consistent NLTE treatment of the radiative line force yields $\dot{M} = 3.3 \times 10^{-7}\ M_\odot/yr$ and $v_\infty = 225$ km/s for these parameters _. This compares well with the observed values of $3 \times 10^{-7}\ M_\odot/yr$ and 127 km/s, respectively (see 2.4.). This means that R 71 in quiescence is obviously no problem for the theory.

P-Cygni, on the other hand, is far more complicated as it lies much closer to the Eddington-limit. This means that its wind properties depend strongly on the ratio $L/L_E \sim L/M$. The detailed and extensive calculations presented by Puls et al. and Pauldrach et al. (both contributed papers in these proceedings) demonstrate, however, that even in this extreme case the theory works well and reproduces $\dot{M}$, v_∞ as well as the very gradual slope of $\rho(r)$ and $v(r)$, inferred from the far IR-energy distribution. Moreover, the detection of the bistability of the wind of P-Cygni due to the sudden change of optical thickness in the Lyman-continuum might provide an interesting tool for the understanding of the variability of LBV-winds!

4. UNIFIED MODEL ATMOSPHERES INCLUDING WINDS AND SPHERICAL EXTENSION

4.1. Motivation

Hydrostatic photospheric models neglect the emission of the surrounding stellar wind envelope. They are therefore unable to produce the observed wind induced strong emission features in the optical spectra of hot stars (H_α, HeII 4686, IR-emission) or the distortion of photospheric absorption profiles (H_β, HeI 5876, HeII 5412).

The normal procedure to account for these effects is to put a wind on top of a hydrostatic photosphere and then to study its influence on the spectrum. This method has crucial limitations as it makes an artificial division between photospheres and winds and introduces unnecessary free parameters which strongly influence the calculations at the borderline between the two regions (for a discussion see Ku 88 or Gabler et al., 1988). The alternative is the concept of "unified model atmospheres" introduced recently by Gabler et al., 1988. This new model atmosphere code takes $\dot{M}$, v(r), ρ(r) from the hydrodynamics of radiation driven winds (see section 2.) and calculates T(r), n_i (H), n_i (He) by a detailed NLTE treatment using radiative equilibrium and correct spherical radiative transfer. These models need only T_{eff}, log g, R_* at the inner atmospheric boundary as the natural free parameters and are in principle sel-fconsistent. They yield the entire sub- and supersonic atmospheric structure, the stellar energy distribution and the hydrogen and helium line spectra. As has been shown by Gabler et al., the surrounding wind can have significant influence on the emergent spectrum even in cases where the wind density is not extremely strong. In the following we restrict the discussion to models for luminous O-stars.

4.2. Unified models for luminous O-stars

Four models have been calculated here with the following parameters:

model	T_{eff} (10^3 K)	log g (cgs)	log $L/L_\odot$	spectral type
A	45	3.6	6.0	O4f supergiant
B	32	3.2	5.8	O9.5 giant
C	32	2.9	5.8	O9 supergiant
D	32	2.9	6.1	O9 supergiant

Note that models B and C have the same luminosity, but different gravity, whereas the opposite is true for models C and D.

For the sake of brevity we discuss only a few effects: Fig. 8 demonstrates how the energy distribution (in particular the EUV and the FIR) is affected by the wind envelope. Fig. 9 shows that for a low temperature O-giant the wind has almost no influence on the line spectrum. However, for a supergiant of the same temperature even H_γ is influenced by the surrounding wind and sphericity (Fig. 9). The corresponding H_α emission is clearly luminosity dependent (emission of model D larger than of model C). This indicates that the unified models and H_α line emission may provide an independent tool for distance determinations! For future extragalactic work this will be an important path to follow. We expect the method of unified model atmospheres to become very important for all objects close to the Eddington-limit.

ACKNOWLEDGEMENTS: This work was supported by the Deutsche Forschungsgemeinschaft under grants Ku 474-11/3 and 474-13/2.

References

Abbott, D.C.: 1979, IAU Symp. 83, p. 237
Abbott, D.C.: 1982, Astrophys. J. 259, 893
Azzopardi M.: 1981, ESO Workshop on "The Most Massive Stars", 227
Bohannan, B., Abbott, D.C., Voels, S.A., Hummer, D.G.: 1986, Astrophys. J. 308, 728
Conti, P.S., Alschuler, W.R.: 1971, Astrophys. J. 170, 325
Conti, P.S., Frost, S.A.: 1977, Astrophys. J. 212, 728
Dubois, P.: 1986, Proc. IAU Symp. 116, 415
Fehrenbach, Ch., Duflot, M.: 1972, Astron. Astrophys. 21, 321
Fitzpatrick, E.L.: 1987, Astrophys. J.312, 596

Gabler, R., Gabler, A., Kudritzki, R.P., Puls, J., Pauldrach, A.: 1988, Astron. Astrophys., in press
Garmany, C.D., Conti, P.S.: 1985, Astron. J. 293, 407
Gehren, T., Husfeld, D., Kudritzki, R.P., Conti, P.S., Hummer, D.G.: 1986, Proc. IAU Symp. 116, p. 413
Hummer, D.G., Abbott, D.C., Voels, S.A., Bohannan, B.: 1988, Astrophys. J. 328, 704
Humphreys, R.M.: 1983, Astrophys. J. 265, 176
Kudritzki, R.P.: 1973, Astron. Astrophys. 28, 103
Kudritzki, R.P.: 1980, Astron. Astrophys. 85, 174
Kudritzki, R.P., Simon, K.P., Hamann, W.R.: 1983, Astron. Astrophys. 118, 245
Kudritzki, R.P., Pauldrach, A., Puls, J.: 1987, Astron. Astrophys. 173, 293
Kudritzki, R.P., Groth, H.G., Butler, K., Husfeld, D., Becker, S. Eber, F., Fitzpatrick, E.: 1987, Proc. ESO Workshop on "SN 1987a", ed. I.J. Danziger, p. 37
Kudritzki, R.P., Groth, H.G., Husfeld, D., Hummer, D.G., Conti, P.S.: 1988b, Astron. Astrophys., in prep.
Kudritzki, R.P., Cabanne, M.L., Husfeld, D., Niemela, V.S., Groth, H.G., Puls, J., Herrero, A.: 1988a, Astron. Astrophys., submitted
Kudritzki, R.P.: 1988, "The Atmospheres of Hot Stars: Modern Theory and Observation", 18th Advanced Course, Swiss Society of Astronomy and Astrophysics, Leysin, March 1988
Lucy, L.B., Solomon, P.: 1970, Astrophys. J. 159, 879

Maeder, M., Meynet, G.: 1987, Astron. Astrophys. 182, 243
Pauldrach, A., Puls, J., Kudritzki, R.P.: 1986, Astronom. Astrophys. 164, 86
Pauldrach, A.: 1987, Astron. Astrophys. 183, 295
Puls, J.: 1987, Astron. Astrophys. 184, 227
Simon, K.P., Jonas, G., Kudritzki, R.P., Rahe, J.: 1983, Astron. Astrophys. 125, 34
Voels, S.A., Bohannan, B., Abbott, D.C., Hummer, D.G.: 1988, Astrophys. J., in press
Walborn, N.R., Panek, R.J.: 1984, Astrophys. J 280, L27
Walborn, N.R., Panek, R.J.: 1984, Astrophys. J 286, 718
Walborn, N.R., Panek, R.J.: 1985, Astrophys. J 291, 806
Walborn, N.R., Nichols-Bohlin, J., Panek, R.J.: 1985, "IUE Atlas of O-Type Spectra from 1200 to 1900 A", NASA Reference Publication 1155
Wolf, B., Appenzeller, I., Stahl, O.: 1981, Astron. Astrophys. 103, 94

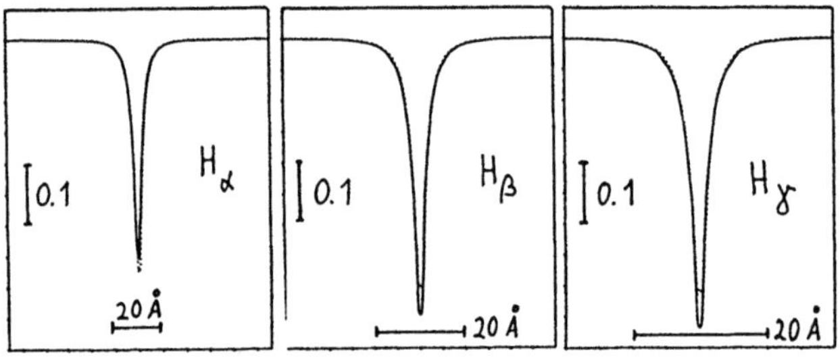

Fig. 9: The Balmer lines of model B (dots: hydrostatic, fully drawn: unified model).

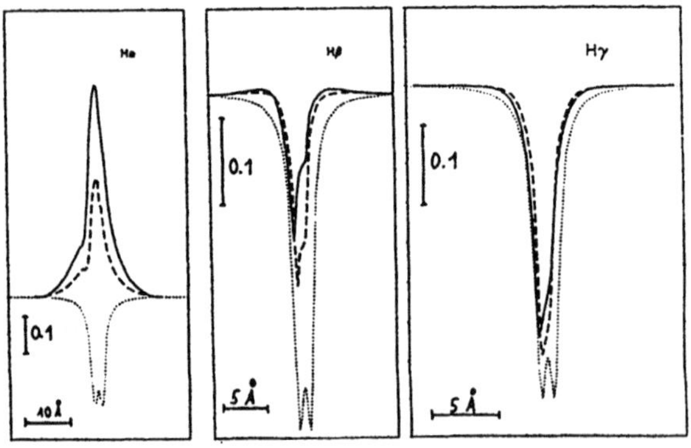

Fig. 10: The Balmer lines of model C (dashed) and model D (fully drawn) compared with the hydrostatic model (dotted).

DISCUSSION

Maeder: Your beautiful results represent an essential constraint for understanding stellar evolution. Surface helium and nitrogen enrichments in models are usually produced by a number of physical effects: convective dredge-up, mass loss, overshooting, intermediate convective zone, *etc.* These effects lead to considerable He and N enrichments for stars in some parts of the H-R diagram (*cf. Astr.& Astrophys.* **173**, 287): very massive O stars, red supergiants, blue supergiants, WR stars. The main question now is: do the observations and evolutionary predictions correspond? If yes, OK; if no, this means that new mixing mechanisms have to be included to reproduce the richness of Nature. In the case of O stars, we already know that in order to account for the data we must assume that a fraction of the O stars evolve close to homogeneity. Shall we have to assume something similar for B supergiants?

Kudritzki: Well, all I can say now is that it looks like this.

Davidson: It seems to me that the most urgent goal for **LBV** studies now is to explain and quantify their eruptive instability; no one can safely claim to understand the upper H-R diagram until this has been done. From this point of view, the most important thing in your talk is the "bistability" that arises in your group's analysis of P Cygni. It would be good to know more about the mutual similarities and differences among the phenomenon that Andre Maeder proposed, Immo Appenzeller's bimodal ideas that he will probably discuss, and the bistability that you have found. But for now, here's an assertion that may seem cryptic but which will probably be specified better sometime later at this meeting (I will do it myself if nobody else does): If we simply take a fixed value of the radiation-force gallows Γ_{crit} defined in a particular way, and suppose that your bistability or something very much like it occurs near that value, then we can reproduce the observed Upper Limit in the H-R diagram. This is a quantitative conjecture, based on some idealized figuring.

Walborn: Janet Drew finds that when the wind temperature is derived from the model rather than adopted or similar to T_{eff}, it is significantly lower and the super-ionization is not reproduced. Do you expect your new united models to resolve this discrepancy?

Kudritzki: Janet's calculations are in radiative equilibrium and include the important effect of cooling by metal lines. However, two crucial simplications are still present in her calculations: an approximate treatment of optically thick continua, and ionization from excited levels. The situation with respect to radiative equilbrium is not clear until the models are improved. In our unified models, which at present treat only hydrogen and helium, the next step to be done by Rudi Gabler will be the correct inclusion of the most important metals, avoiding these simplifications. A check on the radiative equilibrium will then be possible. If radiative equilibrium is not fulfilled, as I expect for the outer wind layers, then this will be a direct proof of the importance of dissipative processes as calculated by Stan Owocki and his collaborators. The only way to investigate this is to do first the full NLTE calculations, including radiative equilibrium correctly. Our present full-NLTE wind calculations for adopted $T_{wind} \sim T_{eff}$, which nicely fit the observed line profiles, can then be used as constraints for the observed temperatures in the wind.

Cassinelli: X-rays can have a major effect on the super-ionization stage. There is evidence for a significant X-ray intensity in ζ Puppis, for instance. Given that, it is surprising to hear you say that the observed super-ionization can be explained without accounting for X-rays. Were you able to produce the strong O VI λ1040 line seen in ζ Pup with your model?

Kudritzki: There is no doubt that the X-rays are there. Whether they affect the ionization equilibrium depends on assumptions about the layers in the wind where they originate. Pauldrach has discussed this (1987; *Astr.&Astrophys.* **183**, 295), finding that if the observed X-ray flux originates from the full wind volume, then the mean X-ray ionization rate is small compared to ionization caused by optically thick radiation shortward of 228 Å (the He II edge), which is treated correctly in our calculations. However, if we assume that the X-rays come from the outer wind layers only, *i.e.* where $v(r) \geq 0.8\,v_\infty$ (which is supported by observations), then they are very important for ionization in these layers. This is a nice coincidence, since Pauldrach can reproduce observable O VI only up to $0.8\,v_\infty$ in the case of ζ Puppis. This means that we have an indication that the correct NLTE treatment is sufficient in the whole wind region, but that in the outermost layers X-ray Auger ionization has to be included in the rate equations. This can be done easily in our code, provided we have the local mean X-ray intensity in the wind. To get this in a more self-consistent way, one has to follow and to extend the impressive work by Owocki/Castor/Rybicki, who by elaborate time-dependent hydro-calculations investigate wind instabilities and the corresponding evolution of dissipative shocks. We will follow this line.

Humphreys: In your talk you mentioned an initial mass around $25\,M_\odot$ for R 71. What is its current "spectroscopic" mass? You also mentioned $23\,M_\odot$ from your spectroscopic analysis of P Cygni. These masses are not inconsistent with evolutionary models, especially when we consider that LBV's are in a rapid-mass-loss phase.

Kudritzki: I agree, $23\,M_\odot$ for P Cyg fits perfectly with Maeder/Meynet at $log\ L/L_\odot \sim 5.8$ and $T_{eff} \sim 19000$ K. I did the same exercise with R 71 and it agrees also, $\sim 10\,M_\odot$ indicated by the $log\ g$, $log\ T_{eff}$ diagram. So we have no disagreement for these stars but we have the "normal" B0 Ia, B1 Ia supergiants; all the stars analyzed by us so far show the enrichment, which disagrees with our expectation that only those returning from the red giant branch should be enriched and not those on their way to the RGB.

Rolf Kudritzki
(reflecting on boundary conditions)

THE H-R DIAGRAMS OF THE LMC AND SMC

C. D. Garmany
Joint Institute for Laboratory Astrophysics
University of Colorado and National Institute of Standards and Technology, Boulder, CO 80309-0440

E. L. Fitzpatrick
Princeton University Observatory, Peyton Hall
Princeton, NJ 08544

ABSTRACT. We present new determinations of the H-R diagrams for luminous, hot stars in the LMC and the SMC. Using all available photometric and spectroscopic data, we discuss the conversion of the observed measurements to temperature and luminosity. The resulting H-R diagrams for normal stars confirm the upper luminosity limit, and also show an interesting "ridge" not noted before.

1. INTRODUCTION

It was first noted by Hutchings (1976) that the maximum luminosity observed for early-type stars decreases with decreasing effective temperature. Humphreys and Davidson (1979) further demonstrated that this decrease stops at an effective temperature of about 10^4 K and that the maximum luminosity is essentially constant for cooler temperatures. More recent estimates of this luminosity limit, often referred to as the "HD limit," are found in Humphreys (1987), Garmany, Conti and Massey (1987), and Humphreys (this volume). It is generally assumed that these observations indicate that stellar evolution tracks reverse their direction, i.e., from rightward evolution in the H-R diagram (HRD) to leftward evolution, at effective temperatures which depend on the initial masses of the stars. Lamers and Fitzpatrick (1988) showed that there is a very good correspondence between the observed upper luminosity limit and the locus of very low stellar effective gravities, as deduced from line blanketed, LTE, plane parallel model atmospheres calculated using Kurucz's (1979) atmosphere program.

In order to understand the physical processes which control the evolution of massive stars, it is necessary that the distribution of stars in the HRD be well-determined. This includes accurate measurements of the location of the HD limit, as well as of the ratio of red-to-blue supergiants, and of the metallicity dependence of the HRD. As has been pointed out frequently, the Large and Small Magellanic Clouds (LMC and SMC) are excellent candidate galaxies for such studies, providing the opportunity to study nearly complete samples of early-

K. Davidson et al. (eds.), Physics of Luminous Blue Variables, 83–89.

type stars. The uniform and relatively well known distances of the stars in these galaxies, and their generally low interstellar reddening offer obvious advantages over their Galactic counterparts. In addition, LMC and SMC studies, when combined with Galactic studies, allow the effects of stellar metallicity differences of up to a factor of 10 or more to be investigated.

In this paper, we present new determinations of the HRD's for the LMC and SMC, emphasizing the M_{bol} vs. T_{eff} distributions of the normal stars, for which the effective temperatures and reddenings can be determined accurately (i.e., we exclude the LBV's and the WR stars). The obvious extension of our work will be to compare our results with the M_{bol} vs. T_{eff} distributions of the more exotic objects, which may lead to some insight into the evolutionary relationships between the various classes of stars. The data for this project have been compiled from numerous sources, including catalogs of UBV photometry, spectral types, and reddenings. The sheer number of stars for which published data exist now allows for a better definition of the upper luminosity limit for normal stars than available previously and brings out other features of the upper HRD.

2. THE LARGE MAGELLANIC CLOUD

The data we have for the LMC stars usually include UBV photometry and spectral types of widely varying accuracy. In compiling these data, we began with a copy of the Rousseau et al. (1978) catalog of 1822 members of the LMC, which contains both new and previously published spectral types and UBV photometry. We added additional photometry from Isserstedt (1982) and slit spectral types determined by Walborn (1983), Crampton (1979), Conti, Garmany, and Massey (1986) and Fitzpatrick (1988). All known WR stars (Breysacher 1981) and H-alpha emitters later than type O (Henize 1956; Bohannan and Epps 1974) were removed from the dataset. Also removed were stars lacking UBV photometry and objects noted by Rousseau et al. as being non-stellar, multiple, or having composite spectra. The final dataset consisted of 1334 "normal" stars with spectral types in the range early-O to late-G and visual brightnesses in the range V = 10-15 mag.

In order to construct an HRD, we first determine T_{eff} (from which we determine the bolometric correction) and E(B-V) (from which we determine the absolute magnitude) for each star. Our sample includes 114 O-type stars which have well-determined spectral types derived from slit spectra. For these stars, we determined T_{eff} from a calibration of T_{eff} vs. spectral type derived from recent non-LTE spectral line analyses of O stars by Simon et al. (1983), Bohannan et al. (1986), and Voels et al. (1989). To derive E(B-V) for the O stars we assumed their intrinsic B-V colors to be -0.30. A LMC distance modulus of 18.3 mag was adopted.

For the rest of the stars, we pieced together a calibration of spectral class (temperature type) vs. effective temperature, bolometric correction, and intrinsic color. At a given spectral type, the intrinsic colors were tabulated for luminosity classes Ia, Ib, and II.

The adopted T_{eff} calibration is independent of the luminosity class. These calibrations were taken from various published sources and are reasonably well-determined and noncontroversial. In a more detailed version of this paper being prepared for publication, we will list all the sources and the final adopted calibrations.

E(B-V) was derived independently of the available spectral type information using the Johnson Q-method and the intrinsic color relations between B-V and U-B. The choice between the Ia, Ib, and II intrinsic colors depended on the absolute magnitudes of the stars. For the stars which have slit spectral types, the effective temperatures were taken from the calibration of spectral type vs. T_{eff}. For the majority of the stars, only poorly determined objective prism spectral types exist. For these, the temperature was derived from a calibration between T_{eff} and intrinsic U-B color.

The final result of all these calibrations is shown in Figure 1, where we plot T_{eff} vs. M_{bol} for the LMC stars. The open circles represent stars for which the photometry indicates E(B-V) > 0.4. These stars are highlighted because we feel that the reddening estimates may be incorrect. In most cases they are located in regions where such high reddening is not expected. Photometry errors may be responsible for the large E(B-V)'s, which result in erroneously large M_V and M_{bol} values. At the cool end of Figure 1 (log T_{eff} < 3.6) we have included about 40 M-type supergiants from Humphreys (1979). It should be realized that the sample of cool stars shown in the figure is by no means complete. We show the M stars to indicate the maximum luminosity observed for such stars in the LMC, NOT to indicate their numbers relative to the hotter stars. The x's indicate the locations of some

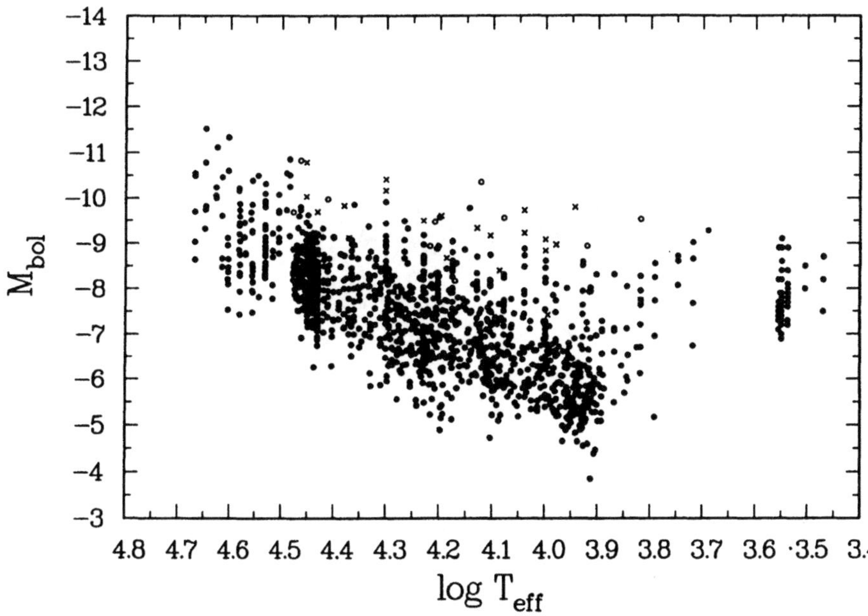

Figure 1. HRD for the LMC. Symbols are described in the text.

of the brightest "semi-normal" emission line stars. These stars exhibit emission only in the Balmer lines and have otherwise normal optical spectra. Note the pileup of stars between log T_{eff} values of 4.43 and 4.48 (i.e., T_{eff} = 27000-30000 K). These are stars for which no slit spectral types exist and which have UBV colors bluer than our calibration for type B0. Using their U-B values, we interpolated these stars onto a temperature grid ranging from 27000 K (the temperature adopted for B0) to 30000 K (the temperature adopted for O9.5). This group undoubtedly contains a large number of O stars which cannot be identified photometrically because of the well-known degeneracy of UBV colors for the O and early B stars. Slit spectral types are required to correctly assign their temperatures. The vertical strings of stars seen at several temperatures in Figure 1 result from T_{eff} being determined from a slit spectral type. These stars are forced into discrete bins in the figure, while the rest of the stars are spread more uniformly because their T_{eff}'s were determined from their UBV photometry.

There are two astrophysically interesting features in this diagram. The first is the upper luminosity limit of the normal supergiants. Figure 2 shows the same HRD with the inclusion of several previous estimates of the upper limit, as well as recent evolutionary tracks by Maeder and Meynet (1987). The distribution of the normal supergiants is consistent with the existing measurements of the limit

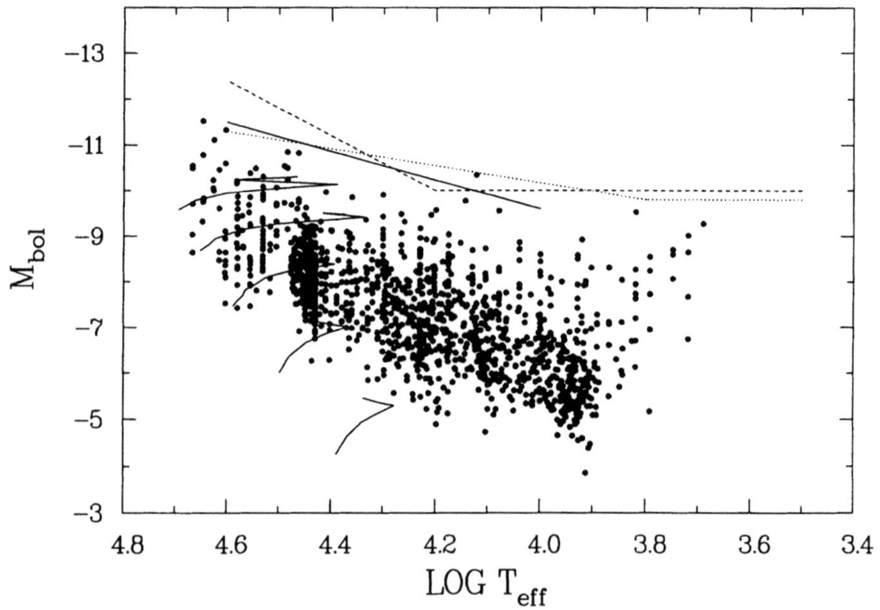

Figure 2. Same as Fig. 1, with the inclusion of evolutionary tracks by Maeder and Meynet for 60, 40, 25, 15, and 9 $M_\odot$ and various estimates of the HD-limits from Garmany, Conti and Massey (1987) (solid line), Humphreys and Davidson (1979) (dashed line) and Humphreys (1987) (dotted line).

and we see no reason to make any additional refinements, given that the exact placement of the line cannot be determined from the relatively sparse data at the brightest M_{bol}'s. A second, and unexpected, feature of the HRD in Figures 1 and 2 is the diagonal "ridge" line running through the data from upper left to lower right. Above this well-defined ridge the density of stars decreases significantly. This line runs roughly parallel to lines of constant absolute magnitude and is located at an M_V of about -6.5 mag. The stars above the line are mostly Ia supergiants. Note that this ridge is nowhere near the end of H burning as defined by Maeder's models except for the 40 and 60 $M_\odot$ tracks. We do not understand the physical significance of this ridge, but it is tempting to ask if it could represent the end of H burning. If so, the main sequence widening discussed by Meylan and Maeder (1982) and others is even greater than assumed. Alternatively, the ridge may indicate the location of some instability occurring in the stellar atmospheres. Obviously, further study is needed.

3. THE SMALL MAGELLANIC CLOUD

The data on SMC stars have been treated in the same manner as for the LMC stars. We started with the catalog by Azzopardi and Vigneau (1982), which consists of 524 stars having UBV photometry, and in some cases, spectral types. To this we added slit spectral types from Walborn (1983), Humphreys (1983) and Garmany, Conti and Massey (1987). Conversion to M_{bol} and T_{eff} was done as for the LMC, with an assumed distance modulus of 18.8. A major uncertainty in producing the SMC HRD is in the temperature calibration. For lack of better information, we adopted the same spectral type vs. temperature and intrinsic color calibrations as for the LMC. However, the lower metallicity of the SMC means that this method is undoubtedly incorrect, although the size of the error is unknown. Better measurements of the physical properties of the SMC stars are certainly required. The results of our procedure are shown in Figure 3. The location of the upper luminosity limit for the SMC may be slightly lower than for the LMC, but this may only be an effect of the smaller number of stars. There is also a suggestion in the SMC data of the same ridge line as seen for the LMC. But, again, the smaller number of stars makes the feature difficult to pinpoint.

This work was supported in part by NSF grant AST88-06594 to the University of Colorado. We thank Elaine Seymour-Kidman for doing the necessary data entry.

REFERENCES

Azzopardi, M. and Vigneau, J. 1982, Astron. Astrophys. Suppl. **50**, 291.
Bohannan, B. and Epps. H. W. 1974, Astron. Astrophys. Suppl. **18**, 47.
Bohannan, B., Abbott, D. C., Voels, S. A. and Hummer, D. G. 1986, Astrophys. J. **308**, 728.

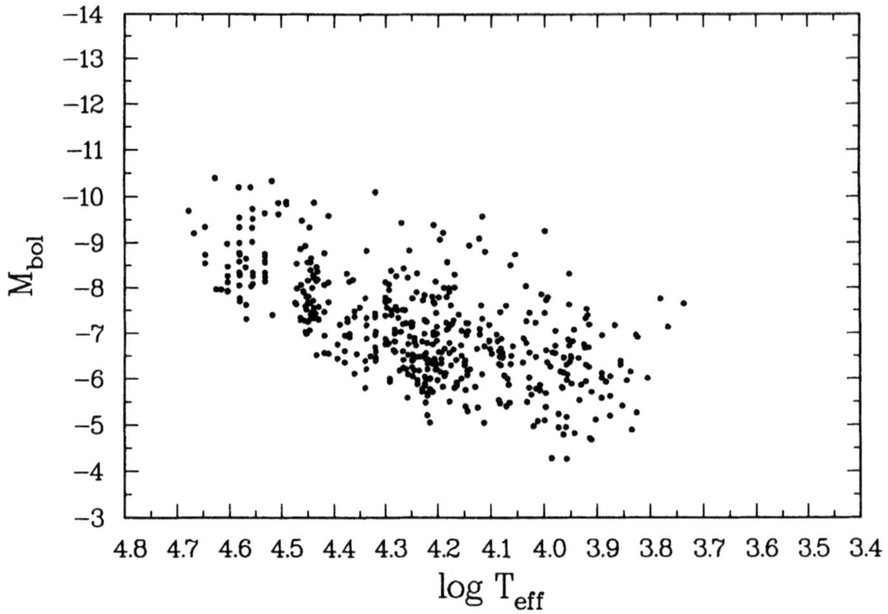

Figure 3. HRD for the SMC.

Breysacher, J. 1981, Astron. Astrophys. Suppl. **43**, 203.

Conti, P. S., Garmany, C. D. and Massey, P. 1986, Astron. J. **92**, 48.

Crampton, D. 1979, Astrophys. J. **230**, 717.

Fitzpatrick, E. L. 1988, Astrophys. J. (Dec. 1, 1988) in press.

Garmany, C. D., Conti, P. S. and Massey, P 1987, Astron. J. **93**, 1070.

Henize, K. G. 1956, Astrophys. J. Suppl. **2**, 315.

Humphreys, R. M. 1979, Astrophys. J. Suppl. **39**, 389.

______. 1983, Astrophys. J. **265**, 176.

______. 1987, in "Instabilities in luminous early-type stars," ed. C. deJager and H. J. G. L. M. Lamers (Dordrecht: Reidel), p. 3.

Humphreys, R. M. and Davidson, K. 1979, Astrophys. J. **232**, 409.

Hutchings, J. P. 1976, Astrophys. J. **203**, 438.

Isserstedt, J. 1982, Astron. Astrophys. Suppl. **50**, 7.

Kurucz, R. L. 1979, Astrophys, J. Suppl. **40**, 1.

Lamers, H. J. G. L. M. and Fitzpatrick, E. L. 1988, Astrophys. J. **324**, 279.

Maeder, A. and Meynet, G. 1987, Astron. Astrophys. **182**, 243.

Meylan, G. and Maeder, A. 1982, Astron. Astrophys. **108**, 148.

Rousseau, J., Martin, N., Prevot, L., Rebeirot, E. and Robin, A. 1978, Astron. Astrophys. Suppl. **31**, 243.

Simon, K. P., Jonas, G., Kudritzki, R. P. and Rahe, J. 1983, Astron. Astrophys. **125**, 34.

Voels, S. A., Bohannan, B., Abbott, D. C. and Hummer, D. G. 1989, Astrophys. J. (in press).

Walborn, N. R. 1983, Astrophys. J. **265**, 716.

DISCUSSION

Sreenivasan: Are you suggesting that LBV's are in or close to the main sequence or hydrogen-burning phase?

Garmany: At least in the LMC, the LBV's are intermingled in the H-R diagram with normal supergiants (luminosity Ia). We are not suggesting that the LBV's are still in the core-hydrogen-burning phase, but the emergence of the ridge in our treatment of the data raises questions about where hydrogen burning stops. This ridge line must have some physical significance.

Maeder: I am very fond of observational constraints, but for a given contraint one unique picture should emerge and that is not the case here. The color-magnitude diagrams of well-studied clusters and associations (*cf.* Mermilliod) do not show the enormous main sequence extension that you suggest. There is only a moderate extension which has been used to calibrate the distance of overshooting. I very much prefer to rely on well-analyzed cluster sequences rather than on a mixture of field stars of different ages.

Garmany: We agree that this effect has not been seen in young clusters. However, a complete data set for a galaxy such as the LMC should represent a steady-state situation. If young clusters and associations expand and disperse on a timescale that is much smaller than the nuclear timescale, as suggested many years ago by Blaauw and others, then one might not expect the older stars that form our ridge to still be identified as cluster or association members. In fact, as we showed, only a third of the supergiants in the LMC are within association boundaries.

Humphreys: Two incidental remarks. (1) It is important to remember that there are many more M-type supergiants in both the LMC and SMC; Katy's diagram shows just a few of these. (2) The rough original upper boundary (Humphreys & Davidson 1979) was based on a hotter temperature scale, which caused it to be steeper than it is in more recent H-R diagrams. (The upper left end was **not** intended to be determined by η Car, by the way.)

Garmany: We added some representative M stars to the diagram just to show where they lie, but all we are really concerned with here are types O to early G.

Gallagher: Could the bunching of stars along the ridge line be due to problems in the conversion between spectra and effective temperatures?

Garmany: I don't think so. We see the effect with both the spectral type -- temperature conversion and the $(U-B)_0$ -- temperature conversion. Temperatures for B stars, the most important subset in the ridge, have been determined by Fitzpatrick (1987) from Kurucz models using lower gravities than in the published models. The use of a different temperature scale will probably not eliminate the ridge.

Vanbeveren: Observers and theoreticians like to compare the data with evolutionary tracks in the H-R diagram. However, when T_{eff} is considered, are you sure that T_{eff}*(observed)* can be compared with T_{eff}*(evolution)*, *i.e.*, that the observed core-hydrogen-burning band can be compared with the theoretically predicted one?

Garmany: Well, one hopes that observers and theoreticians are talking about the same temperature scale. It seems unlikely that there is a large difference between them -- at least the ZAMS is agrees!

Kudritzki: No, Danny [Vanbeveren], I do not think this effect is large enough (factor of 1.5 to 2 in T_{eff}) to explain Katy's results.

Garmany: Binaries would raise the ridge by some amount, and a starburst could also be considered --

Gallagher: No, I do not think a starburst could produce the ridge line; the time scale would have to be too short for stars so widely dispersed across the LMC.

Vanbeveren,
Fitzpatrick, Groth

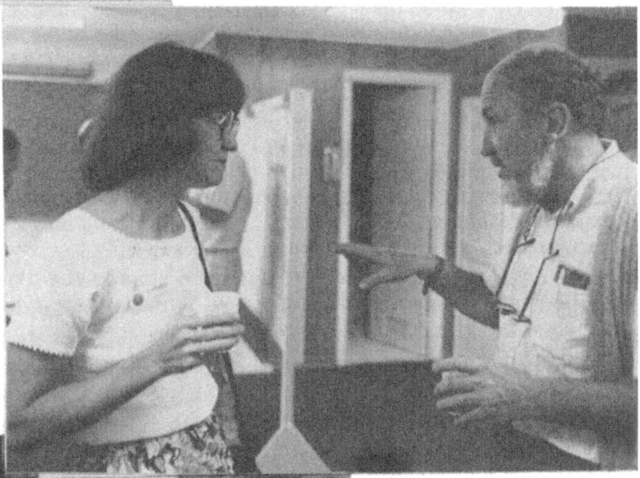

Garmany
and
Conti

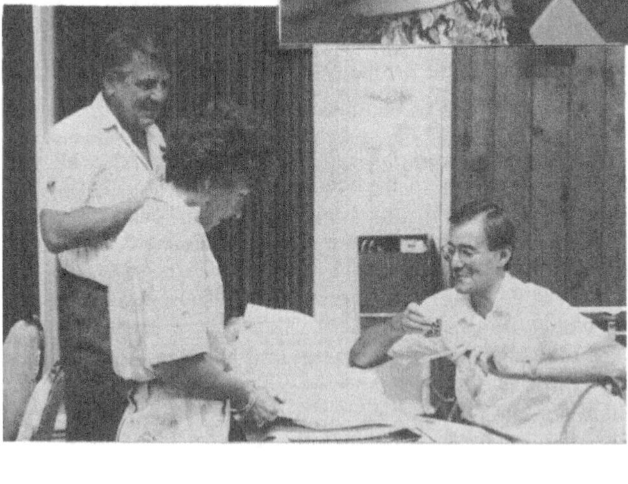

Wolf,
Rossi,
Leitherer

"NORMAL" LBV ERUPTIONS A LA S DORADUS

B. WOLF
Landessternwarte Königstuhl
D-6900 Heidelberg 1

ABSTRACT. The main characteristics of the LBVs with "normal" S Dor-type outbursts are reviewed. At quiescence they are luminous ($M_{Bol} \approx$ -9 to -10) OB supergiants (14 000 K $\lesssim T_{eff} \lesssim$ 30 000 K) occupying an inclined instability strip in the HRD. The most luminous S Dor variables are the hottest ones. During outburst dense envelopes are expelled of equivalent spectral type middle A to early F. S Dor variables occupy at maximum a vertical strip in the HRD at T≈8000 K. The observed amplitudes in the visual range increase from 1 mag (e.g. R71; T_{eff} = 14500 K at quiescence) to 2.5 mag (e.g. R127; T_{eff} = 33000 K at quiescence). The existence of an amplitude-luminosity relation is suggested which can be used to derive extragalactic distances.

1 Introduction

Among the LBVs the S Dor type variables are particularly distinguished. S Dor type varibles or Hubble-Sandage variables are (apart from supernovae) during outburst the visually brightest stars in the Universe. Due to their characteristic line spectra and their irregular variability of typically one to two magnitudes in the visual in timescales of years to decades they can be easily recognised in extragalactic systems. For these reasons they belong potentially to the most powerful extragalactic distance indicators (cf. Wolf,1988).

During the past few years S Dor variables became particularly important in connection with current theories of the evolution of very massive (M $\gtrsim$ 50 $M_\odot$)stars. S Dor type variables are supposed to represent a short-lived ($\sim 10^4$ yrs) unstable phase before presumably becoming Wolf-Rayet stars (cf. Wolf et al. 1981a, Maeder, 1982, 1983).

During the past few years from detailed studies of some selected S Dor type variables (particularly of the LMC) our knowledge of their nature and outburst properties has considerably improved (for recent reviews see Wolf, 1986, Lamers, 1986). An important result of such investigation was the finding that although the visual brightness during outburst increases typically by more than one magnitude the bolometric luminosity remains essentially unchanged.This refers, however, to the more modest type of eruptions with amplitudes $\Delta V \lesssim 3$ mag only, but not to the most extreme cases like η Car, which seem to erupt with some extra brightening. For these "plinian" type of LBV eruptions see Davidson (1988; this volume). We call the type of outbursts occurring under the condition M_{Bol} = const "normal" LBV eruptions à la S Dor. The main properties of these objects are discussed here.

K. Davidson et al. (eds.), Physics of Luminous Blue Variables, 91–99.

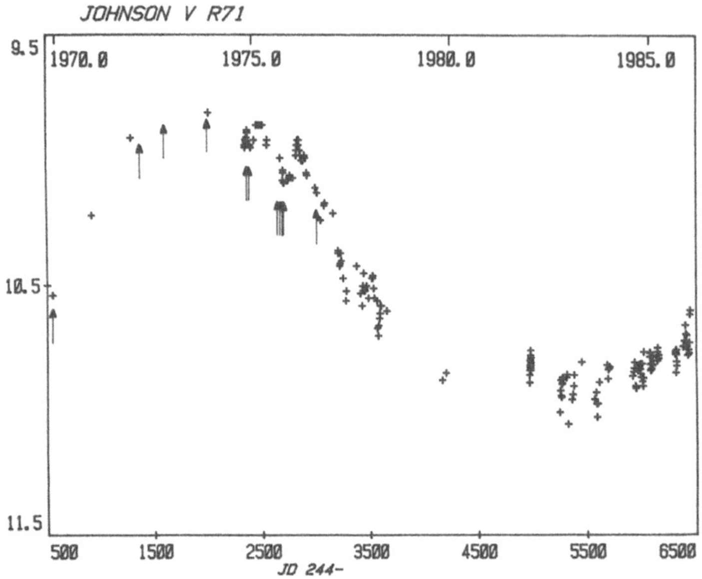

Figure 1: Light curve of the LMC-S Dor variable R71. The observations prior to 1978 were taken from van Genderen (1978) and the sources quoted by van Genderen. Observations after 1982 are from the "Long-term photometry of variables" group, initiated by C.Sterken. Arrows denote epochs at which coudé spectrograms were taken

2 General characteristics of S Dor variables

S Dor type variables are as bright as $M_v \approx -9$ to - 11. Members of this class have been identified in the Galaxy, in irregular and spiral galaxies of the local group and due to their extreme brightness in the galaxies M81, NGC 2403 and M101 outside the local group (cf. e.g. Wolf, 1988). They show irregular photometric variations in the visual of 1 to 2.5 mag in timescales of years or longer. From surrounding material from previous episodes of mass ejection a kinematic age of the order of several 10^4 years has been estimated (cf. Stahl, 1988, this volume) showing that S Dor variables represent indeed a very short-lived evolutionary phase of very massive stars.

Historically, R 71 of the LMC was the first case of a normal S Dor variable which has been observed both during minimum and maximum by means of high dispersion spectroscopy and is in the following regarded as a paradigmatic case for sketching the main properties of this group of LBVs. R 71 is now the best investigated S Dor variable. Apart from forbidden lines (mainly [FeII]) R 71 looks at quiescence like a rather normal B-type supergiant (Wolf et al. 1981) with a temperature of 14500 K. From a recent model atmosphere analysis (Kudritzki, 1988, this volume) the mass of this very luminous star ($M_{Bol} \approx -9$) was found to be surprisingly low ($M \approx 15 M_{\odot}$) and its He-abundance to be high (at least 20 %). Both findings are in agreement with the widely accepted scenario, that S Dor type variables represent a very late evolutionary phase prior to becoming WR-stars. R 71 is being continuously monitored by van Genderen's (1988) group and by Sterken's (1988, this volume) LTPV group. Fig. 1 shows the light curve of R 71 from the beginning of the maximum which R 71 had in the seventies to the present minimum state.

Apart from the large amplitude of the eruption of more that 1 mag between maximum and

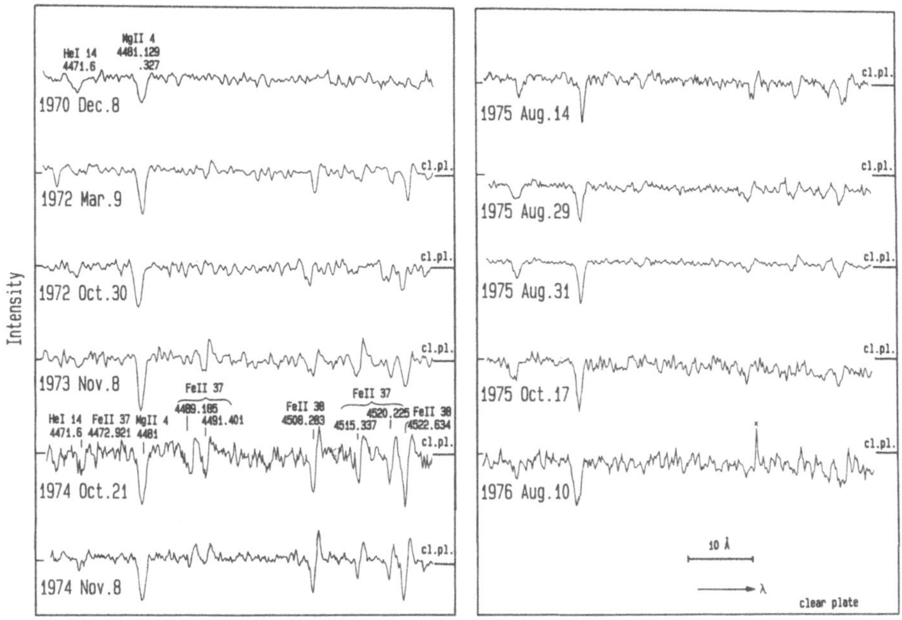

Figure 2: Intensity tracings of a section of some spectrograms taken between 1970 and 1976. For details see text

minimum there are superimposed microvariations occurring with timescales of the order of 23.5 days (1983 - 1985) and 14.3 days (1986 - 1987)(van Genderen et al., 1988). R 71's outburst in the seventies was monitored by Maart de Groot and the author with the ESO coudé spectrograph. Sections of this sequence of photographic spectrograms of high resolution are shown in Fig.2. The early spectrum (1970, Dec.8) clearly shows the minimum characteristics (e.g.HeI 4471), the spectra observed about maximum (e.g. 1974, Oct.21) shows typical cool A-type envelope spectrum (dominated by P Cygni type lines of singly ionised metals) of S Dor variables at outburst. From Fig. 2 it can be easily seen that even small photometric variations are paralleled by significant spectroscopic variations in the sense that the spectrum becomes later when the star becomes visually brighter. From Fig. 2 also the slow expansion velocity of the envelopes is evident which is typically about 100 to 200 $km\,s^{-1}$ for S Dor variables. It is this low velocity and the high density (typically $N \approx 10^{11}$ cm^{-3} for S Dor variables at maximum) of the envelope which even allows dust formation around this early B-type star (Wolf and Zickgraf, 1986). The mass loss rate of R 71 during maximum phase was estimated to $\dot{M} = 5 \cdot 10^{-5}$ $M_\odot$ and a factor 10 to 100 lower during minimum phase (Wolf et al. 1981a). These are again typical values for S Dor variables. It is the variable mass loss which causes by a variable flux redistribution the brightness variations in the visual whilst the bolometric luminosity remains practically unchanged. This earlier hypothesis (Wolf et al. 1981, Appenzeller and Wolf, 1982) has been confirmed in the case of AG Car by direct flux integration (IUE and ground-based) during various phases of its light curve (Viotti et al. 1984).

The variability is suggested to be a consequence of a strong line driven wind which can become dynamically unstable (Appenzeller 1986, Lamers 1986 and Appenzeller 1988, this volume).

3 S Dor Variables at Minimum - Instability Strip

Another example of a well investigated(both during minimum and maximum) S Dor Variable is R 127 of the LMC. This star has been detected as an S Dor variable by Stahl et al.(1983). During its present outburst it was for a short time the visually brightest star of the LMC before it was surpassed by SN1987A. Its outburst characteristics have been described in several papers (Stahl et al. 1983, Stahl and Wolf, 1986, Appenseller et al. 1987, Wolf et al. 1988). R 127 has at minimum been classified by Walborn (1977, 1982, 1988, this volume) as a transition Ofpe/WN9 star. Hence this very luminous ($M_{Bol} \approx$ -10.3) star is at minimum much hotter (T$\approx$ 33000 K) than R 71. Ofpe/ WN9 spectra are characterised by the simultaneous presence of HeII-and NIII - emission and HI-, HeI- and NII-lines. Two objects (S 61 and R 84 of the LMC) of this class which are anticipated to represent also the main properties of R 127 during minimum have been studied by means of high dispersion spectroscopy(IUE and CASPEC) by Wolf et al. (1987). Although quite normal mass loss rates were derived for these peculiar O-stars (for R 84 see also Schmutz et al. 1988, this volume),particularly the analysis of the UV spectra revealed the surprising result, that all UV-absorption lines (i.e. the lines originating from excited levels) are violet-shifted by about 200 $\mathrm{km\,s^{-1}}$ indicating high column-densities due to an unusually slowly accelerated stellar wind in the deeper layers. In this respect the transition type Of/WN stars resemble the peculiar B-supergiant P Cygni (Lamers, 1986). It was therefore suggested (cf. Stahl et al. 1985, Wolf et al. 1987) that the Of/WN transition type stars are the hotter counterparts of the early B-type P Cygni stars.

Quite recently a detailed multi-frequency study has been carried out of Var C of M33 by Humphreys et al(1988). It was convincingly shown by these authors that Var C resembles in many details S Dor of the LMC. Var C showed at quiescence a spectrum not unlike P Cygni (Humphreys, 1975) and T = 22000 - 25000 K and $M_{Bol} \approx$ - 9.8 were estimated by Humphreys et al..

The location of R 71, R 127 and Var C at quiescence in the HRD are shown in Fig. 5. These objects occupy an inclined strip (hatched in Fig. 5). S Dor and AG Car (with its new distance of Humphreys, 1988) and P Cygni are also located in this strip which is called S Dor-instability strip (Wolf, 1988). It means that the temperature of S Dor type variables at quiescence increases with the absolute luminosity of the star.

Also located in this strip is R81 of the LMC which was previously shown to resemble P Cyg in many respects (Wolf et al. 1981b) and was recently established as an eclipsing binary (Stahl et al. 1987). From the lightcurve a mass of $\approx 25 M_{\odot}$ was estimated by these authors for this luminous star. 22 $M_{\odot}$ were derived for P Cyg and 15 $M_{\odot}$ have been estimated for R71 (Kudritski, 1988, this volume) from model atmosphere analyses. Although these masses are low they are expected from evolutionary models (Maeder and Meynet, 1987) with mass loss for very evolved initially massive stars. Since there are other "normal" blue supergiants in the S Dor instability strip it is suggestive that only the very evolved low mass stars in the strip become S Dor- type unstable.

4 S Dor types at outburst - Opaque-wind limit

Whereas normal S Dor variables at quiescence occupy a wide temperature range (14000 to 35000 K) the outburst spectra are very much alike for all S Dor variables. This is well demonstrated in Fig. 3 which shows recent high resolution and high S/N ratio spectra (taken with CASPEC) of S Dor and R 127. From such observations it has been shown that during maximum S Dor variables are surrounded by cool (Te $\approx$ 8000 K), dense (N_e = 10^{11} $\mathrm{cm^{-3}}$), slowly expanding ($V_{exp} \approx$ 100 - 200 $\mathrm{km\,s^{-1}}$) envelopes. As shown in Fig. 5 the S Dor variables are at maximum located in the shaded vertical area at T $\approx$ 8000 K in the HRD. This observationally well established vertical limit has recently got its theoretical interpretation. On the basis of his opaque wind model Davidson (1987) found that the temperature of the envelope of LBVs cannot fall far below 7500 K even if the mass-loss rate is enormous. A major role in this model plays the low speed ($v \lesssim 10\,\mathrm{km\,s^{-1}}$) pseudo photosphere which is observed in maximum spectra of normal S Dor variables. The main characteristics of this false photosphere which is much more extended than the hot core visible during minimum have been described in the

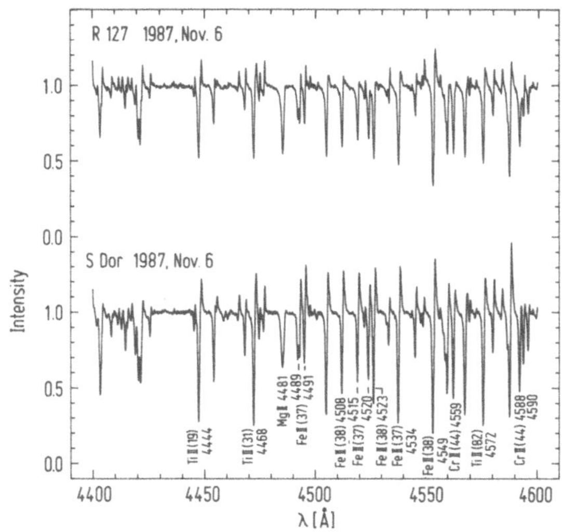

Figure 3: Section of the spectra of S Dor and of R127 of the LMC. Both stars are presently in the maximum state. The striking similarity of the spectra proves the very similar physical conditions in the maximum state envelopes

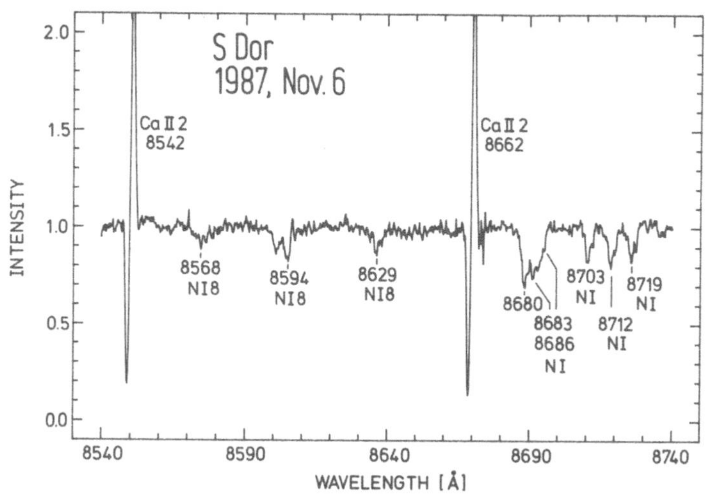

Figure 4: Section of a near infrared CASPEC spectrum of S Dor. Apart from the P Cygni-type profiles of the CaII triplet numerous absorption lines of NI and OI are conspicuous. These lines are unshifted with respect to the system velocity and are formed in the pseudo-photosphere of S Dor

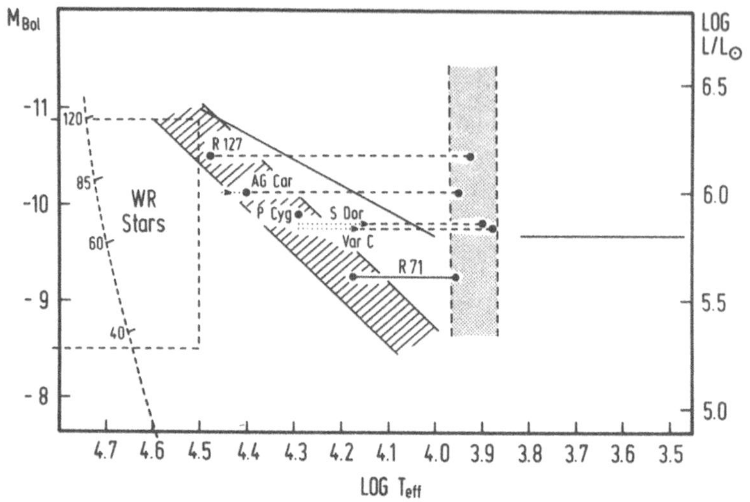

Figure 5: HRD showing the position of well studied S Dor variables and their range of variations. P Cyg is also included. At quiescence the S Dor variables are located in the inclined hatched area called S Dor instability strip. At maximum state they occupy the vertical shaded area called opaque wind limit. also included is the ZAMS (broken line). The solid line indicates the upper limit of stellar luminosities (see Humphreys and Davidson, 1979)

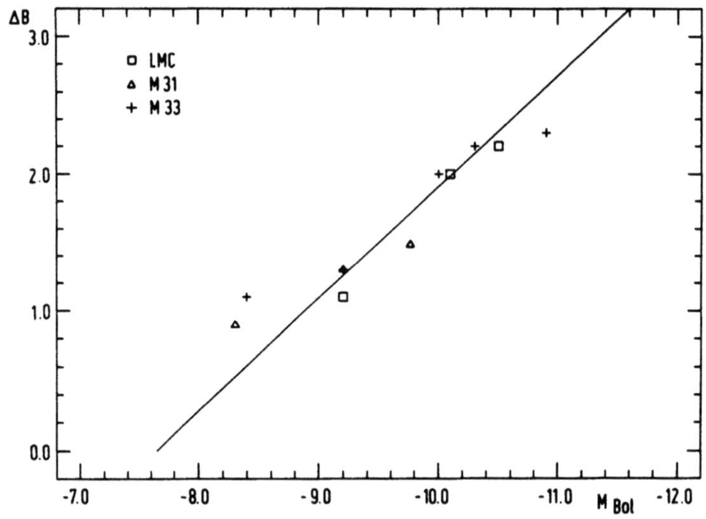

Figure 6: Composite amplitude-luminosity relation for S Dor variables derived from the sources indicated at the upper left (Wolf,1988)

case of S Dor by Leitherer et al.(1985). It is evidenced by high excitation lines like MgII 4481, HeI lines and SiII lines, which are essentially unshifted (with respect to the systemic velocity) except for small radial velocity variations with amplitudes of the order of 10 km s^{-1}. In addition to the few well known lines in the photographic wavelength range quite recently a considerable number of pseudo photospheric lines particularly of NI and OI have been identified in the near infrared range (cf. Fig. 4). On the basis of this large number of lines now available it should be possible in the future to derive a detailed physical model of the pseudo-photosphere which obviously plays a crucial role in the understanding of S Dor type outbursts.The finding that S Dor variables at quiescence are located in an inclined instability strip (hatched in Fig. 5) combined with the observationally well established existence of the vertical opaque wind limit(shaded in Fig. 5) for S Dor type variables at outburst has an interesting consequence. Since normal LBV eruptions à la S Dor occur under the condition $M_{bol} \approx$ const, this implies that the variation of the bolometric correction or likewise the photometric amplitude in the visual and blue range from minimum to maximum increases with the absolute luminosity of the star. The existence of an amplitude- luminosity relation is indicated. In fact such a relation has recently been derived by Wolf(1988) by combining published photometric data of the S Dor variables of the LMC and of the Hubble-Sandage variables of M31 and M33 (see Fig. 6). This relation provides a more physical basis for the determination of absolute luminosities of S Dor variables and hence can considerably improve the usefulness of S Dor variables for the extragalactic distance scale.

5 Conclusions

Since a previous review of S Dor variables by Sharov (1975) considerable technical advances allowed us to extend the observations of S Dor variables particularly of the LMC to the much wider wavelength range from the satellite UV to the infrared. On the basis of these new data our knowledge of the nature of S Dor variables and their outburst mechanisms has been considerably improved.Yet, many basic problems are still unanswered. Although it seems now well established that radiation pressure causes the S Dor type eruptions there is no quantitative theory predicting the photometric amplitudes of the eruptions or the amplitude-luminosity relation indicated by the observations. What role plays the pseudo-photosphere for the outburst mechanism? What is the duration of the S Dor phase and how do the characteristics depend on abundances? However, as the example of Var C of M33 has shown, with the advanced technical facilities rather detailed studies can be carried out for more distant objects. On the basis of such a more extended set of data further progress in understanding the nature of LBV eruptions à la S Dorados can be expected.

References

Appenzeller, I.: 1986, *IAU Symposium* **116**, p. 139
Appenzeller, I.: 1988, this volume
Appenzeller, I., Wolf, B.: 1982 *ESO Workshop* "The Most Massive Stars", p. 131
Appenzeller, I., Wolf, B., Stahl, O.: 1987, "Instabilities in Luminous Early Type Stars", p. 241
Davidson, K.: 1987, *Astrophys. J.* **317**, 760
Davidson, K.: 1988, this volume
Humphreys, R.M.: 1975, *Astrophys. J.* **200**, 426
Humphreys, R.M.: 1988, this volume
Humphreys, R.M., Davidson, K.: 1979, *Astrophys. J.* **232**, 409
Humphreys, R.M., Leitherer, C., Stahl, O., Wolf, B.: 1988a, *Astron. Astrophys.* **203**, 306
Humphreys, R.M., Stahl, O., Davidson, K., Wolf, B., Zickgraf, F.-J.: 1988, this volume Kudritzki, R.: 1988, this volume
Lamers, H.J.C.L.M.: 1986, "Instabilities in Luminous Early-Type Stars", p. 99
Leitherer, C., Appenzeller, I., Klare, G., Lamers, H.J.C.L.M., Stahl, O., Waters, L.B.F.M., Wolf, B.: 1985, *Astron. Astrophys.* **153**, 168

Maeder, A.: 1982, *Astron. Astrophys.* **105**, 149
Maeder, A.: 1983, *Astron. Astrophys.* **120**, 113
Maeder, A., Meynet, G.: 1987, *Astron. Astrophys.* **182**, 243
Schmutz, W., Leitherer, C., Torres-Dodgen, A.V., Vogel, M., Conti, P.S., Hamann, W.-R., Wessolowski, U.: 1988, this volume
Sharov, A.S.: 1975, *IAU Symposium* **67**, p. 275
Stahl, O.: 1986, *Astron. Astrophys.* **164**, 321
Stahl, O.: 1988, this volume
Stahl, O., Wolf, B.: 1982, *Astron. Astrophys.* **110**, 272
Stahl, O., Wolf, B.: 1986, *Astron. Astrophys.* **158**, 371
Stahl, O., Wolf, B., de Groot, M., Leitherer, C.: 1985, *Astron. Astrophys. Suppl. Ser.* **61**, 237
Stahl, O., Wolf, B., Klare, G., Cassatella, A., Krautter, J., Persi, P., Ferrari-Toniolo, M.: 1983, *Astron. Astrophys.* **127**, 49
Stahl, O., Wolf, B., de Groot, M., Leitherer, C.: 1985, *Astron. Astrophys. Suppl. Ser.* **61**, 237
Stahl, O., Wolf, B., Zickgraf, F.-J.: 1987, *Astron. Astrophys.* **184**, 193
Sterken, C., 1988, this volume
van Genderen, A.M., Thé, P.S., Augusteijn, Th., Engelsmann, E.C., van der Grift, E.W., Prein, J.J., Remijn, R.L., Steeman, F.W.M., van Weeren, N.: 1988, *Astron.Astrophys. Suppl. Ser.* **74**, 453
Viotti, R., Altamore, A., Barylak, M., Cassatella, A., Gilmozzi, R., Rossi, C.: 1984, *NASA* Conf. Publ. 2349, p. 231
Walborn, N.R.: 1977, *Astrophys. J.* **215**, 53
Walborn, N.R.: 1982, *Astrophys. J.* **256**, 452
Walborn, N.R.: 1988, this volume
Wolf, B.: 1986, *IAU Symp.* **122**, p. 409
Wolf, B.: 1988, *Astron. Astrophys.* submitted
Wolf, B., Appenzeller, I., Cassatella, A.: 1980, *Astron. Astrophys.* **88**, 15
Wolf, B., Appenzeller, I., Stahl, O.: 1981a, *Astron. Astrophys.* **103**, 94
Wolf, B., Stahl, O., de Groot, M.J.H., Sterken, C.: 1981b, *Astron. Astrophys.* **99**, 351
Wolf, B., Stahl, O., Seifert, W.: 1987, *Astron. Astrophys.* **186**, 182
Wolf, B., Stahl, O., Smolinski, J., Cassatella, A.: 1988, *Astron. Astrophys. Suppl. Ser.* **74**, 239
Wolf, B., Zickgraf, F.-J.: 1986, *Astron. Astrophys.* **164**, 435

Bernhard Wolf

DISCUSSION

Bohannan: From measurement of terminal velocities for many of the Of/WN stars in the LMC with low-resolution IUE spectra, I would not want to conclude that **all** of the Of/WN stars have low-velocity winds compared with normal O-type supergiants.

Wolf: I agree with you. I referred to those transition types originally introduced in 1977 and 1982 by Walborn, which are very closely related to R 127 at minimum. In fact R 127 is a member of the group discussed in that paper.

Leitherer: I think you can derive only an upper limit to the expansion velocity of the photosphere of R84 from radial velocities of absorption lines. An absorption line is by definition formed outside the photosphere. The question is, how far outside? This is immediately related to the shape of v(r), which should be derived from self-consistent, extended model atmospheres applied to the star -- not from plane-parallel hydrostatic models.

Wolf: I agree.

Sreenivasan: I find the suggestion of an instability strip for S-Dor stars interesting. These are thought to be non-radial pulsators. Where do the β Cephei stars lie with respect to the S Dor instability strip, if you extend it downward? (They are thought by those who study them to lie in a strip also.)

Wolf: It is widely believed that radiation pressure plays the important role as a physical mechanism for 'normal' S Dor eruptions. I do not know if the β Cep non-radial pulsation mechanism is important in connection with S Dor stars.

Walborn: I think the possibility remains that the true minimum states of P Cyg and S Dor have not yet been observed spectroscopically, and that they are actually Ofpe/WN9 objects like R 127 and AG Car. The light curve of P Cyg shows that it has been in an intermediate state throughout modern times, and its spectrum is identical to the intermediate-B-type spectrum of AG Car, while the spectrum of S Dor is identical to the maximum spectrum of R 127. Also, the detailed definition of an inclined instability strip for these stars may be an artifact. I do have a minimum-state echellogram of R 127 which can be subjected to further quantitative analysis.

Wolf: The S Dor instability strip suggested here is a conjecture to define the range in the H-R diagram where known LBV's are located. We do not know whether S Dor is an Ofpe/WN9 star. The spectrum taken by Thackeray during S Dor's deep minimum in the 1960's is certainly different from Ofpe/WN9; but it might well be that the star was not in an absolute minimum then, as you surmise. I would at least expect that R 71 has a later spectral type during minimum.

De Groot: I still see a lot of virtue in the S Dor instability strip. Its location in the H-R diagram seems to coincide with the ridge of luminous stars shown by Garmany earlier this morning. I do not know what this means, but I think we should not lose sight of this interesting coincidence.

Lamers, Moffat, Stahl, Davidson

PLINIAN ERUPTIONS À LA ETA CARINAE

Kris Davidson
University of Minnesota
Astronomy Department
Minneapolis, MN 55455
U.S.A.

Eta Carinae is at once our *most notable* LBV but also a *doubtful* LBV, because its Great Eruption observed 150 years ago was so unique. Since that event exceeded other major LBV outbursts by orders of magnitude in ejected mass and energy, we are not certain that it really was an example of S Dor, P Cyg, R127-type behavior. But even if it turns out to be a freak, η Car has provided either the original example or one of the earliest hints for several facets of the LBV phenomenon, *e.g.*:

-- Visual-wavelength brightening with little change in luminosity during each major mass-loss event;

-- The idea of a modified Eddington limit causing instability in a critical range of surface temperature;

-- CNO-processed surface material;

-- Bipolar structure in the ejecta.

The first two of these were recognized nearly 20 years ago for this star (Westphal and Neugebauer 1969, Davidson 1971). So Eta has given us clues to the topic even if it turns out not to be a proper LBV! But I think that it probably is one, and that its extreme nature makes it more interesting as a physical puzzle than the other famous LBV's.

Since many references to the history and nature of η Car can be found in (*e.g.*) van Genderen and Thé (1985), Davidson *et al.* (1986, hereinafter "DDWG"), Meaburn *et al.* (1987), and Davidson (1987a, "D87"), I shall omit some primary references in this brief commentary. Some points made at the Lunteren meeting in 1986 (D87) are worth repeating here -- they are pertinent especially to Maeder's, Appenzeller's, and de Jager's discussions in these proceedings -- but I'll try to add a few other points and to emphasize those that seem most interesting now. First, though, some essential facts:

The light curve (see figure on following page) shows why 19th-century commentators from John Herschel to Agnes Clerke ascribed such great importance to η Car. It was a first-magnitude star for about 20 years, briefly flickering brighter than Canopus during 1843 -- an impressive performance for a star more than 2 kpc away.[1] Astronomers became less aware of Eta during most of the 20th century and vaguely supposed it to be a weird nova, or an unusual supernova, or a strange red supergiant. Then Westphal and

[1]In those days they hadn't yet decided what to call it: η Navis, η Argus, even η Roburis. In the old map shown on the front of the American Astronomical Society's Membership Directory in recent years, Eta appears in "Karlseiche" = Robur Carolinum = Charles' Oak, a constellation invented by Halley around 1680 to commemorate a hiding place that Charles II of England had once found useful. It's a pity that we can't name constellations to appeal to potential funding agencies any more.

K. Davidson et al. (eds.), Physics of Luminous Blue Variables, 101–108.

Neugebauer (1969) found it to be the brightest extra-solar-system object in the sky at IR wavelengths around 20 μm, which eventually inspired us to recognize that it is a very luminous **blue** star surrounded by dusty ejecta. Infrared observations of η Car have been crucial in two or three ways (here I assume that $D = 2.5$ kpc):

(1) They indicate the present luminosity, about $10^{6.6} L_\odot$, requiring $M >$ about $90 M_\odot$ according to the classical Eddington limit. (A naïve Zanstra estimate gives the same result but depends on more assumptions.) *Caveat:* If there are holes in the dusty envelope, say in the polar directions, then the luminosity just quoted is an underestimate -- though a large error seems unlikely.

(2) The mass of the IR-emitting dust is of the order of $0.01 M_\odot$ (see, *e.g.*, Robinson *et al.* 1987, DDWG, Hackwell *et al.* 1986), showing that at least $1 M_\odot$ of gas has been ejected in less than 200 years. If the dust/gas ratio is lower than "normal" because carbon and oxygen are relatively scarce (most of the CNO is in the form of N), then 2 or even $3 M_\odot$ seems a plausible guess for the mass ejected in the Great Eruption -- implying a temporary mass-loss rate of the order of $0.1 M_\odot/y$. The present rate is not known but is probably less than $10^{-3} M_\odot/y$, possibly far less. (Any larger value must be confined near the equatorial plane -- see DDWG and Davidson 1987b.)

(3) Infrared data are also important in showing more clearly that bipolar structure or axial symmetry is present in the ejecta (see, *e.g.*, Meaburn *et al.* 1987, Hackwell *et al.* 1986, and refs. therein). This may be quite important, and personally I hope that the axis is defined by rotation rather than by a close binary orbit, because a binary scenario for the Great Eruption seems ill-adapted to explaining recurrent eruptions (see below) -- although a close companion might help by inducing rapid rotation.

UV and visual-wavelength observations have also been essential, giving us indications of the surface composition (He-rich, N > C, N > O) and of the star's effective temperature; see DDWG and refs. therein. T_{eff} is most likely in the 24000 -- 32000 K range, close to the hypothetical modified Eddington limit instability as mentioned earlier. With a bold surmise one could have *predicted* the IR brightness in 1967 by a Zanstra argument based on visual-wavelength data then available; so in a sense the UV and visual wavelengths are more essential here than the infrared -- I say this to foster a sense of proportion on the part of IR astronomers!

Weigelt and Ebersberger (1986) used speckle observations at $\lambda \sim 8500$ Å to discover that the central object in η Car appears multiple, one dominant object plus 3 fainter companions. We are still not certain whether the secondary components are stars or nebular blobs; either alternative presents difficulties (Davidson and Humphreys 1986), which, however, have been reduced somewhat by Hofmann and Weigelt's (1988) revision of the brightness ratios. At red wavelengths the brightest object is roughly 12 times as bright as each of its companions. A nebular blob model is difficult because the fainter components seem too small to intercept enough energy to explain their brightnesses. One problem with a multiple-star model is that the apparent separations are all comparable, making it an unstable "trapezium" system unless it really consists of two pairs almost superimposed along the line of sight.

If the three fainter components are stars, they are probably 30-to-50 $M_\odot$ O-type stars with separations of a few hundred a.u. -- too far apart for much interaction. This model would give the primary a luminosity close to $10^{6.4} L_\odot$ (abs. bol. mag. ~ -11.2, $M_{Edd} \sim 55 M_\odot$) and an initial mass around $120 M_\odot$; but obviously we need more speckle results and some observations with a space telescope. At least tentatively, we can use Ockham's Razor to suppose that the brightest component was the site of the Great Eruption; any alternative model would be less straightforward.

The foregoing page is an outline of the probable present condition of η Car. With this in mind, now let me mention a few points that I think are interesting, including some possible clues to the eruptive phenomenon. Let's begin with the light curve.

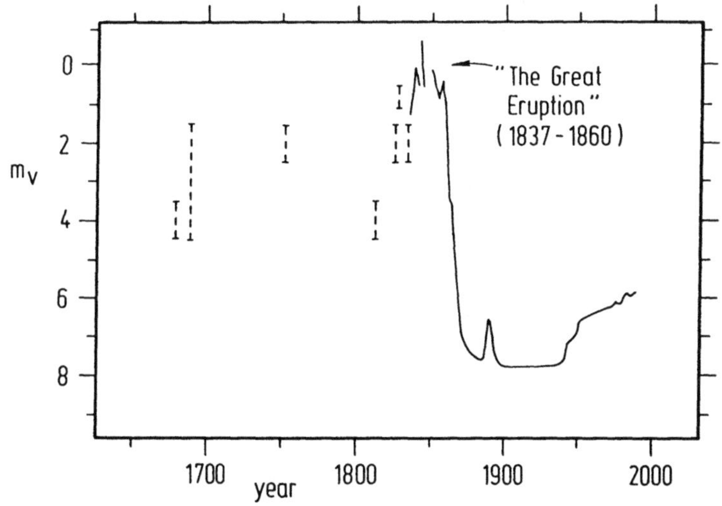

Historical visual-wavelength brightness of η Carinae. See D87 for refs.

Since 1860 the brightness has been reduced by circumstellar dust that began to form around that time. Without that circumstellar extinction, the apparent magnitude would now be about $m_V \sim 4.0 - - 4.5$, comparable to the low state sometimes seen before 1820. The light curve has the following points of interest -- these seem obvious when they are pointed out, but I emphasize them here because they are important and because they have never been used in a genuine physical model of the system:

(1) Before 1820 the star was noticeably unstable; the 2-magnitude discrepancies among various observers were too large to be mistakes. But then *the Great Eruption apparently relieved the immediate cause of instability* and the star has not been so erratic during this century. (This is why I like the term "eruption"; the light curve is reminiscent of a volcano or maybe a geyser.) However, the outburst around 1890, which was indeed a mass-loss event (see Walborn and Liller 1977), seems mildly inconsistent from this point of view and needs some explanation.

(2) *The star was behaving like a 2-mode or 2-state system before 1820.* Sometimes it was a 2nd-magnitude star, sometimes 4th, but seldom 3rd magnitude. Presumably the fainter state resembled its present condition, a hot luminous star. In the brighter state there was a dense wind that looked like a cooler photosphere and therefore was brighter at visual wavelengths, the now-familiar LBV style of brightening. Since any mass-loss rate above about $10^{-3.6} M_{\odot}/y$ will give about the same visual appearance (Davidson 1987b), "on" and "off" are adequate descriptions of the two states to explain the brightness changes semi-quantitatively.

(3) *The luminosity must have increased during the big event.* So far as I know, van Genderen and Thé (1985) were the first to clearly note this ironic point, ironic because the idea of an LBV visual-wavelength outburst at near-constant luminosity originated in connection with this event, as I mentioned at the beginning of this discussion. The brightest visual magnitude that can be attained at the present luminosity, even without

any circumstellar dust, is roughly in the range $m_V \approx 1.5 -- 2.0$, which was approached or attained in the pre-1820 maxima. From 1840 to 1860 the average brightness was about a magnitude brighter than this; in 1843 it briefly reached $m_V \approx -1$, corresponding to absolute magnitude < -14! The average total luminosity during the 20-year event was probably about $10^7\ L_\odot$, almost certainly exceeding the classical Eddington limit. A luminosity change large enough to be noticed may be a trait that distinguishes giant LBV eruptions from the more normal variety (*cf.* Wolf, these proceedings).

(4) The luminosity fluctuated during the Great Eruption. The photosphere was then located in the dense wind, at $r \sim 10^{14.2}$ cm $\sim 2000\ R_\odot$, $T \sim 7000$ K, with bolometric correction near zero. Therefore the observed fluctuations in visual brightness, *e.g.* in 1843, were not merely variations in bolometric correction. They cannot all have been errors, either; the observers described η Car relative to specific bright stars that are well known today, including Canopus, $m_V \approx -0.7$.

Having seen the light curve, we can ponder the energy budget. The "extra" energy radiated between 1840 and 1860 was roughly $10^{49.2}$ ergs -- about 60 years' worth of radiation at the normal luminosity. (The eruption was synergistic: Radiation caused the gas to escape, which in turn allowed radiation to escape.) If the mass of the ejected gas was around $2\ M_\odot$, then each of the other contributions to the energy budget is also about 10^{49} ergs. The kinetic energy of $2\ M_\odot$ is 10^{49} ergs at a velocity of 700 km/s, close to the observed expansion speed. If the star's mass was (*e.g.*) $80\ M_\odot$ before the eruption, then a simplified calculation shows that its outer $2\,M_\odot$ extended quite deeply (to $r \sim 0.7\,R_*$, $T \sim 10^{6.3}$ K) and had thermal energy $\approx 10^{49}$ ergs, gravitational potential energy $\approx -10^{49}$ ergs. I suppose that the readjustment energy of the remaining star after removal of the outer layers would have been similar. Anyway, the conjecture that the mechanical and radiative energies released in the event were roughly equal is consistent with the idea that the ejected mass was of the order of $2\,M_\odot$.

Regarding energy balance, and assuming that it was a single-star event, the eruption was strangely reminiscent of autoionization in a doubly-excited atom. The initial state looked like a hot star, the final state looks like a hot star plus ejecta plus radiation, and somehow the initial state had as much energy as the final state but was bound. As in the atom, characteristic energies are much larger in the stellar core but we suspect that the core is not directly relevant to the process. This analogy, in addition to the suspected bipolar morphology, is why I fear that rotation may prove crucial even if we dislike the extra parameters involved (see Sreenivasan's discussion, these proceedings). As in an atomic transition, the initial and final states might differ mainly in their angular momentum distributions. In any case we need a model of the "before-and-after" energy balance.

Next let's consider timescales, particularly the duration of a major outburst and the recurrence time between outbursts. The *dynamical* time for the outer layers of η Car is ~10 days for the basic star, or 30 to 100 days for the pseudo-photosphere during an eruption. Large fluctuations in brightness did occur this quickly between 1820 and 1860, and this fact may eventually prove significant for modeling the instability, but the characteristic duration of a major event appears to be longer. The *thermal* timescale for just the outer layers, containing roughly the mass that was ejected, is obviously comparable to the duration of the Great Eruption (*cf.* remarks by Maeder and by Appenzeller, these proceedings). However, the pre-1820 outbursts probably involved much less mass but did not have proportionally shorter durations. The characteristic duration of an outburst for η Car as well as other LBV's seems to be about equal to the thermal timescale for the outer 0.1 to 1 percent of the mass; maybe this is a hint that the eruptive process has a deep-seated cause, or maybe not! A calculative experiment proposed in D87 is pertinent here: Instantaneously remove outer layers Δm from an equilibrium stellar model, and then give it enough time to return to dynamical equilibrium but not thermal equilibrium. Both the luminosity and the radius are temporarily changed; then, as a function of Δm, does the star become more stable or less stable according to the modified Eddington limit or any other hypothetical criterion? Maeder and also Appenzeller (these proceedings) and Heisler and Alcock (1986) have discussed almost, but not quite, this experiment, and the results seem ambiguous to me.

A few outlying nebular blobs around η Car have proper motions indicating ages of several hundred years (Walborn *et al.* 1978); therefore some of the pre-1820 outbursts must have ejected significant amounts of mass. Perhaps the recurrence time between major eruptions is again related to a thermal timescale, but on the other hand, it may instead be related to a nuclear/evolutionary timescale -- the time needed to evolve just a tiny bit, to become unstable again. I wish that we knew whether any giant outbursts of this star were seen thousands of years ago; pertinent but unrecognized records may exist somewhere (see D87, DDWG). Finally, regarding timescales: If rotation is crucial, then the characteristic recurrence time between outbursts may be related to a fourth physical timescale, that for diffusion of angular momentum (see Sreenivasan, these proceedings).

We do not have any well-studied extragalactic examples of LBV eruptions as large as that of η Car. A giant outburst of variable V12 in NGC2403 was observed during the 1950's (see Tammann and Sandage 1968); it probably attained an absolute magnitude of -11 or brighter, and, like η Car, it fluctuated wildly just before the outburst and faded quickly afterward, perhaps as dust formed in the ejecta. However, the duration of the event was much briefer than in the case of η Car. We don't know much about the present state of V12; maybe it is a worthwhile IR object. An explosion called SN1961V, in NGC1058, was likened by Zwicky (1964) to η Car, but it was extremely luminous and may have been a supernova. In terms of observational technique, looking for giant LBV eruptions in nearby galaxies is an extension of the idea of seeking low-luminosity supernovae like SN1987A.

Now, at last, for my main point. Eta Carinae is a beautiful, tantalizing physical system; we have some clues to its causes of instability, though we cannot say exactly what caused its eruptions; we suspect that this is essentially the same phenomenon that occurs in other LBV's; but *no serious attempt has yet been made to model the instability.* A number of people have calculated evolutionary tracks or detailed atmospheres and winds for this part of the H-R diagram, but we need something different, namely a stability analysis of a **simplified** equilibrium star. Simplification is essential, I think, not merely to reduce the calculative effort, but more importantly to eliminate distracting details that are not crucial to the eruptive process.

With Ockham's Razor in mind, what do we expect to be necessary in an idealized theoretical configuration to be examined for LBV-like behavior? (1) The L/M ratio must be within shouting distance of the classical Eddington Limit. (2) The surface temperature should be LBV-like. (3) A CNO-processed chemical composition, with helium abundance around $Y \sim 0.4$, is most likely adequate for the outer layers. (4) Only the outer layers of the star, perhaps the outer 10% or so of the mass, seem crucial at the outset. This means that there is no obvious need for evolutionary calculations of the interior structure; suitable inner boundary conditions can probably substitute. (5) There is also no obvious need for a fancy non-LTE atmosphere model; even if the instability is triggered in the photosphere and involves opacity dependences, a simplified atmosphere is likely to include the the essential physics of the outburst phenomenon. (6) Realistic hydrodynamics, on the other hand, may be essential (see de Jager, these proceedings). (7) Initially the star should be non-rotating. If such a star refuses to behave like an LBV, then the next step may be to introduce rotation; whether or not rotation is essential is one of the main questions here. As for binary scenarios, I suspect that a rotational model would be able to simulate the most likely effects of a close companion. (8) If the program outlined above unexpectedly fails to produce eruptions, then one can justifiably devise more complicated models!

Some cleverly simple form of analysis might reveal the desired instability. More likely, a careful plasma-physics-style linearized analysis is appropriate. But it is also conceivable that the phenomenon is intrinsically hydrodynamic and non-linear. An obvious approach would be to construct equilibrium outer layers for a star, put this equilibrium configuration in a supercomputer with a hydrodynamic code (keeping the radiative transfer simple), and see what happens! It is remarkable that none of these approaches has yet been undertaken by anybody, so far as I know. By now we are fairly sure that a simple zeroth-order modified Eddington limit does not suffice to explain LBV outbursts, and we suspect, variously, that the true explanation is a more subtle instability involving either a modified Eddington limit (Maeder's as well as Appenzeller's discussions in these proceedings fall in this category), or de Jager's turbulent pressure, or rotation as discussed by Sreenivasan... Which of these ideas actually work? Which are *sufficient* and which are *necessary* for LBV's?

(*Addenda* -- A few recent papers are notable in connection with η Car:)

(1) Bandiera *et al.* (these proceedings) have found a dramatic change in the profile of the He I λ5876 emission line. Ruiz *et al.* (1984) previously described rapid changes in the Hα profile. The interesting point is that small velocity dispersions of the order of 30 km/s are involved. The speeds and timescales suggest that the relevant gas is in a region only a few a.u. across. This compact gas with a small velocity dispersion may be a circumstellar disk and may be evidence *against* the presence of a companion star close enough to affect the primary star; some special trick would seem necessary to allow circumstellar gas to have a small velocity dispersion in a binary system with orbital velocities of hundreds of km/s.

(2) Walborn and Blanco (1988) report evidence for deceleration of the outlying ejecta around η Car. They suggest that pre-1820 ejection dates deduced from proper motions of the outermost visible blobs are invalid. However, the proposed rates of deceleration seem implausible on energetic grounds.

(3) Goodrich *et al.* (1989) have obtained spectra of the site of "SN 1961V" in NGC 1058, mentioned above. They believe that this object was indeed an LBV outburst rather than a supernova, and that it was more extreme than η Car.

REFERENCES

Davidson, K. 1971, *Monthly Not. Roy. Astron. Soc.* **154**, 415.

Davidson, K. 1987a (D87), in *Instabilities in Luminous Early Type Stars* (ed. Lamers & de Loore), p. 127.

Davidson, K. 1987b, *Astrophys. J.* **317**, 760.

Davidson, K., Dufour, R.J., Walborn, N.R., & Gull, T.R. 1986 (DDWG), *Astrophys. J.* **305**, 867.

Davidson, K., & Humphreys, R.M. 1986, *Astron. Astrophys.* **164**, L7.

Goodrich, R.W., Stringfellow, G.S., Penrod, G.D., & Filippenko, A.V. 1989, *Astrophys. J.*, in press (July 15 issue?)

Hackwell, J.A., Gehrz, R.D., & Grasdalen, G.L. 1986, *Astrophys. J.* **311**, 380.

Heisler, J., & Alcock, C. 1986, *Astrophys. J.* **306**, 166.

Hofmann, K.-H., & Weigelt, G. 1988, *Astron. Astrophys.* **203**, L21.

Meaburn, J., Wolstencroft, R.D., & Walsh, J.R. 1987, *Astron. Astrophys.* **181**, 342.

Robinson, G. Mitchell, R.M., Aitken, D.K., Briggs, G.P., & Roche, P.F. 1987, *Monthly Not. Roy. Astron. Soc.* **227**, 535.

Ruiz, M.T., Melnick, J., & Ortiz, P. 1984, *Astrophys. J. Letters* **285**, L19.

Tammann, G.A., & Sandage, A. 1968, *Astrophys. J.* **151**, 840.

Walborn, N.R., & Blanco, B.M. 1988, *Publ. Astron. Soc. Pacific* **100**, 797.

Walborn, N.R., Blanco, B.M., & Thackeray, A.D. 1978, *Astrophys. J.* 219, 498.

Walborn, N.R., & Liller, M.H. 1977, *Astrophys. J.* **211**, 181.

Weigelt, G., & Ebersberger, J. 1986, *Astron. Astrophys.* **163**, L5.

Westphal, J.A., & Neugebauer, G. 1969, *Astrophys. J. Letters* **156**, L45.

van Genderen, A..M., & Thé, P.S. 1985, *Space Sci. Rev.* **39**, 317.

Zwicky, F. 1964, in *Stars & Stellar Systems, Vol. 8, Stellar Structure* (ed. Aller & McLaughlin; Univ. of Chicago Press), pp. 410-412.

DISCUSSION

Zickgraf: Were the brightness variations of η Car before the Great Eruption similar to the variations of S Dor and other LBV's? The amplitude seems to be of the same order of magnitude. This could mean that LBV's are in a pre-η Car phase.

Davidson: Indeed the pre-1820 behavior may have been a series of "normal" LBV outbursts. Of course we do not know whether LBV behavior normally leads to a dramatic catharsis à la η Car; but I suspect that for LBV's that are significantly less luminous, it does **not**. The similarity in photometric amplitude among various "normal" LBV eruptions is merely a consequence of just 3 circumstances: (1) the stars are fairly hot when quiescent ($T_{eff} \sim$ 15000--30000 K?), (2) their luminosities don't change much during eruptions, and (3) their coolest pseudo-photospheric temperatures (giving maximum visual brightness) are around 7000 K for the same reasons as in supernovae (see *Astrophys. J.* **317**, 760). So, unfortunately, eruptions caused by various different mechanisms might look photometrically alike.

De Groot: It is interesting to see that the oscillations of V 12 may be similar to the to the pre-1830 oscillations of η Carinae. Do you have an explanation for the difference in time-scale?

Davidson: I can offer only the usual vague generalities. Many of us suspect that the thermal time-scale for some outer portion of the star is relevant, but we don't really know whether this outer portion consists of the ejected layers, or a critical depth for the instability mechanism, or something else. And the rate of interior evolution may also be pertinent: conceivably each ejection event temporarily stabilizes the star until just a little bit of further evolution de-stabilizes it again. It seems intuitively plausible that for a given star the recurrence time should be correlated with the violence of the most recent event (this is another sense in which the term "eruption" evokes colorful geological analogies). Comparisons between stars with different L and M are not safe, but one obvious idea is that the typical recurrence time may be longer for lower luminosities.

Humphreys: (1) Regarding V 12 in NGC 2403, I have deep CCD frames of that field, showing stars of 22nd magnitude or fainter. V 12 is not present on the B and V frames but is there in R and I. So it is red today, maybe because of dust in the ejecta. (2) You mentioned a change in the structure of the H and He lines in η Car. The narrow component was absent between 1984 and 1987, but had reappeared in 1988. This time-scale seems very short.

Davidson: Maybe the short time-scale for this change indicates that the narrow-line gas is photoionized. A few years ago Zanella, Wolf, and Stahl proposed that a temporary disappearance of [Ne III] emission was due to an outburst that temporarily reduced the far-UV luminosity (*Astr. Astrophys.* **137**, 79). Admittedly this scenario seems inconsistent with the simplest stellar interpretation of Weigelt's faint objects, wherein the bright component is the LBV but most of the far UV is likely to come from the other components.

CHANGES IN THE ATMOSPHERIC STRUCTURE OF LBV'S DURING ERUPTIONS

Claus Leitherer, David C. Abbott, and Werner Schmutz
Joint Institute for Laboratory Astrophysics
University of Colorado and National Bureau of Standards
Boulder, CO 80309-0440

I. INTRODUCTION

Some LBV's undergo brightness variations in the visual of about 1-2 magnitudes on time-scales of decades. These outbursts occur at approximately constant bolometric luminosity with the stellar radius and the temperature changing correspondingly. This subgroup of the LBV's is generally called Hubble-Sandage variables or S Doradus stars and will be the focus of this talk. We present self-consistent radiation-hydrodynamic calculations to model the observations from minimum to maximum states.

II. THE ATMOSPHERIC MODELS

Previous analyses of LBV's were based on models which made use of hydrostatic, plane-parallel atmospheres representing the stellar photosphere and, in addition, an expanding envelope of a specified density structure to account for the stellar wind. A recent review of results based on this model has been given e.g., by Wolf (1987). While these approaches can be justified for typical O-stars, it is obvious that in LBV's several assumptions inherent in these models are not valid. Most importantly, the continuum forming region of LBV's is clearly extended and the hydrodynamic and radiative properties of the wind and the photosphere may not be separated. Keeping in mind these considerations, we will not review previous attempts to model LBV's semi-theoretically but report on new methods now in progress.

Our models make use of a spherically extended, dynamical NLTE atmosphere with 27 levels of H and He. The radiative transfer is calculated in the comoving frame. This code has been described in some detail by Hamann and Schmutz (1987). The effects of photospheric and wind blanketing due to metallic lines are accounted for empirically. We obtain the number of line scatterings relative to electron scatterings as a function of wind density from a Monte Carlo calculation following a photon on its way through the atmosphere. For a given pre-specified set of stellar parameters including the mass-loss rate

K. Davidson et al. (eds.), Physics of Luminous Blue Variables, 109–115.

and the velocity law we obtain line profiles of H and He and the continuum energy distribution which are then compared with the observations.

The hydrodynamic structure of the wind is computed with a modified CAK code which does not make use of the radial streaming approximation. The radiative acceleration is calculated from Abbott's (1982) line list which is essentially complete up to Z = 30. The ionization and excitation in the wind is treated as in Abbott and Lucy (1985). Multi-line effects (multiple scatterings and line-overlap) are empirically accounted for by a Monte Carlo simulation as described above. The equation of motion of the flow is solved for a set of stellar parameters which in turn leads to a prediction for the mass-loss rate and the velocity field v(r).

By combining the two codes for the radiative transfer and the hydrodynamics we developed an iterative scheme to find a self-consistent solution. The photospheric parameters, which are input for the CAK code, are derived from the dynamical NLTE atmosphere as $R_{2/3}$ and $T_{2/3}$, where 2/3 refers to the region in the atmosphere where Rosseland optical depth 2/3 occurs. Using these photospheric parameters, we obtain a hydrodynamic solution for the wind with a new mass-loss rate and v(r) which in turn leads to a new estimate for the multi-line effects and can also be compared with the wind parameters derived from the line-profile fitting.

The results presented here are based on a large number (~50) of converged models obtained from the iteration scheme outlined above.

III. THE IONIZATION STATE OF THE ATMOSPHERE

A very striking property of LBV's from the point of view of a radiatively driven wind is the non-variable bolometric luminosity at different phases. Since a radiatively driven wind is primarily sensitive to L, the huge $\dot{M}$ variations observed in LBV's between maximum and minimum states are at first glance not obvious to understand in terms of radiation pressure.

Figure 1 shows the observed energy distributions of R71 in its minimum state and of R88 in its maximum state. Although the spectrum of R71 shortward of ~3000 Å has not been observed during its latest maximum (around 1974), the close similarity in their stellar parameters implies that the IUE region of R71 at maximum should resemble R88 at maximum. Notice the absence of the Balmer discontinuity in the maximum as well as in the minimum phase emphasizing a very low gravity and the importance of sphericity effects. The discontinuities visible in the UV spectra are caused by line blanketing. The minimum spectrum of R71 is blanketed by lines of (mostly) photospheric doubly ionized iron-group elements. On the other hand, the spectrum of R88 at maximum shows very strong wind blanketing due to lines of singly ionized iron-group elements. It is obvious that a self-consistent model of the expanding atmospheres of R71 and R88 must account for these dramatic changes in ionization and excitation.

Figure 2 shows the dependence of $\dot{M}$ and the terminal velocity v_∞ on the ionization state of the wind expressed by T_{ion}. The stellar

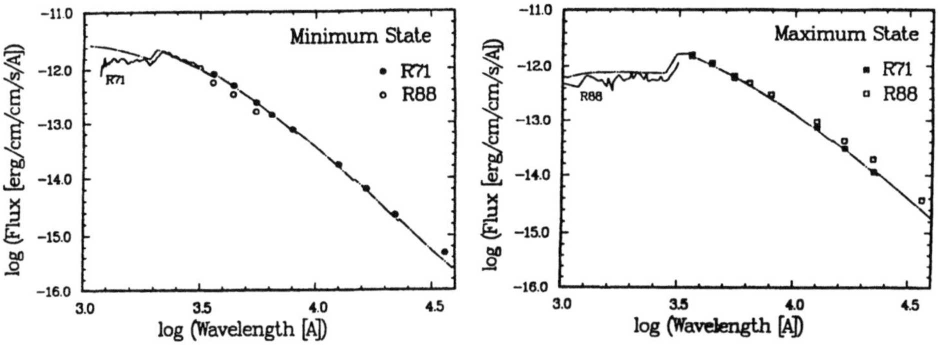

Fig. 1. Observed and computed energy distributions of R71 and R88 at minimum and maximum states.

parameters refer to our final model of R71 in the minimum states (see Table 1). In the right-hand section of the two figures the important metallic lines driving the wind are doubly ionized, whereas on the left side they are singly ionized. The strong sensitivity of $\dot{M}$ and v_∞ can be explained by the different term schemes of singly and doubly ionized iron-group elements. The important lines, which provide most of the wind acceleration, are resonance lines, lines originating from metastable levels, or other low-excitation lines, which are located longward of 911 Å. (The stellar flux shortward of 911 Å is too small for an efficient momentum transfer.) Doubly ionized iron-group elements have most of their strong, low-lying transitions shortward of 911 Å, whereas longward of 911 Å most lines are due to transitions with lower excitation potentials above ~5 eV. These lines can only be excited collisionally. They are strong in the photosphere (cf. Figure 1) but not in the wind around the critical point. In contrast, singly ionized iron-group elements possess most of their strongest resonance

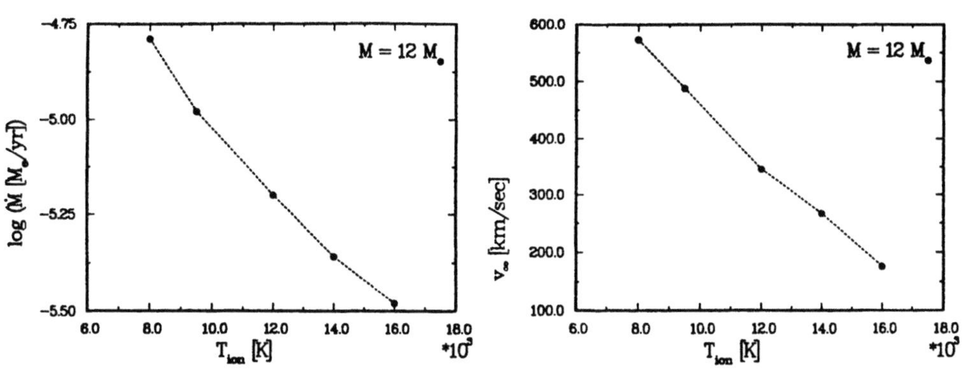

Fig. 2 Mass-loss rate and terminal velocity as a function of the ionization state of the wind.

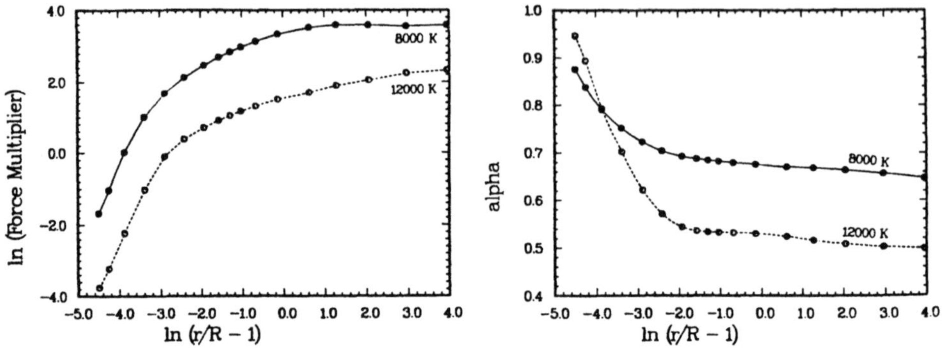

Fig. 3 Run of the force multiplier and α for two models with different values of T_{ion}.

lines longward of 911 Å in the Balmer continuum, which provides an efficient momentum transfer from the stellar radiation field to the stellar wind.

Figure 3 shows the behavior of the force multiplier (= the ratio of acceleration due to lines and due to continuum) and α (= relative fraction of optically thick [α=1] and thin [α=0] lines) in two representative wind models. The model with T_{ion} = 8000 K has a higher line acceleration and thus a higher mass-loss rate and a higher fraction of optically thick lines producing a higher terminal velocity since $v_\infty \sim \alpha/1-\alpha$ (cf. Kudritzki et al. 1987).

For reasons of clarity, the above discussion assumed identical photospheric parameters for the maximum and minimum states, only T_{ion} has been varied. This will now be relaxed, and we determine $R_{2/3}$ and $T_{2/3}$ from an atmospheric fine analysis with our dynamical NLTE atmosphere. During all phases we have $R^2(2/3)\ T^4(2/3)$ = const. The results derived above for $\dot{M}$ are virtually unaffected if one adopts the photospheric parameters appropriate for a given T_{ion}. This is because the stellar energy distribution in the Balmer continuum is not very sensitive to a change in $T_{2/3}$ between 16000 K and 8000 K if L is constant. On the other hand, log g is much lower in the maximum state, and so is the surface escape velocity. Since the terminal velocity of the wind scales with v_{esc}, we find that v_∞ is essentially unchanged between the minimum and maximum states due to the counteracting influence of α and v_{esc}.

Table 1 presents the most important wind parameters of R71 in maximum and minimum states derived from our hydrodynamic analysis. The results are in good agreement with what is derived from a detailed line profile and continuum analysis without making specific assumptions about the acceleration mechanism (see the poster by Leitherer et al., these proceedings). This provides support for the assumption that radiation pressure accelerates the stellar wind outside the optically thick photosphere.

Table 1. Converged hydrodynamic wind models for R71

	$T_{2/3}$	$R_{2/3}$	log $\dot{M}$	v_{∞}	∞
Minimum	15500	88	-5.92	234	0.97
Maximum	8100	332	-4.79	259	1.1

IV. THE VARIATION OF THE PHOTOSPHERIC RADIUS

In our models the effective temperature and the stellar radius are defined by

$$4\pi R_c^2 \ \sigma \ T_{eff}^4 = L$$

where the expansion velocity of the atmosphere is negligible at R_c. Models which are very extended and/or have high mass-loss rates have $T_{2/3} \neq T_{eff}$ and $R_{2/3} \neq R_c$. It appears attractive to interpret the observed changes in $R_{2/3}$ and $T_{2/3}$ in terms of opacity effects, where $R_{2/3}$ occurs further out in the atmosphere due to the higher wind densities and R_c remains fixed. However, given the functional dependence of T(r), v(r), and $\kappa\rho(r)$, we find that the difference between R_c and $R_{2/3}$ is by far less than the observed factor-of-4 variation, even if $\dot{M}$ is varied by a factor of 100. This conclusion is derived from our line-profile and continuum analysis as well as from our hydrodynamic calculations.

Davidson (1987) discussed the relation between apparent temperature, the velocity of the photosphere, and $\dot{M}$ in LBV eruptions. Lamers (1987) used this relation to derive a formula for the V-amplitude between maximum and minimum states of an LBV and the difference in $\dot{M}$. Our results disagree with this relation. While we agree with Davidson's result that $T_{2/3}$ of R71 and R88 at outburst is around 8000 K, we disagree in that we can not simultaneously reproduce the quiescent state of these stars with the same value of R_c; in other words, these stars do not move along one of the graphs in Lamers' (1987) Figure 2.

Table 2 summarizes the photospheric parameters found for R71. Notice the low gravity even in the minimum state and the low mass. The IR energy distribution is severely affected by the low log g and sphericity effects (as is the Balmer jump). The good agreement between observed and theoretical energy distribution (cf. Figure 1) is due to the use of unified, extended model atmospheres. The "IR excess" which arises when plane-parallel atmospheres without stellar winds are compared to observations is accounted for self-consistently.

Table 2. Photospheric parameters of R71

	log L	T_{eff}	R_c	$T_{2/3}$	$R_{2/3}$	log g	M
Minimum	5.5	16000	84	15500	88	1.3	12
Maximum	5.5	10000	230	8100	332	0.0	12

We find a relatively extended subsonic region of v(r) with a very gradual acceleration and a steep velocity law in the wind region. This is in contrast to what is usually adopted for LBV's, namely a very shallow wind law (Leitherer et al. 1985, Waters and Wesselius 1986). Despite these differences, $\dot{M}$ derived here is not too different from what is published in the literature. This is due to the fact that a core-halo model with a linear v(r) happens to give the same answer when deriving $\dot{M}$ as does a unified, extended model with a hydrostatic density stratification plus a steep wind law.

V. CONCLUSIONS

Current spherical, dynamic models, which treat the radiation-hydrodynamics in the expanding atmosphere self-consistently are in good agreement with the observations of R71 (and R88) at maximum and minimum states. The mass derived for R71 (~12 $M_{\odot}$) clearly hints at this LBV being a post-RSG. The same may be true for R88.

Radiation pressure is able to account for the $\dot{M}$ variations between the minimum and maximum states due to the recombination of lines of doubly ionized iron-group elements in the relevant temperature regime. However, it is important to realize that this is no instability mechanism. Rather, it gradually operates between 16000 K and 8000 K. The drastic change in the photospheric radius observed in connection with the change of $\dot{M}$ can not be accounted for self-consistently in our models. Although recombination and ionization effects may be responsible for the so-called microvariability, which occurs in time-scales of weeks, it is hard to understand why these effects can account for variations on time-scales of decades. Probably the triggering mechanism has its origin in deeper, subphotospheric layers.

This work has been supported by NSF grants AST88-02937 and AST88-06594 through the University of Colorado. C. L. acknowledges receipt of a Feodor-Lynen Fellowship of the Alexander-von-Humboldt Foundation.

References

Abbott, D. C. 1982, Ap. J., **259**, 282.
Abbott, D. C., and Lucy, L. B., 1985, Ap. J., **288**, 679.
Davidson, K. 1987, Ap. J., **317**, 760.
Hamann, W.-R., and Schmutz, W. 1987, Astr. Ap., **174**, 173.
Kudritzki, R. P., Pauldrach, A., and Puls, J. 1987, Astr. Ap., **173**, 293.
Lamers, H. J. G. L. M. 1987, in Instabilities in Luminous Early-Type Stars, ed. H. Lamers and C. deLoore (Dordrecht: Reidel), p. 99.
Leitherer, C., Appenzeller, I., Klare, G., Lamers, H. J. G. L. M., Stahl, O., Waters, L. B. F. M., and Wolf, B. 1985, Astr. Ap., **153**, 168.
Waters, L. B. F. M., and Wesselius, P. R. 1986, Astr. Ap., **155**, 104.
Wolf, B. 1987, in IAU Symposium 122, Circumstellar Matter, ed. I. Appenzeller, C. Jordan (Dordrecht: Reidel), p. 409.

DISCUSSION

Appenzeller: Did I understand correctly that your minimum and maximum models result simply from using different core radii and that you can also produce (hypothetical) intermediate models by choosing an intermediate core radius?

Leitherer: Yes, we varied the core radius in the range implied by observations. We were not able to reproduce the observed energy distribution and line profiles with a fixed core radius and variable $R(\tau=2/3)$ only.

Cassinelli: Stellar interior theorists tell us that stars that lose mass and become quasi-homogeneous evolve to the left, toward the homogeneous-He-burning main sequence. Why does your star stop its leftward evolution around 16000 K?

Leitherer: Our models only try to reproduce the constraints given by the observations, *i.e.*, effective temperature between about 16000 K and 8000 K. The question of why the LBV phenomenon occurs just in this temperature region and what prevents the star from going leftward should be addressed to stellar-evolution theorists.

Langer: What relation is there between your proposed mass-loss-enhancing mechanism due to Fe recombination and the Munich group's model wherein LBV mass loss is provoked by the recombination of hydrogen?

Leitherer: Both mechanisms involve changes in the ionization conditions in the wind. The Fe II/III mechanism that I described in my talk, however, generates no instability or outburst. It operates relatively continuously in the relevant temperature region. We encountered instabilities several times in our series of model calculations with our core-halo CAK code, but they turned out to be artifacts when we tried to confirm them with our self-consistent Monte Carlo code.

Schmutz: It is important to realize that the mechanisms of S Dor type outbursts proposed by the Munich group (Pauldrach *et al.*, this meeting) and by us (Leitherer *et al.*, this meeting) are identical. The high mass-loss rate is due to Fe II lines, the low state occurs if iron is Fe III. Therefore it is only a question of the iron ionization balance in the wind whether the mass-loss rate is high or low. We achieved the ionization shift by adopting the observed temperatures of S Dor and R 71 in maximum and minimum states, respectively. The Munich group gets the ionization shift at constant temperature by increasing the mass-loss rate, *i.e.* by increasing the density in the wind.

Williams, Balona, van der Hucht, Sterken

THE B[e] STARS

F.-J. ZICKGRAF
Landessternwarte Königstuhl
D-6900 Heidelberg 1

ABSTRACT. B[e] supergiants show evidence for a non-spherical two-component stellar wind. The general appearance and the physical properties of the suggested disk-like configuration are discussed. The high mass-loss rates, the surprisingly large number and the location in the H-R diagram make these stars important for the understanding of the post-main-sequence evolution of massive stars.

1. Introduction

In the Magellanic Clouds (MCs) presently 10 B[e] supergiants are known: three in the SMC and 7 in the LMC (Stahl et al. 1983, 1984, 1988, Zickgraf et al. 1985, 1986, 1988). Apart from several individual characteristics they show the following common properties

- optical spectra dominated by strong Balmer emission lines (some stars with P Cygni profiles) with Hα equivalent widths of the order of 10^2 to 10^3 Å;
- narrow, low-excitation metallic emission lines like FeII, [FeII] and [OI];
- strong infrared excesses due to thermal reradiation of circumstellar dust ($T_{dust} \approx$ 1000 K, $R_{dust} \approx 300\ R_*$).

In contrast e.g. to the S Dor variables the B[e] supergiants show *no* striking photometric variations (at most $0.^m1$ to $0.^m2$). Spectroscopic variations are also very weak with the exception of Hen S 18 in the SMC which shows a strongly variable HeIIλ4686 emission line (cf. Shore et al. 1987, Zickgraf et al. 1988).

2. The two-component wind model

R 126 in the LMC can be regarded as the proto-type of the B[e] supergiants. Its spectrum has been described by Zickgraf et al. (1985) as *hybrid*, meaning the simultanous presence of broad UV resonance absorption lines of CIV, SiIV, and NV indicating expansion velocities of about 1600 km s^{-1} and very narrow FeII, [FeII]

K. Davidson et al. (eds.), Physics of Luminous Blue Variables, 117–120.

and [OI] emission lines with typical line widths of several 10 $km\,s^{-1}$. Hen S 18 in the SMC is another example of this kind. Sharp emission lines of FeI, FeII, [FeII], and [OI] (FWHM $\approx$ 20 to 60 $km\,s^{-1}$) are contrasted by broad P Cygni absorption components of the Balmer lines ($v_\infty \approx$ -750 $km\,s^{-1}$). Obviously two different environments of line formation are required to produce such spectra. Zickgraf et al. 1985 suggested a two-component wind model consisting of a hot and fast line-driven (Castor et al. 1975) normal OB star wind in the polar region and a dense , cool and slowly expanding wind in the equator zone forming an *excretion* disk. The disk is the site of dust condensation and the formation of the low excitation lines. The molecular band emission of TiO and/or CO observed in some B[e] supergiants (Zickgraf et al.1988, McGregor et al. 1988) is also supposed to originate in the disk.

Individual characteristics of the B[e] supergiants are explained by taking into account different angles of aspect. Whereas R 126 is an example of a pole-on seen star, R 50 in the SMC represents the edge-on case showing no evidence for a high velocity wind component. Its line spectrum exhibits properties similar to a shell star (cf. Zickgraf et al. 1986) with Beals type III P Cygni profiles of the Balmer lines and sharp absorption features of singly ionized metals, both indicating a low expansion velocity.

The fast wind component is visible in seven out of ten B[e] supergiants either in the UV or in the visual wavelength region. Only in two stars (R 50 in the SMC and R 82 in the LMC) no evidence of a fast wind is found. One star is an intermediate case (R 4 in the SMC). With the assumption of the same overall geometry for all B[e] supergiants and randomly distributed inclination angles this small statistics allow to estimate the solid angle of the disk to about 20 to 30 % of the full sphere corresponding to a disk opening angle of 20° to 40°. Note, that White and Becker (1985) derived a disk opening angle of this size for the galactic B[e] supergiant MWC 349 from VLA observations.

Near IR excess in the J-band and Hα emission-line strength allow to derive the emission measure *EM*. It was found to be of the order of 10^{62} cm^{-3} for all objects of the sample (cf. Zickgraf et al. 1988). With a disk geometry as suggested above and a disk radius $R_{disk} \approx 300\ R_*$ densities of several 10^9 cm^{-3} were derived. A similar result was independently obtained by McGregor et al. (1988) from CO emission.

The ratio of the mass flux densities f_{disk} and f_{pole} in the different wind components can be estimated as follows. Assuming for the excretion disk the above given density, a mean radius of $10^2\ R_*$, and $v_{exp} = 20$ $km\,s^{-1}$ and taking the polar wind to be a normal OB star wind with mass loss rate and flux density given by the relations of Lamers (1981) values of f_{disk}/f_{pole} of the order of 50 were estimated. With a ratio of the expansion velocities of $v_{pole}/v_{disk} \approx 20$ the density ratio between disk and polar wind is typically 10^2 to 10^3 (cf. Zickgraf et al. 1988).

The total mass-loss rate $\dot{M}_{tot}$ is determined by the mass outflow in the excretion disk. For Hen S 18 in the SMC e.g. mass-loss rates of $\dot{M}_{disk} \approx 4\ 10^{-5}\ M_\odot\ yr^{-1}$ and $\dot{M}_{pole} \approx 4\ 10^{-6}\ M_\odot\ yr^{-1}$ for the two wind components were obtained, i.e. $\dot{M}_{tot} \approx \dot{M}_{disk}$. $\dot{M}_{tot}$ is comparable to that of LBVs, however, it appears not to be variable but rather stable.

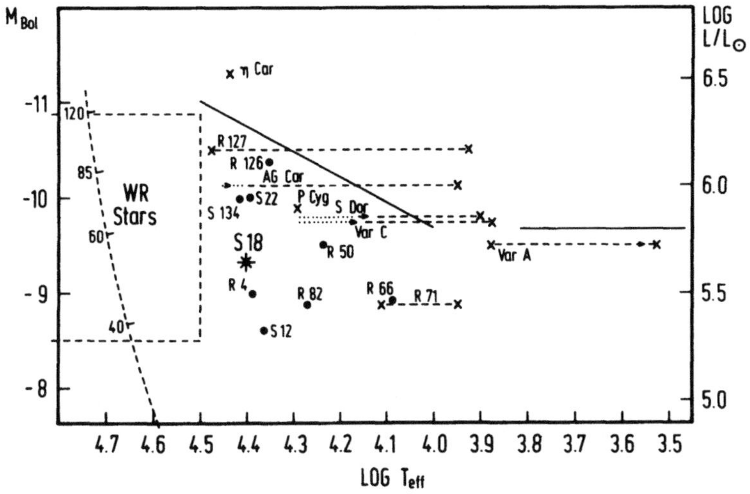

Figure 1: H-R diagram showing the location of the B[e] supergiants and the LBVs (×)

In the H-R diagram the B[e] supergiants and S Dor variables are located nearly in the same region (see Figure 1). Note the large number of B[e] supergiants compared to the S Dor type stars. In the SMC e.g. no S Dor variables are known, whereas already three B[e] supergiants have been found in this galaxy. Due to the surprisingly large number, the high mass-loss rates and the location in the H-R diagram the B[e] supergiants are certainly an important evolutionary stage of massive stars.

References

Castor, J.I., Abbott, D.C., Klein, R.I.: 1975, *Astrophys. J.* **195**, 157

Lamers, H.J.G.L.M.: 1981, *Astron. Astrophys.* **245**, 593

McGregor, P.J., Hillier,D.J., Hyland, A.R.: 1988, *Astrophys. J.*, in press

Shore,S.N., Sanduleak, N., Allen, D.A.: 1987, *Astron. Astrophys.* **176** 59

Stahl, O., Wolf, B, Zickgraf, F.-J., Bastian, U., de Groot, M.J.H., Leitherer, C.: 1983, *Astron. Astrophys.* **120**, 287

Stahl, O., Leitherer, C., Wolf, B., Zickgraf, F.-J.: 1984, *Astron. Astrophys.* **131**, L5

Stahl, O., Smolinski, J., Wolf, B., Zickgraf, F.-J.: 1988, IAU Coll. **113** "Physics of Luminous Blue Variables"

Zickgraf, F.-J., Wolf, B., Stahl, O., Leitherer, C., Klare, G.: 1985, *Astron. Astrophys.* **143**, 421

Zickgraf, F.-J., Wolf, B., Stahl, O., Leitherer, C., Appenzeller, I.: 1986, *Astron. Astrophys.* **163**, 119

Zickgraf, F.-J., Wolf, B., Stahl, O., Humphreys, R.M.: 1988, *Astron. Astrophys.*, submitted

DISCUSSION

Shore: (1) It is gratifying to see that the mass-loss results that you are deriving from optical data agree with results from the ultraviolet. We find C III] and Si III] both giving densities of 10^8--10^{10} cm^{-3} in some of the emission-line stars. HD 38489, S 18/SMC, and S 131/LMC are good examples of this. (2) Here is a crazy, toy, model. There must be massive systems with large mass ratios between components. What happens if one star is evolving through the supergiant region while the other is still on its way to the main sequence? Ablation, making the companion look something like a comet, should produce many of the same effects one sees in B[e] stars. This is not the same as the model we proposed for S 18/SMC (Shore, Sanduleak, & Allen 1987, *Astr.&Astrophys.* **176**, 59) nor that which I have suggested for R Aqr with an accreting companion sitting in the wind of a massive star, perhaps in an elliptical orbit. We might ask: What are the environmental effects of massive stars when less massive stars are formed nearby?

Walborn: Conti *et al.* have argued observationally that mass ratios near unity are favored for massive stars, and that very large ratios are rare.

Shore: So are LBV's.

Henrichs: What supports the excretion disk against gravity in your picture?

Zickgraf: It is supported by radiation pressure, probably due to opacity in the lines of ionized metals, similar to the mechanism suggested for S Dor winds.

Friedjung: There are good reasons for supposing that disks exist around these stars, but their nature is uncertain. A few years ago we suggested that S 22 has an excretion disk (Bensammar *et al.* 1983: *Astr.&Astrophys.* **126**, 427). Though this interpretation is less likely now, we must remember that we do not understand very well the disks supposed to exist around normal Be stars. Winds can come from both the central star and the disk. Moreover, some disk regions may be optically thick and such disks may be heated by radiation from the central star; so your density may not represent the whole disk.

Lamers: The spectral energy distributions of B[e] stars are very similar to those of classical Be stars. In classical Be stars, the IR excess is due to free-free emission (see, *e.g.*, Waters 1986, *Astr.&Astrophys.* **162**, 121). How do you know that the far-IR excess of B[e] stars is due to dust?

Zickgraf: The far-IR energy distributions can be fit well by black-body curves with temperatures around 1000 K. At wavelengths greater than 2 or 3 μm they are clearly different from free-free emission.

Conti: Does any observer have any information on rotational velocities of B[e] stars? Are they indeed rapid?

Zickgraf: The lack of photospheric absorption lines in B[e] supergiants makes it practically impossible to determine their $v \sin i$.

MODELING MASS LOSS FROM B[e] STARS

J.P. CASSINELLI, R.E. SCHULTE-LADBECK, C.H. POE*, and M. ABBOTT

Department of Astronomy
University of Wisconsin-Madison
475 N. Charter Street
Madison, WI 53706

**High Altitude Observatory*
NCAR
P.O. Box 3000
Boulder, CO 80307-3000

1. Introduction

It has been suggested by Zickgraf et al. (1986) that the outer atmosphere of some B[e] stars has a two-component structure: a fast, radiation-driven wind from the pole, and a dense, slow outflow from the equator. This is also the basic picture investigated by Poe, Friend, and Cassinelli (1989) (hereafter PFC) to explain the momentum problem associated with Wolf-Rayet stars. Here we apply the PFC multiforce wind theory to model the B[e] outflow phenomenon.

The theory requires that the B[e] stars rotate rapidly, i.e., $\omega = v/v_{crit}$ greater than about 0.5, where

$$v_{crit} = \sqrt{\frac{GM\,(1-\Gamma)}{R}} \tag{1}$$

M and R are the stellar mass and radius and Γ is the ratio of the luminosity to the Eddington maximum luminosity; $\Gamma = k_F\, L/(4\pi c\, GM)$ with k_F being the flux mean opacity.

We have investigated two general questions. 1) Can B[e] stars be rotating near critical speed, and if so, 2) what constraints can be placed on the parameters that determine the two-component flow structure?

2. Near-Critical Rotation of B[e] Stars

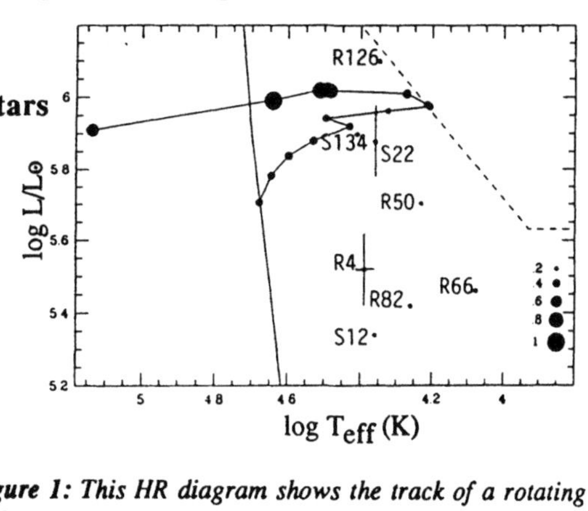

Figure 1: This HR diagram shows the track of a rotating 60 $M_\odot$ star and the positions of observed B[e] stars.

We have used information regarding the density distribution of non-rotating stars to investigate the possibility that $\omega > 0.5$ for luminous post-main sequence stars. The results are shown in Figure 1, where the size of the dots is proportional to ω. Our main conclusion is this: high ω values can only occur after the star has undergone substantial mass loss. Stars evolve back to the left in the HR diagram from the Humphreys -

K. Davidson et al. (eds.), Physics of Luminous Blue Variables, 121–125.

Davidson limit after they have lost at least 40% of their initial mass. In this lowered mass, overly luminous state, v_{crit} is reduced, and ω can be increased as is shown in Figure 1.

We infer the following: a) B[e] stars are in the post Humphreys-Davidson part of their evolutionary track, and are moving to the left in the HR diagram, b) the B[e] stars are possibly the precursors of rapidly rotating Wolf-Rayet stars, and c) the B[e] stars are not an extrapolation, to high luminosity, of classical Be stars, but require a pre-existing major mass-loss phase.

3. Theoretical Constraints on the B[e] Winds

We have applied the PFC theory and used limits on the surface magnetic field as derived from the analysis of Maheswaran and Cassinelli (1988).

The wind theory of PFC accounts for three forces to drive the equatorial flow: the line-radiation force of the Castor, Abbott, and Klein (1975, CAK) theory, the effects of a finite-sized disk, and the centrifugal and magnetic forces. The effects of all of these forces can best be illustrated in a diagram of mass-loss rate ($\dot{M}$) vs. terminal wind speed (v_∞), as is shown in Figure 2a. The solid lines show the dependence of $\dot{M}$ and v_∞ on rotation rate and surface magnetic fields, B, ranging from 0 to 500 Gauss. Several interesting limits and boundaries can be discerned. There is a minimal mass-loss rate that is set by the Friend-Abbott (1986, FA) line-driven wind theory; increasing ω leads to larger $\dot{M}$. The asymptotic behavior of $\dot{M}$ vs. v_∞ for the magnetic cases can be rather well described by the Michel velocity, v_M, of "fast magnetic rotator" (FMR) theory (Hartmann and MacGregor 1982),

$$v_\infty \approx v_M = \left(\frac{(R^2B_o)^2\,\Omega^2}{\dot{M}}\right)^{1/3} \tag{2}$$

where B_o is the radial component of the magnetic field at R, and $\Omega^2 = GM(1-\Gamma)/R^3$. Lines satisfying equation (2) are shown in Figure 2b and demonstrate that the radiation and magnetic fields needed to explain the wind properties can be derived from rather simple considerations.

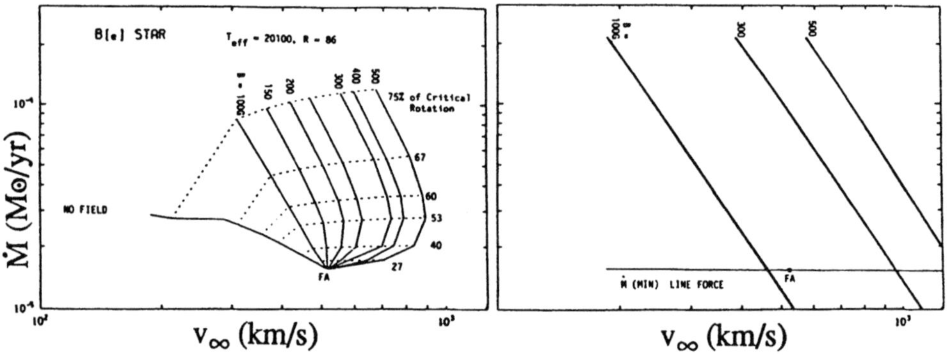

Figure 2a and b: Mass-loss rate versus terminal wind speed for different rotation rates and magnetic field strengths calculated numerically (left side) and using equation (2) (right side).

Constraints on the magnetic fields in the winds of rotating stars have recently been described by Maheswaran and Cassinelli (1988). They point out that rapid rotation leads to circulation currents in the stellar-interior envelope. If the circulation currents are faster than the Alfvén speed, the magnetic field becomes submerged below the stellar surface. This means that either the surface field is essentially zero or is greater than a certain lower limit. There is also an upper limit on the field beyond which hydrostatic structure is drastically modified. The allowed non-zero values for B_0 are shown in Figure 3. Combining these limits with the results of Figure 2a gives rise to Figure 4.

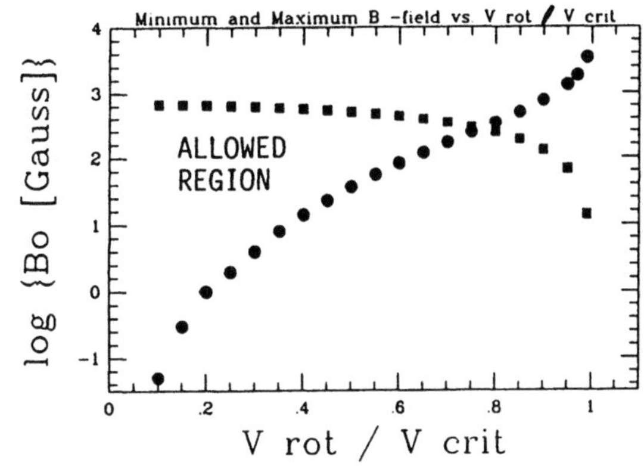

***Figure 3:** The logarithm of the magnetic field strength as a function of the rotation rate. The allowed region of values is indicated.*

We conclude that there are two allowed types of equatorial flow:

1) $B_0 = 0$ and one has the no-field rotationally-driven outflow that was first analyzed by Castor (1979) in conjunction with CAK line forces. Note that for this type of flow v_∞ decreases as $\omega = v_0/v_{crit}$ increases. Or,

2) B is in the range allowed by Figure 3 and the equatorial flow is determined by the PFC analysis.

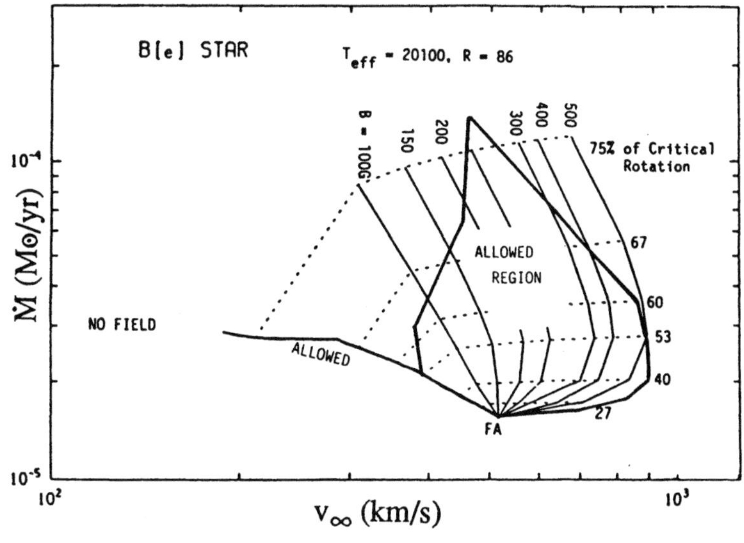

***Figure 4:** The numerical solutions for mass-loss rate versus terminal wind speed are shown for the same models as in Figure 2a, with the allowed values of the magnetic field indicated.*

The density and velocity structure for our model with $T_{eff} = 20{,}100$ K and $R = 86\ R_\odot$ is shown in Figure 5a. Note that the polar terminal speed is not as high as the observed about 1500 km s^{-1} in R126. The speed is increased for $T_{eff} = 40{,}000$ K (Figure 5b). This temperature is higher than indicated by the continuum of R126. However, we think that it is quite plausible that the effective temperatures of B[e] stars have been underestimated because of extended atmosphere effects. The extended equatorial region may well provide the radiation that flattens the continuum, giving rise to a lower temperature estimate.

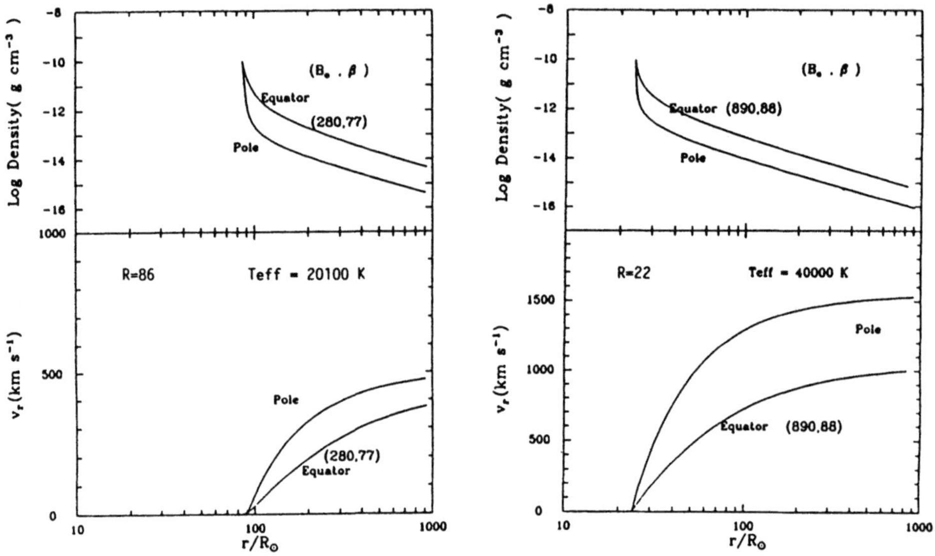

Figure 5a and b: The logarithmic density and velocity structure of two models are shown. The outflow from the cooler star (left panel) is slower than the one from the hotter star (right panel).

We should also comment on the evidence for 900 K dust in R126 (Zickgraf et al.). Using the Mathis, Rumpl and Nordsieck (1976) grain model, we find that even for the 20,100 K model the grain temperature exceeds 900 K all the way to 2000 R, where the density shown in Figure 5a has fallen below the minimal value of 10^{-17} gm cm^{-3} needed for grain growth. The fact that grain growth occurs requires that the outflowing material be optically thick. This will only happen if there is a strong density enhancement as suggested for the equatorial zone of a two-component model. We consider the ongoing grain formation to be added evidence that the flow from B[e] stars is not spherically symmetric.

Acknowledgements: R.E.S.-L. gratefully acknowledges financial support by NASA Contract NAS 5-26777, J.P.C. and M.A. were supported by NASA Grant NAGW-422.

References

Castor, J.I. (1978), in: IAU Symp. **83**, eds. P.S. Conti and C.W.H. de Loore (Dordrecht:Reidel), p.175.
Castor, J.I., Abbott, D.C., and Klein, R.I. (1975), Ap. J., **195**, 157.
Friend, D.B. and Abbott, D.C. (1986), Ap. J., **311**, 701.
Hartmann, L. and MacGregor, K.B. (1982), Ap. J., **259**, 180.
Maheswaran, M, and Cassinelli, J.P. (1988), Ap. J., **335**.
Mathis, J., Rumpl, , and Nordsieck, K.H. (1976), Ap. J., **217**, 425.
Poe, C.H., Friend, D.B., and Cassinelli, J.P. (1989), Ap. J., **336**.
Zickgraf, F.-J., Wolf, B., Stahl, O., Leitherer, C., and Appenzeller, I. (1986), Astron. Ap., **163**, 119.

DISCUSSION

Henrichs: What is the thickness of the disk in your model?

Cassinelli: We are solving the wind momentum equation in the equatorial plane and along the polar axis. Since we are not doing a 2-dimensional solution, I cannot answer your question from theoretical considerations. However, recall that Zickgraf estimated the equatorial zone in R 126 to have a half-angle of 15°.

Shore: If the envelope is differentially rotating and torus-like, then all bets are off concerning gravity darkening. It is generally true, however, that the measured rotational velocity ($v \sin i$) will be quite a bit less than the true velocity.

Vanbeveren: When a rotating star is losing mass as a consequence of angular momentum loss it will slow down -- even a contracting star. Then what is the mechanism that brings the star to break-up velocity in your model?

Cassinelli: The stars lose about 40% of their mass while in the neighborhood of the Humphreys-Davidson limit, before they enter the B[e] phase. I did not say that this pre-B[e] mass loss was caused by angular-momentum-driven winds. However, the mass loss has the effect of increasing the ratio v_ϕ/v_{crit} during the final evolutionary movement to the left in the H-R diagram. This is because $v_{crit} = \sqrt{GM(1-\Gamma)/R}$ is reduced by a factor of about 2, because M is smaller and Γ is larger than when the star was evolving to the right in its early post-main-sequence evolution.

Schulte-Ladbeck, Zickgraf, Cassinelli, Appenzeller

Bohannan, de Jager, Henrichs, Fitzpatrick, Garmany

LUMINOUS BLUE VARIABLES NEED NOT BE BLUE

Cornelis de Jager and Arnout M. van Genderen
Laboratory for Space Research Utrecht and Leiden Observatory

ABSTRACT. A number of yellow and red super- and hypergiants show phenomena that are similar to those shown by Luminous Blue Variables. The LBV phenomenon may not be restricted to the blue part of the Hertzsprung-Russell diagram and the conventional name 'S Dor variables' seems more appropriate.

1. CHARACTERISTICS OF LUMINOUS BLUE VARIABLES

It is convenient to distinguish between the LBV-characteristics and the LBV-phenomenon. LBV's are usually defined as blue stars that occasionally undergo mass ejections of such a degree that the level where the continuous optical depth reaches unity moves so far outward that the surface brightness, effective temperature and spectral type change considerably while also the visual apparent brightness changes - because the bolometric correction does so. If so defined, it appears that they occupy an area in the Hertzsprung-Russell diagram as shown in Figure 1 (after Van Genderen, 1988). These stars thus form a subclass of the Alpha Cygni variables. Van Genderen (1988) showed that the maximum light amplitude of the LBV's, in his paper called S Dor variables, are definitely larger that those of the other Alpha Cygni variables, and also larger than those of the hypergiants (the Ia+ type luminosity class). We refer to Figure 1 in Van Genderen's paper (1988).
LBV's must be pulsating stars. Their region of occupancy in the HR diagram is part of that of the (quasi-regularly pulsating) supergiants (Burki, 1978; Maeder, 1980). Van Genderen (1988) remarks that their light variation lacks the high-frequency components that are so often found in those of the non-LBV blue supergiants. This may mean that the average densities of LBV's (and hence also their masses) are smaller than those of non-LBV's at the same position in the Hertzsprung-Russell diagram. In comparison with such stars the LBV's must therefore be more evolved.
It is remarkable that LBV's do hardly occur at the Humphreys-Davidson limit (with the exeption of Eta Car); they are rather found somewhat below it.

2. THE LBV-PHENOMENON ALSO SHOWN BY COOLER STARS

A number of cool super- and hypergiants also show the LBV-phenomenon. We

K. Davidson et al. (eds.), Physics of Luminous Blue Variables, 127–129.

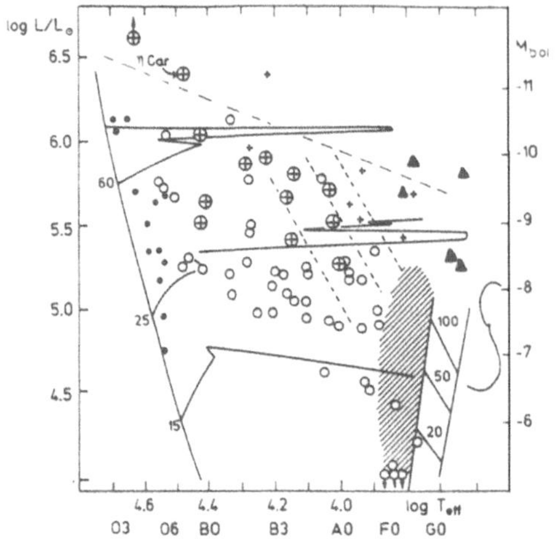

Fig.1: The S Dor Variables in the Hertzsprung-Russell Diagram

Dots: type V stars
Circles: supergiants
Plusses: hypergiants
Circles with plusses: LBV's (blue S Dor stars)
Triangles: cool S Dor stars

describe a few of them; we refer also to a description in De Jager (1980) ch. 3, particularly section 3.8.
Rho Cas (HD 224 014) was classified in 1943 as F8Ia, but during the years 1946-49 it had M-type characteristics, related to a mass ejection. Another ejection started in 1957; it lasted till 1960.
HR 8752 (V509 Cas, HD 217 476) varied between G0Ia to K5Ia and back between 1950 and now, due to a mass ejection between 1968 and 1980 (Piters et al., 1988; Smolinski, priv. comm.).
HR 5171A (V766 Cen. HD 119 796) is at present showing the effects of an increased rate of mass loss that occurred before or since 1953. Since that time the colour (B-V) increased from ~ 1.95 to ~ 2.5 in 1985, which shows the gradual building up of a dust envelope. The extinction at 3200 Å increased by a factor 30 between 1971 and 1987!
Other stars that behave similarly are Var A in M33 (Humphreys et al., 1987) and the infrared source IRC 10420 (F8Ia, Fix and Cobb, 1987).

We conclude that the LBV phenomenon is not restricted to early-type stars. Therefore it is more appropriate to use the name "S Dor stars" to describe the stars described above, in accordance with the definition of the General Catalogue of Variable Stars (Kukarkin et al., 1974; Kholopov et al., 1985).

REFERENCES:

Burki, G.: 1978, Astron. Astrophys. **65**, 357.
De Jager, C.: 1980. The Brightest Stars, Reidel, Dordrecht.
Fix, J.D., Cobb, M.L.: 1987, Astrophys, J. **312**, 290.
Humphreys, R.M., Jones, T.J., Gehrz, R.D.: 1987, Astron, J. **94**, 315.
Kholopov, P.N. et al.: 1985, Gen. Cat. of Var. Stars, 4th ed., Nauka, Moscow.
Kukarkin, B.V., et al.: 1974, Gen. Cat. of Var. Stars, 3rd ed., Nauka, Moscow.
Maeder, A.: 1980, Highlights of Astron. **5**, 473.
Piters, A., De Jager, C., Nieuwenhuijzen, H.: 1988, Astron. Aph. **196**, 115.
Spaan, F., De Jager, C., Nieuwenhuijzen, H.: 1989, Astron. Aph.; in prep.
Van Genderen, A.M.: 1988, Astron. Astrophys.; in press.

DISCUSSION

Sterken: The light curves that you showed for AG Car and R71 were collected with the same sampling frequency. We expect a shorter period for a less luminous supergiant, and if a short-period light curve is observed with the same time frequency as a longer-period variation, one may automatically obtain a bumpier light curve. Some of the supergiants may have this phenomenon in their observed light curves.

De Jager: I agree. The bumpy character of the light curves of normal supergiants (non-LBV's) may indicate that their periods are shorter than those of LBV's.

Humphreys: IRC +10420 is an OH/IR source. Recent observations show that the strength and structure of of the OH emission is changing. The changes are consistent with dissipation of the dust shell. IRC+10420 is a good candidate for post-red-supergiant evolution. I doubt that all of the cool hypergiants are post-red supergiants.

Walborn: (1) Is a spectral type range observed for HR 5171? (2) Another remarkable point of similarity between HR 8752 and LBV's, which you have discussed previously, is the [NII] emission which may indicate circumstellar processed material.

De Jager: I have no data on the spectral type variation of HR 5171. But the temperatures derived in the past 20 years decreased with the increasing extinction. This may suggest an expanding pseudo-photosphere.

Maeder: Do you also have indications on radial velocity variations for these yellow and red supergiants, in addition to the photometric monitoring?

De Jager: The only star for which a good series of v_{rad}-measurements is available over a sufficiently long period (several years) is α Cygni, which was observed by Paddock around 1930. Lucy has used these measurements for a Fourier analysis and found 16 frequencies with periods from 7 to 100 days (1976, *Astrophys. J.* **206**, 499). It is high time that such measurements be done for other supergiants.

de Jager and Smolinski

MASS EJECTIONS FROM THE G-TYPE HYPERGIANT HR 8752

J.Smoliński[1], J.L. Climenhaga[2] and J.M. Fletcher[3]

[1]*N. Copernicus Astronomical Center, Toruń, Poland*
[2]*Department of Physics and Astronomy, University of Victoria, Victoria, BC, Canada*
[3]*Dominion Astrophysical Observatory, Victoria, BC, Canada*

One of the most luminous stars in our Galaxy (Humphreys 1988) HR 8752 = HD 217476 (G0 Ia), is similar in some respects to Luminous Blue Variables. Studies by several authors (Sargent 1965; Smoliński 1971; Smoliński et al. 1977; Stickland and Harmer 1978; Lambert and Luck 1978; Lambert et al. 1981; Smoliński et al. 1986; Sheffer and Lambert 1987) have contributed to the understanding of activities in the atmosphere of this star, but its nature is still not well understood.

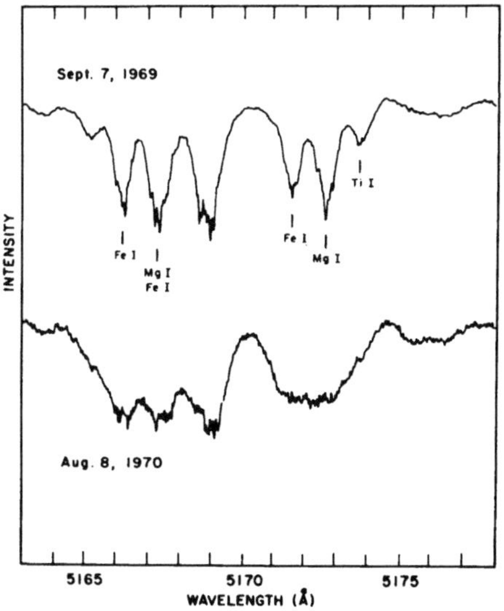

Fig. 1. Comparison of two spectra of HR8752 showing an increase in line widths during 1970.

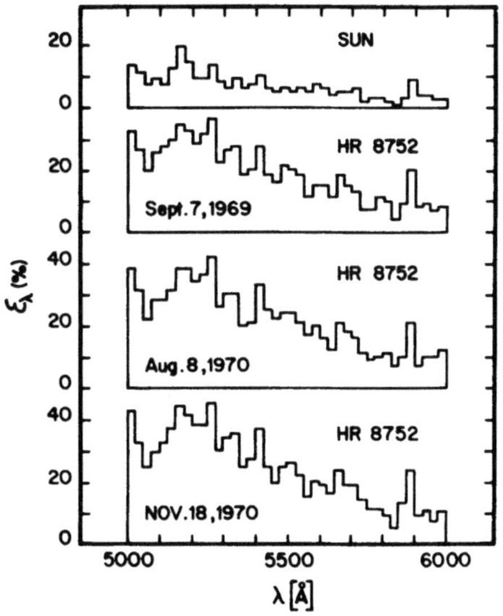

Fig. 2. Line absorption values ϵ_λ for HR 8752 during 1969 and 1970 compared with ϵ_λ for the Sun.

K. Davidson et al. (eds.), Physics of Luminous Blue Variables, 131–134.

Since 1969, we have made spectroscopic observations of this star with the same equipment at the Dominion Astrophysical Observatory, Victoria, BC, Canada. During this period various changes occured in the spectra: Changes in the broadening of absorption lines, splitting of some absorption lines into several components and the appearance of emission components in certain lines.

From our observations we are able to distinguish clearly at least 3 periods of time or episodes when spectral features are different.

(i) *The 1970 episode*: During this episode, which lasted at least 8 months, the absorption lines were at their broadest. The widths of the absorption lines increased by a large amount (by about a factor of 2) from 1969 to 1970. This increase can be seen clearly in Fig.1 which shows two spectra of HR 8752 obtained on September 7, 1969 and August 8, 1970. Line absorption values, ϵ_λ, in 1970 were found to be about 25% higher than in 1969 (Fig.2). Such a large increase should change the colour of the star considerably. In 1970, when the lines were very strong and broad it is possible to distinguish 3 blue-shifted absorption components of neutral metallic lines and of some ionized lines, with radial velocities with respect to the star system of about 35, 54 and 84 km/s.

(ii) *The 1979 episode*: During this period inverse P Cygni profiles were observed in the metallic lines, especially in the Fe I lines (Smoliński et al. 1979). The blue emission components were visible for at least 7 months with variable intensities (Fig.3) and with very little change in radial velocities.

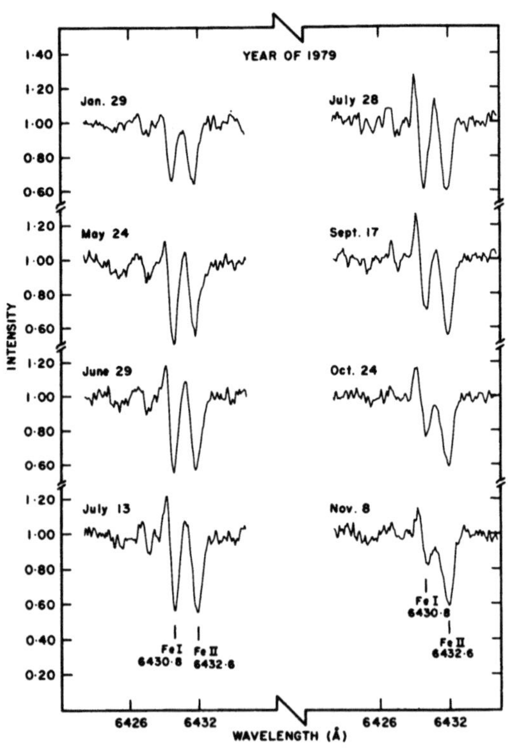

Fig. 3. A series of spectra of HR 8752 showing a blue shifted emission component of variable intensity in the Fe I line during 1979.

(iii) *The 1982 episode*: During this episode a blue-shifted absorption component with velocity of about -100 km/s was strong and clearly distinguished in neutral metallic lines, particularly in the Fe I lines (Fig. 4) and in the Ti II lines (Smoliński et al. 1982).

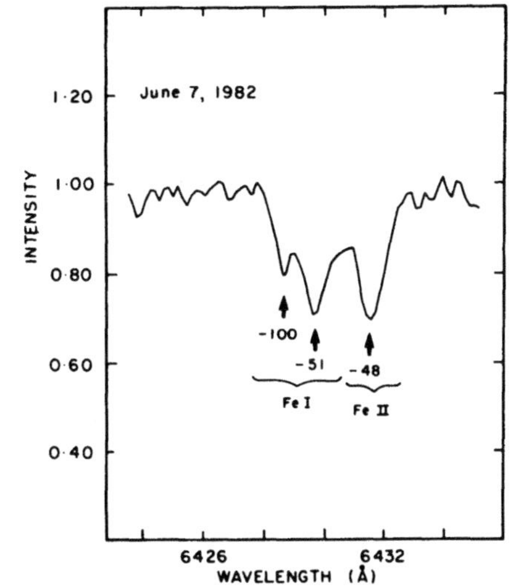

Fig. 4. In this June, 1982, spectrum of HR 8752, a blue shifted absorption component in the Fe I line is present and has a velocity of -100 km/s.

We interpret the changes during these episodes to be the result of mass ejections and the existence of an expanding circumstellar envelope.

Our results are in agreement with the model proposed by us (Smoliński, Climenhaga and Fletcher, 1986) suggesting that HR 8752 is a binary system with a period of 620 days contained in a common expanding envelope and surrounded by an H II region.

REFERENCES:

Humphreys, R.M.: 1988, in press.

Lambert, D.L., Hinkle, K.H., Hall, D.N.B.: 1981, *Astrophys. J.*, **248**, 638.

Lambert, D.L. and Luck, R.E.: 1978, *Mon. Not. R. Astr. Soc.*, **184**, 405

Sargent, W.L.W.: 1965, *Observatory*, **85**, 33.

Sheffer, Y. and Lambert, D.L.: 1987, *Pub. A.S.P.*, **99**, 1277

Smoliński, J.: 1971, in *Colloquium on Supergiant Stars*, ed. M. Hack, 68

Smoliński, J., Climenhaga, J.L., Funakawa, H. and Fletcher, J.M.: 1979, *I.A.U. Circ*, No. 3382.

Smoliński, J., Climenhaga, J.L., Funakawa, H. and Fletcher, J.M.: 1982, *Inf. Bull. Var. Stars*, No. 2229.

Smoliński, J., Climenhaga, J.L., Fletcher, J.M.: 1986, *Symposium on Luminous Stars and Associations in Galaxies, ed. De Loore, C.W.H., Willis, A.J., Laskarides*, P., 89.

Smoliński, J., Feldman, P.A., Higgs, L.A.: 1977, *Astron. Astrophys.*, **60**, 277.

Stickland, D.J., Harmer, D.L.: 1978, *Astron. Astrophys.*, **70**, L53.

DISCUSSION

Gallagher: (1) What is the period for this system? (2) Is it a super-symbiotic star?

Smolinski: (1) We have determined the orbital period to be 620 days. (2) This is an interesting suggestion. However, the orbital separation in Mira-type symbiotic stars is hundreds of AU, while our estimate of the separation for HR 8752 is only several AU. It does have some similarity to ζ Aurigae stars, but it is possibly different from both types.

Walborn: Did you suggest that the apparent change in spectral type is due to the appearance of extra blue-shifted components in the neutral lines, rather than a temperature change?

Smolinski: Yes, we suggest such a possibility. The presence of blue components in the Fe I lines as we observed in our 1970 high-dispersion spectra would affect the estimate of spectral class in low-dispersion spectra even without any temperature change. However, we do not rule out either possiblity.

Zickgraf: Where in your model are the inverse P Cygni profiles of the Fe I lines coming from?

Smolinski: We believe that the blue emission components of the inverse P Cygni profiles of the Fe I lines are produced in a shell surrounding the system and expanding with a velocity of about 35 km/s. During 1979 the emission components were visible for at least 7 months with continuously changing intensities.

McGregor: Lambert, Hinkle, and Hall found CO emission in HR 8752 (1981; *Astrophys. J.* **248**, 638). The CO lines at high resolution showed P Cygni profiles and later inverse P Cygni profiles. They interpreted this as a shell which was ejected and which then fell back onto the star.

Smolinski: Their infrared CO observations are interesting. However, our observations do not indicate the shell falling back onto the star.

MASS LOSS FROM LUMINOUS BLUE VARIABLES

Henny J. G. L. M. Lamers[1]
Joint Institute for Laboratory Astrophysics
University of Colorado and National Institute of Standards and Technology
Boulder, Colorado, 80309-0440 USA

1. INTRODUCTION

Luminous Blue Variables (LBV's) are losing mass at a rate which is higher than in normal stars of the same luminosity. This high mass loss is evident from the occurrence of P Cygni profiles in the visual spectrum, the large numbers of UV lines which are Doppler shifted or show P Cygni profiles and the large IR excess or radio free-free emission. Mass loss from LBV's is strongly variable on a wide range of timescales from months to centuries and possibly even millenia. During these variations the mass loss may vary from values as low as 10^{-6} to 10^{-5} $M_\odot$/yr, when the star is quiet, to outbursts of the type observed in P Cygni in AD 1600 and η Car in 1837 (for reviews see Davidson, 1987; Lamers, 1986, 1987).

The variability of mass loss is related to the photometric variability, in that $\dot{M}$ is higher when the star is visually brighter at lower T_{eff}. Since the observations of a small number of LBV's suggest that the photometric variations are due to changes in T_{eff} and R_* at constant L, we may conclude that $\dot{M}$ is highest when the photospheric radius is largest and the gravity is smallest.

The size of the variations in R_* and T_{eff} can be very substantial; e.g., the star S Dor changed from $T_{eff} \simeq 22000$ K and $R_* \simeq 44\ R_\odot$ during minimum V = 11.3 in 1965 to $T_{eff} \simeq 8000$ K and $R_* \simeq 330\ R_\odot$ during maximum V = 9.3 in 1983. The mass loss rate was 5×10^{-5} $M_\odot$/yr at maximum; the mass loss rate at minimum is unknown (Leitherer et al., 1985; Lamers, 1987). (It is worth noting that the values of T_{eff} and R_* are ill-defined in extended atmospheres because $R(\tau_\lambda \simeq 1)$ depends on λ. The values of T_{eff} given in this paper are some kind of spectroscopic or photometric temperature and R_* is the corresponding value derived from $L = 4\pi R_*^2\ T_{eff}^4$).

What causes the variations in R_*, T_{eff}, and $\dot{M}$? This is a chicken and egg problem: 1) does R_* increase because some subphotospheric

[1]On leave from SRON Laboratory for Space Research and The Astronomical Institute at Utrecht, The Netherlands

K. Davidson et al. (eds.), Physics of Luminous Blue Variables, 135–147.

mechanism forces the star to a higher mass loss rate so that $R(\tau \simeq 1)$ moves outward or 2) does $\dot{M}$ increase because some subphotospheric mechanism increases R_* and decreases g? The study of R71 by Leitherer et al. (1989) suggests that 2) is the more likely possibility. This replaces the question about the origin of the variable $\dot{M}$ by the question about the mechanism for the variations in R_*.

We should emphasize that the situation regarding the large eruptions is not clear at all. The fact that as much as ~1 $M_\odot$ can be ejected in a major eruption suggests an instability located deep below the photosphere (Davidson, 1989).

In this paper I will discuss the observational evidence for mass loss from LBV's and its variability. I will try to derive "time-averaged" mass loss rates which are important because of their evolutionary consequences. I must apologize that the values which I will derive are rather uncertain for the simple reason that the number of LBV's for which mass loss has been estimated is very small. The estimates about the variability of mass loss are even more scarce and the information about large eruptions is limited to two or three stars only. So the "time-averaged mass loss rates" depend on low number statistics. Unfortunately we cannot do better at the moment.

2. VARIABILITY AND TERMINOLOGY

A typical LBV light curve is shown in Fig. 1 for AG Car between 1969 and 1985 (Viotti, 1988). The maximum amplitude is $\Delta V = 2\overset{m}{.}6$, but the typical "ups and downs" have a range of $0\overset{m}{.}5 \lesssim \Delta V \lesssim 2\overset{m}{.}0$. I will call these the <u>moderate variations</u>. They are the characteristic LBV variations for which the mechanism is unknown. They seem to occur at constant L, and so ΔV is due to variations in the Bolometric Correction. The variations are not periodic but occur on a peak-to-peak timescale of years (in AG Car) to two decades (in S Dor). The maxima should not be called "outbursts," although this name is often used in the literature, because the observations show that they are extended periods of high mass loss which last considerably longer than the expected dynamical timescales for real outbursts. I will refer to "maxima" and "minima."

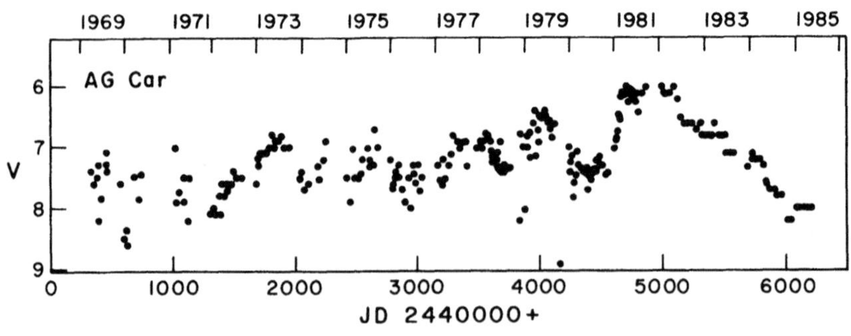

Fig. 1. The light curve of AG Car as an example of a typical LBV light curve (Viotti, 1988).

In addition to the moderate variations, the LBV's which have been studied in detail show photometric micro-variations of $\Delta V \lesssim 0\overset{m}{.}2$ on timescales of months to weeks (Van Genderen, 1986; Lamers, 1987). They are probably due to non-radial pulsations.

Some LBV's, such as P Cyg, η Car and V12 in NGC 2403 have shown major brightening of $\Delta V \gtrsim 3^m$ connected with large mass ejection in the recorded history. I will call these eruptions, although their real nature is not known. Other LBV's, such as AG Car and R127 are surrounded by rings or shells which indicate that large amounts of mass were ejected in the past.

Some LBV's, such as P Cyg and possibly η Car, go through extended periods of quiescence, which may last for a century, in which the moderate variations are absent or diminished to $\Delta V \lesssim 0\overset{m}{.}3$. It is possible that the quiescence is due to the occurrence of eruptions previously. It is not known whether quiescence should be considered as an extended minimum or an extended maximum.

I will discuss the mass loss rates at the various phases separately. In reality there is probably a continuous transition between minima and maxima and quiescence.

3. PRESENT MASS LOSS RATES OF LBV'S

3.1. Mass Loss During Quiescence or at Minimum

The mass loss rate during quiescence or at minimum has been determined for a few LBV's only (Table 1).

P Cygni: The most accurate estimate of $\dot{M}$ is based on the free-free radio emission (Abbott et al., 1981; White and Becker, 1982). This $\dot{M}$ depends only on the distance and the parameters of the star, for which we adopted the values of Lamers et al. (1983).

η Car: Van Genderen and Thé (1985) and Davidson et al. (1986) derived upper limits for the present mass loss rate of $\dot{M} \lesssim 10^{-3}$ $M_\odot$/yr. Davidson et al. argued that the temperature of the pseudo-photosphere would be less than the observed value of $T \simeq 30000$ K if $\dot{M}$ was larger than this value. The much higher rate of 10^{-2} - 10^{-1} $M_\odot$/yr derived from the amount of dust in the homunculus (Andriesse et al., 1978; Hyland et al., 1979) does not represent the present $\dot{M}$, but some average over the last centuries.

R71: The parameters of this star at minimum in 1980 and its mass loss rate of $\dot{M} = -6 \times 10^{-7}$ $M_\odot$/yr have been determined by Leitherer et al. (1989) from a study of the H and He lines and the energy distribution, using non-LTE model atmospheres and self-consistent wind models.

3.2. Mass Loss During Maximum

AG Car: We adopt the stellar parameters and the new distance determination of 7.1 kpc derived from the radial velocity and from the studies of normal stars close to AG Car (Humphreys et al., 1989; Lamers et al., 1989). The mass loss rate is highly variable: during the maximum of 1981 (V=6.1) the visual spectrum showed many strong P Cygni profiles, whereas during minimum in 1975 (V=7.4) it showed

Table 1. Mass loss rates of LBV's

Galaxy	Star	Phase	Type	V	log d (pc)	M_{bol}	T_{eff}	R_* ($R_\odot$)	v_∞ (km/s)	Ref	log $\dot{M}$ ($M_\odot$/yr)	Method	Ref	$\frac{\dot{M} v_\infty}{L/c}$
MW	P Cyg	Quies	B1 Ia$^+$	4.8	3.26	- 9.9	19300	76	210	1,2	-5.0	Radio	3,4	0.14
	η Car	Quies	--	6.0	3.40	-11.8	30000	76	700	5	<-3.0	Phot.	5,6	<8
	AG Car	Max	B5 Ia	6.1	3.85	-11.3	13700	290	166	7,8	-4.6	HI	9	0.08
LMC	R71	Min	B2.5 Iep	10.9	4.66	-8.8	16000	88	170	10	-6.2	HI, IR	10	0.02
		Max	A1 I	9.9	4.66		10000	330	110	10	-4.7	HI, IR	10	0.41
	S Dor	Max	Aeq	9.5	4.66	-9.5	10000	430	110	10,11	-4.5	HI, IR	10	0.34
	R127	Max	B5 I	10.4	4.66	-10.6	17000	140	200	12	-4.2:	HI, IR	12	0.45:
M31	Var A1	?	--	16.6	5.82	-10.3	15000	150	150	13,14	-5.1	HI	14	0.06
	Var 15	Max	--	17.3	5.82	-9.2	9000	250	200	13,14	-4.3	HI	14	1.30
	AF And	Min:	--	16.4	5.82	-10.0	28000	38	300	13,14	-4.8	HI	14	0.30
M33	Var B	Min:	--	16.2	5.78	-10.5	25000	60	150	13,14	-5.2	HI	14	0.04
	Var C	Max:	--	17.2	5.78	-9.8	7800	440	175	13,14	-4.4	HI	14	0.52
	Var 83	Min:	--	16.2	5.78	-10.8	20000	110	120	13,14	-5.1	HI	14	0.03

References

1. Lamers et al., 1983
2. Lamers et al., 1985
3. Abbott et al., 1981
4. White and Becker, 1982
5. Davidson et al., 1986
6. van Genderen and Thé, 1985
7. Humphreys et al., 1989
8. Lamers et al., 1989
9. Wolf and Stahl, 1982
10. Leitherer et al., 1989
11. Leitherer et al., 1985
12. Stahl et al., 1983
13. Humphreys et al. 1984
14. Humphreys, 1989

emission lines of H I, He I and [Fe II] (Wolf and Stahl, 1982; Whitelock et al., 1983). The mass loss rate has been derived for the maximum only from the extended Balmer emission wings, which are attributed to electron scattering by Wolf and Stahl (1982) who found $\dot{M} \approx 3 \times 10^{-5}\ M_{\odot}/yr$. These authors assumed $R = 130\ R_{\odot}$. Correcting for the larger radius of 290 $R_{\odot}$ adopted here, we find $\dot{M} \approx 7 \times 10^{-5}\ M_{\odot}/yr$. This may be an overestimate since part of the emission wings may be due to non-LTE effects (Hubeny and Leitherer, 1989). On the other hand, the spectra of AG Car at the 1981-maximum is very similar to that of S Dor during its 1986-maximum (Wolf and Stahl, 1982) for which Leitherer et al. (1989) derived $3 \times 10^{-5}\ M_{\odot}/yr$. We conclude that the mass loss rate of AG Car during maximum is $\dot{M} \approx 2 - 7 \times 10^{-5}\ M_{\odot}/yr$.

R71 and S Dor: Leitherer et al. (1989) derived the mass loss rate during the 1983-maximum of S Dor at V = 9.5 and during the 1974-maximum of R71 at V = 9.9 from the H I profiles and from the energy distributions using non-LTE models and self-consistent wind calculations. They found $\dot{M} = 2 \times 10^{-5}\ M_{\odot}/yr$ for both stars, if they assume that both stars have $L = 3 \times 10^5\ L_{\odot}$. In reality the luminosities are $L = 5 \times 10^5\ L_{\odot}$ for S Dor and $L \approx 3 \times 10^5$ for R71 if a distance modulus of 18.3 and E(B-V) = 0.15 is adopted. Consequently we adopt 3×10^{-5} $M_{\odot}/yr$ for S Dor after correction for the larger radius of this star.

R127: The variability of this star has been described by Walborn (1977, 1982), Stahl et al. (1983), and Stahl and Wolf (1986). The mass loss rate of $6 \times 10^{-5}\ M_{\odot}/yr$ during the 1982-maximum (V=10.4) was estimated by Stahl et al. (1983) from the strength of the Balmer emission and from the IR excess. They adopted a ballistic velocity law, which is certainly wrong. So this mass loss rate is uncertain by about a factor of 3.

3.3. Mass Loss From LBV's in M31 and M33

Humphreys et al. (1984), Kenyon and Gallagher (1985), and Humphreys (1989) have studied the LBV's in M31 and M33. The mass loss rates were derived from the Balmer profiles and from the IR excess. The stellar parameters and the mass loss rates are listed in Table 1. For most of these stars, we do not known whether the derived rates represent those at minimum, at maximum, or inbetween, except for Var 15 in M31 which was observed at maximum. Judging from the temperatures we would suspect that Var C in M33 was at maximum because T ≈ 7500 - 8000 K, and that AF And, Var B and Var 83 in M33 with T ≳ 20000 K were probably close to minimum. The phase of Var A1 in M31 is unknown.

3.4. Conclusions

The mass loss rates of the LBV's are plotted versus M_{bol} in Fig. 2. The mean relation for O-stars (Garmany and Conti, 1984) is shown for comparison. We can draw the following conclusions.

1. The mass loss rates at maximum are about the same for all LBV's: $\log \dot{M} = -4.43 \pm 0.15$. The maxima used here are not absolute maxima, so the differences in the range of $-4.6 \le \log \dot{M}_{max} \le -4.3$ are probably real.

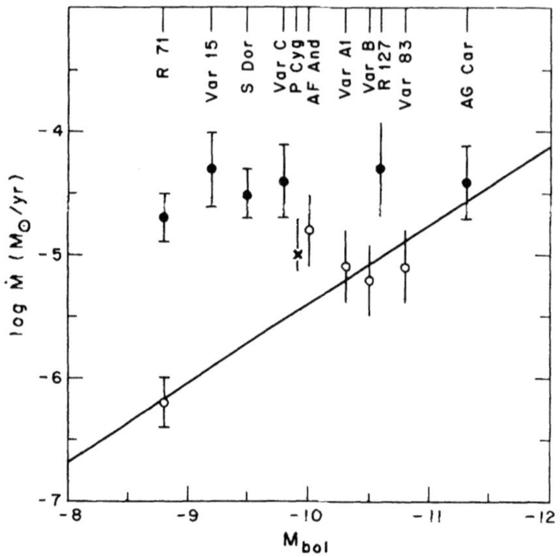

Fig. 2. Mass loss versus M_{bol} for LBV's at different phases: filled circles = maximum; open circles = minimum; cross = quiescence. The line is the $\dot{M}$ - M_{bol} relation for normal stars. The mass loss of LBV's at maximum is about constant at $\dot{M} \simeq 3 \times 10^{-5}$ $M_\odot$/yr. The mass loss at minimum is similar to that of normal stars.

2. The mass loss rates at minimum seem to follow the relation for normal stars. The only exception is AF And in M31. The star P Cygni during quiescence has a 3 times higher mass loss than the mean $\dot{M}$-L relation for normal stars. This does not necessarily imply that $\dot{M}$ is higher than predicted by the radiation driven wind theory. Pauldrach et al. (1986) have shown that the mass loss of P Cygni in quiescence agrees with the predictions, because P Cyg has a lower mass than normal stars of the same luminosity. Unfortunately we do not know the parameters of the LBV's accurately enough to make this comparison between predicted and observed $\dot{M}$ for each star. However, the fact that the observable and predicted rates agree for the two LBV's with reasonably well determined parameters, i.e. P Cyg and R71 at minimum, strongly suggests that the mass loss of LBV's at minimum and quiescence is due to radiation pressure.

3. The ratio $\dot{M}\, v_\infty c/L$ between the momenta of the wind and the radiation is given in Table 1. Its accuracy is about a factor 2 except for R127. This ratio is less than 1, within the errors, for all stars. The mean values are

$$\log \dot{M}\, v_\infty c/L \simeq -1.2 \pm 0.5 \text{ at minimum}$$

$$\log \dot{M}\, v_\infty c/L \simeq -0.4 \pm 0.4 \text{ at maximum}$$

This implies that the acceleration of the wind can be provided by radiation pressure.

4. MULTIPLE SHELL EJECTIONS

In addition to the steady but variable mass loss described in the previous section, several (and possibly all) LBV's eject shells on timescales of the order of months. This is evident from the occurrence of multiple narrow absorption components in the UV and visual absorption lines. Such components have been observed in the spectra of AG Car at -60, -151 and -210 km/s (Wolf and Stahl, 1982), and R81 at -70, -150 and -200 km/s (Stahl et al., 1987).

The velocity variations of the multiple narrow components of the star P Cyg have been studied in detail (de Groot, 1969; Markova and Kolka, 1988; and Lamers et al., 1985). In general three or more violet displaced absorption components can be seen at any moment, sometimes even up to six. The components occur first at low velocities, $-50 \lesssim V \lesssim -100$ km/s, and shift in time to more negative velocities up to -220 km/s. In the v(t) diagrams each component produces a more or less straight line with a slope indicating an acceleration of 0.2 - 0.6 cm/s^2. Each component can be followed during a period of 0.5 - 1.5 yr, during which time the shells travel a distance of 2 - 7 $\times$ 10^{14} cm or 40 - 140 R_*. The repetition time of the shell ejections is 3 - 5 months for P Cygni. Cassatella et al. (1979) estimated that the column density of the shells is typically $N_H \simeq 10^{20}$ cm^{-2} when v $\simeq$ 150 km/s. Assuming a typical distance of 10^2 R_* from the star we find that the mass of a shell is 10^{-7} - 10^{-6} $M_\odot$. If two to four shells are ejected per year, the total mass lost in the shell ejections is smaller than the average mass loss rate, which is 1 $\times$ 10^{-5} $M_\odot$/yr for P Cygni.

Such a detailed analysis has been done for P Cygni only. It suggests that mass loss by multiple shell ejections plays a minor role to the overall mass loss of the LBV's.

5. MASS LOSS DURING ERUPTIONS

η Car: The eruption of η Car in 1837-1860 produced a circumstellar nebula. The core of this nebula, the homunculus, contains 2 $\times$ 10^{-2} $M_\odot$ of dust, which suggest that a total of about 2 - 3 $M_\odot$ was ejected at a velocity of about 700 km/s (see review of Davidson, 1989). The star η Car is also surrounded by "old" ejecta at a distance and velocity which does not agree with an ejection in the previous century. This suggests that η Car has gone through more than one eruption at intervals of several hundred years (Walborn et al., 1978; Davidson et al., 1986).

AG Car: The ring around AG Car suggests the occurrence of one or more ourbursts about 10^4 yr ago with a total mass of about 2 $M_\odot$ (Section 6.1). Previous suggestions that the ring may be due to the interaction between slow matter ejected at a red supergiant phase and the faster wind of the LBV phase are untenable because the new distance determination of 7.1 kpc implies that the star is too luminous to have been a red supergiant. The almost empty region inside the nebula (Paresce, 1989) suggests that no major outbursts have occurred over the last 400 yr if we assume a typical ejection velocity of 700 km/s during an eruption.

P Cyg: The seventeenth century eruption did not leave a visible remnant. The only indication of a remnant is the IR dust emission (6.2) and a radio-arc at a distance of about 0.3 pc (Wendker, 1982; Baars and Wendker, 1987). If the gas-to-dust ratio is 10^2, the total amount of mass ejected in the outburst is on the order of 10^{-2} $M_{\odot}$.

The evidence for mass ejections in recorded eruptions or in previous major outbursts is summarized in Table 2.

Table 2. Evidence for mass-ejection in eruptions or major outbursts

Star	Remnant	Total Mass ($M_{\odot}$)	V_{exp} (km/s)	kinematic age (yr)	eruption (AD)
η Car	Homunculus	2-3	700	1.5×10^2	1837-1860
	Old ejecta	--	$\gtrsim 50$	$\lesssim 2\times10^3$	--
AG Car	Ring nebula	2	50	1×10^4	--
P Cyg	30 K dust	10^{-2}	800	4×10^2	1600-1660
R71	140 K dust	3×10^{-2}	20	4×10^2	--

6. THE MASS LOSS HISTORY OF LBV'S DERIVED FROM RING NEBULAE OR CIRCUMSTELLAR MATTER

The presence of circumstellar matter around LBV's, detectable as ring nebulae, extended Hα or [N II] emission or as cold dust may be used to derive the average mass loss rates of these LBV's over the last 10^3 - 10^4 yr.

6.1 Nebulae and Extended Emission

P Cyg: Leitherer and Zickgraf (1987) detected extended Hα and [N II] 6587 emission and showed that this can be explained by a continuous mass loss of 4×10^{-4} $M_{\odot}$/yr or by multiple shell ejection with $2 \times 10^{-5} < \dot{M} < 1 \times 10^{-4}$ $M_{\odot}$/yr over the last 300 yr. However, Zickgraf and Stahl (1989) have questioned the reality of the extended [N II] emission around P Cygni, so the results are uncertain.

R127: Stahl (1987) has measured the extended Hα and [N II] 6584 emission (Fig. 3). The emission is roughly circular with a diameter of about 4" = 0.9 pc (Fig. 3). Adopting the measured nebular velocity of 28 km/s and assuming a density of 10^3 cm^{-3} he derived a kinematic age of ~ 1.7×10^4 yr and a total mass of 2.9 $M_{\odot}$. This corresponds to a mean rate of 1.7×10^{-4} $M_{\odot}$/yr. Figure 2 suggests that the rate is ~10^{-5} $M_{\odot}$/yr at minimum and 4×10^{-5} $M_{\odot}$/yr at maximum. So either the estimated density of 10^3 cm^{-3} is an overestimate or the clumping is more severe than assumed or the mean mass loss rate over the last 10^4 yr has been larger than at present.

AG Car: The ring nebula around AG Car has been studied by many authors (Thackeray, 1956; Johnson, 1976; Stahl, 1987; McGregor et al., 1988; Paresce, 1989). Its average radius is 15" ≈ 0.52 pc, its expansion velocity is 50 km/s and its kinematic age is 1×10^4 yr. Assuming the old distance of 2 kpc most authors derived a nebular mass of 0.2 - 0.3 $M_\odot$. Adopting the new distance of 7.1 kpc (Table 1) the mass estimate increases to ~2 $M_\odot$ (Stahl, 1989). This corresponds to a mean rate of $\dot{M} \sim 2 \times 10^{-4}$ $M_\odot$/yr over the last 10^4 yr.

6.2 Circumstellar Dust

R71: Wolf and Zickgraf (1986) detected cold dust around R71 with a temperature of 140 K. This temperature indicates that the dust is at a distance of 8×10^3 R_* = 4.5 10^{16} cm. The observed expansion velocity of the circumstellar gas around R71 is 20 km/s, which indicates a kinematic age of the dust region of 400 yr. The dust mass is 3×10^{-4} $M_\odot$. Assuming the canonical gas-to-dust ratio of 10^2, the total circumstellar mass is 3×10^{-2} $M_\odot$ and the mean mass loss rate is 7×10^{-5} $M_\odot$/yr, which is about a factor 3 higher than the rate derived at maximum (Table 1).

P Cyg: Waters and Wesselius (1986) showed that the IRAS observations of P Cyg at 60 and 100 μm indicate the presence of cold dust at T ≈ 30 K. The temperature indicates a distance of 2×10^5 R_*. The dust mass is 10^{-4} $M_\odot$ and the total mass is on the order of 10^{-2} $M_\odot$. If this mass was ejected during the eruption of AD 1600, the velocity must have been 800 km/s. This velocity is higher than the typical velocity of 20 - 50 km/s in typical LBV shells, but it is similar to the ejection velocity of the η Car outburst. On the other hand if v ≈ 50 km/s, as for the AG Car nebulae, then the kinematic age is 7×10^3 yr and the mean mass loss rate is 1×10^{-6} $M_\odot$/yr. This is smaller than the quiescent mass loss rate. So the dust is probably

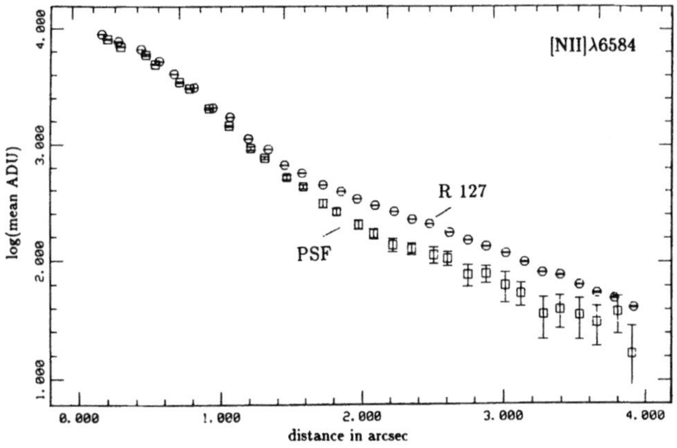

Fig. 3. The radial image profile of the [N II] 6584 line of R127 in digital output units of the CCD versus radial distance from the center. The radial distribution is flatter than that of a point-source (PSF), indicating extended emission around R127, with a mass of 2.9 $M_\odot$ (Stahl, 1987).

formed by the seventeenth century eruption with a total mass of $\Delta M \simeq 10^{-2}$ $M_\odot$ if the gas-to-dust ratio is 10^2.

6.3 Conclusion

We conclude that the time-averaged mass loss rates derived from the circumstellar matter around AG Car, R71 and R127 is in the range of 0.7×10^{-4} to 2×10^{-4} $M_\odot$/yr. The kinematic age of the circumstellar gas is 1×10^4 yr for AG Car and 1.7×10^4 yr for R127.

7. INTEGRATED MASS LOSS AND THE CONSEQUENCES FOR THE EVOLUTION OF MASSIVE STARS

7.1 Time-averaged Mass Loss Rates

The "normal" mass loss rate of the LBV's varies between the values at minimum, $\log \dot{M} \simeq 1.6 \log L - 14.9 \pm 0.5$, and at maximum, $\log \dot{M} \simeq -4.4 \pm 0.2$. Assuming for simplicity that the star spends half the time at minimum and the other half at maximum, we find an average normal mass loss rate of

$$\log \dot{M}_{normal} \simeq -4.7 \pm 0.3 \qquad (1)$$

for stars with $\log L \lesssim 6.3$. The multipole shell ejection does not seem to increase this value significantly.

The mass ejected during eruptions is in the range of 10^{-2} to 2 $M_\odot$. The presence of old ejecta around η Car suggests that eruptions may occur more than once in the LBV phase. The frequency of the eruptions is unknown. The recurrence time is probably larger than 400 yr because P Cygni had its last eruption in 1600-1660. This estimate agrees with the fact that the inner region of the AG Car ring nebula is almost empty (see Section 5) and with the fact that none of the five LBV's in the LMC had an eruption in the last 100 yr. The recurrence time is probably not much longer than 10^3 yrs because the old ejecta around η Car suggest that previous eruptions have occurred several hundred years ago. A recurrence time as large as 10^4 yr is unlikely because of the handful of known LBV's in our galaxy two have suffered eruptions in the last 400 yr. (But, beware of selection effects!) I conclude that the recurrence time of eruptions is probably on the order of 10^3 yr with an uncertainty of a factor 3.

The estimated frequency of the eruptions together with the mass ejected in large eruptions implies an average mass loss rate due to eruptions of

$$-5 \lesssim \log \dot{M}\ (\mathrm{erupt}) \lesssim -3 \qquad (2)$$

This range is consistent with the average mass loss rates derived from study of the circumstellar matter around LBV's (Section 6), which indicates a total average mass loss rate (eruptions plus normal mass loss) of

$$-4.1 \lesssim \log \dot{M}\ (\mathrm{total}) \lesssim -3.7 \qquad (3)$$

for AG Car, R71 and R127. The average mass loss rates may be larger if large eruptions with $\Delta M \sim 1$ $M_\odot$ play an important role.

7.2 Evolutionary Consequences

Massive stars with $50 \leq M_i \lesssim 100\ M_\odot$ have to get rid of 5 to 15 $M_\odot$ after the core H-burning before they become a WR star (Maeder and Meynet, 1987). This mass has to be lost in the LBV phase. Adopting the time-average mass loss rate of the LBV's of $\langle\dot{M}\rangle \simeq 10^{-4}\ M_\odot$/yr we find that the duration of the LBV phase is $\tau_{LBV} \simeq 10^5$ yr. The duration may be shorter by a factor 10 if the large eruptions with $\Delta M \sim 1\ M_\odot$ play a dominant role.

This age-estimate can be compared with independent indicators of the duration of the LBV phase. From the ratio between the numbers of LBV and WR stars of $N(LBV)/N(WR) \simeq 10^{-2} - 10^{-1}$ and from the duration of the WR phase of about 5×10^5 yr (Maeder and Meynet, 1987) we find $\tau_{LBV} \simeq 5 \times 10^3 - 5 \times 10^4$ yr. This is lower than the value derived from $\langle\dot{M}\rangle$, but the uncertainties are large.

A more accurate value of τ_{LBV} mimght be derived from the upper limit for the luminosity of the Red Supergiants of log L = 5.8. To prevent stars just above this limit from evolving to the red requires a minimum $\langle\dot{M}\rangle_{LBV}$ which can be transformed into a maximum value of τ_{LBV}.

8. SUMMARY

1. The mass loss rates of LBV's at minimum are in agreement with the predictions for radiation driven winds. The mass loss rates at maximum are about the same for all LBV's at $\dot{M} \simeq 4 \times 10^{-5}\ M_\odot$/yr.

2. Most LBV's are ejecting shells at intervals of months.

3. During eruptions the LBV's eject $10^{-2} - 10^0\ M_\odot$. The recurrence time of the eruptions is uncertain, but it is probably on the order of 10^3 yr.

4. The time averaged mass loss of a few LBV's derived from their circumstellar matter is $10^{-4}\ M_\odot$/yr. The real time-averaged mass loss may be larger if large eruptions with $\Delta M \simeq 1\ M_\odot$ play a dominant role.

5. The duration of the LBV-phase derived from $\langle\dot{M}\rangle$ and the amount of mass to be lost is 10^5 yr or shorter.

Acknowledgments. I thank P. Conti for hospitality at JILA in the summer of 1988 and the JILA Scientific Reports Office for preparing the manuscript. This study was supported by National Science Foundation grant AST88-06594 to the University of Colorado.

References

Abbott, D.C., Bieging, J.H., Churchwell, E.: 1981, Ap. J. **250**, 645.
Andriesse, C.D., Donn, B.D., Viotti, R.: 1978, MNRAS **185**, 771.
Baars, J.W.M., Wendker, H.J.: 1987, Astr. Ap. **181**, 210.
Cassatella, A. et al.: 1979, Astr. Ap. **79**, 223.
Davidson, K.D.: 1987, in Instabilities in Luminous Early Type Stars, eds. Lamers and de Loore (Reidel), p. 127.

Davidson, K.D.: 1989, these proceedings.
Davidson, K.D. et al.: 1986, Ap. J. **305**, 867.
de Groot, M.: 1969, Bull. Astron. Inst. Netherlands **20**, 235.
Garmany, C.D., Conti, P.S.: 1984, Ap. J. **284**, 705.
Hubeny, I., Leitherer, C.: 1989, these proceedings.
Humphreys, R.M.: 1989, these proceedings.
Humphreys, R.M. et al.: 1984, Ap. J. **278**, 124.
Humphreys, R.M., Lamers, H.J.G.L.M., Hoekzema, N.: 1989, in preparation.
Hyland, A.R. et al.: 1979, Ap. J. **223**, 145.
Johnson, H.M.: 1976, Ap. J. **206**, 469.
Kenyon, S.J., Gallagher, J.S.: 1985, Ap. J. **290**, 542.
Lamers, H.J.G.L.M.: 1986, in Luminous Stars and Associations in Galaxies, eds. de Loore et al. (Reidel), p. 157.
Lamers, H.J.G.L.M.: 1987, in Instabilities in Luminous Early Type Stars, eds. Lamers and de Loore (Reidel), p. 99.
Lamers, H.J.G.L.M., de Groot, M., Cassatella, A.: 1983, Astr. Ap. **128**, 299.
Lamers, H.J.G.L.M. et al.: 1989, these proceedings.
Lamers, H.J.G.L.M., Korevaar, P., Cassatella, A.: 1985, Astr. Ap. **149**, 29.
Leitherer, C. et al.: 1985, Astr. Ap. **153**, 168.
Leitherer, C. et al.: 1989, these proceedings.
Leitherer, C., Zickgraf, F.J.: 1987, Astr. Ap. **174**, 103.
Maeder, A., Meynet, G.: 1987, Astr. Ap. **182**, 243.
Markova, N., Kolka, I.: 1988, Astroph. Sp. Sci. **141**, 45.
McGregor, P. J. et al.: 1988, Ap. J. **329**, 874.
Paresce, F.: 1989, these proceedings.
Pauldrach, A., Puls, J., Kudritzki, R.P.: 1986, Astr. Ap. **164**, 86.
Stahl, O.: 1987, Astr. Ap. **182**, 229.
Stahl, O.: 1989, these proceedings.
Stahl, O., Wolf, B.: 1986, Astr. Ap. **154**, 243.
Stahl, O., Wolf, B., Zickgraf, F.J.: 1987, Astr. Ap. **184**, 193.
Stahl, O. et al.: 1983, Astr. Ap. **127**, 49.
Thackeray, A.D.: 1956, Vistas Astron. **2**, 1380.
van Genderen, A., Thé, P.S.: 1985, Space Sci. Reviews **39**, 317.
van Genderen, A.: 1986, Astr. Ap. **157**, 163.
Viotti, R.: 1988, private communication.
Walborn N.R.: 1977, Ap. J. **215**, 53.
Walborn N.R.: 1982, Ap. J. **256**, 452.
Walborn N.R., Blanco, B.M., Thackeray, A.D.: 1978, Ap. J. **219**, 498.
Waters, L.B.F.M., Wesselius, P.R.: 1986, Astr. Ap. **155**, 104.
Wendker, H.J.: 1982, Astr. Ap. **116**, L5.
White, R.L., Becker, R.H.: 1982, Ap. J. **262**, 657.
Whitelock, P.A. et al.: 1983, MNRAS **205**, 577.
Wolf, B., Stahl, O.: 1982, Astr. Ap. **112**, 111.
Wolf, B., Zickgraf, F.J.: 1986, Astr. Ap. **164**, 435.
Zickgraf, F.J., Stahl, O.: 1989, these proceedings.

DISCUSSION

Sasselov: As I understand it, you interpret the ring nebulae around AG Carinae as actual shells of matter ejected from the star at some time in the past. How are you confident that they are not shock fronts in circumstellar material due to previous continuous mass loss?

Lamers: The new distance estimate for AG Car sets its luminosity well above the upper limit for red supergiants. So AG Car should not have had a previous low-velocity mass-loss phase.

Kahn: Small fluctuations in wind speed can lead to the formation of inhomogeneities at later times. Then *average(density²)* > *(average density)²*. Ignoring this would lead to an overestimate of $\dot{M}$.

Lamers: This effect was roughly taken into account with a filling factor consistent with Leitherer and Zickgraf's study of P Cygni's Hα emission. The rather large uncertainty in $<\dot{M}>$ derived by Leitherer results from the uncertain clumping and filling factors.

Vanbeveren: In view of the timescales, mass-loss rates, and the apparent enhancement of nitrogen, I think that one has to consider the possibility that an LBV is a normal core-hydrogen-burning star that has gone through an accretion phase (to explain the nitrogen overabundance), with a neutron star or a black hole spiralling into its atmosphere. The few computations of such an event that have been done, do show LBV characteristics.

Davidson: Statistics! We would have to be very lucky or unlucky to know so many examples within historical memory, among such a limited sample of luminous stars.

Henrichs: Regarding the binary scenario that Vanbeveren just mentioned, I notice that if the eclipsing binary R81 should undergo an eruption like R71 for instance, then it would become a Roche-lobe-overflowing system. It seems, however, to be difficult to find the binary nature of LBV's.

Walborn: Concerning η Car and whether it has undergone more than one major ejection event in recent centuries -- Betty Blanco and I have recently discussed third-epoch measurements of proper motions in the outer shell (July 1988: *Publ. Astr.Soc.Pacific* **100**, 797). We detected possible decelerations, which, if confirmed, may indicate that all of the material was ejected during the nineteenth-century event. Earlier ejection times have been based on the analysis of plates from just two epochs, necessarily assuming uniform motions.

de Groot, Lamers, Kahn

Walborn, McGregor, Stahl

CIRCUMSTELLAR EJECTA AROUND LBV'S

O. STAHL
Landessternwarte Königstuhl
D-6900 Heidelberg 1

ABSTRACT. The properties of circumstellar shells around LBV's are reviewed. Some LBV's are surrounded by very extended nebulosities, similar to classical planetary nebulae, with AG Car as the prototype. The masses of these nebulae are of the order of a solar mass and their kinematic ages are about 10^4 years. The origin of these nebula is still controversial, but they probably have been ejected in some major outburst in the past. It is then remarkable that the historical outburst of P Cyg has left no bright remnant. There is evidence for CNO processed matter in the ejecta. LBV's with no detectable extended emission also commonly have nebular emission lines. Some of these nebulae are probably related to the resolved nebulae, but some certainly are very much smaller and have much higher densities.

1 Introduction

The observations of circumstellar shells around LBV's are important for a number of reasons.

- The analysis of the spectra of the circumstellar ejecta is much easier than the analysis of the complicated stellar spectra, which is now just becoming feasible. Especially, the chemical abundances of the ejecta can be studied with standard techniques from the nebular lines.
- The circumstellar ejecta allow us to study the mass-loss history of the central stars and also the integrated mass-loss rates, in particular the importance of major eruptions, which are rarely directly observable.

Nevertheless, so far only limited information is available for the circumstellar ejecta around LBV's. This information will be reviewed here.

K. Davidson et al. (eds.), Physics of Luminous Blue Variables, 149–157.

2 Galactic objects

The analysis of spatially resolved objects allows to extract more information than is possible with unresolved objects. Obviously, the galactic objects are most easily resolved. Unfortunately, few LBV's are known in the Galaxy. Most people agree that η Car, AG Car and P Cyg are galactic LBV's. Probably HR Car should be added to this list, but it is a very little studied star. A number of direct images in narrow band filters (Stahl, unpublished) did not reveal any nebulosity around HR Car.

2.1 η Car

The nebula around η Car is of course very well known. Both η Car and its nebula are unique objects and thus it is dangerous to draw general conclusions from η Car. It should suffice here to say that the η Car nebula has a mass of a few solar masses and is expanding with a velocity of about 700 $\mathrm{km\,s^{-1}}$. (For more details see the paper by Davidson in this volume.)

2.2 AG Car and He3-519

The ring nebula around AG Car has been discovered by Thackeray (1950). It is roughly elliptical with a size of about $40 \times 30''$. Milne and Aller (1975) detected the nebula in the radio continuum. Johnson (1976) observed the nebula spectroscopically with a Fabry-Perot instrument. He derived a density and a mass for the nebula. Long-slit spectra of the nebula have been discussed by Thackeray (1977). He found a complicated velocity field which we do not want to discuss here in detail. The typical expansion velocity is ≈ 50 $\mathrm{km\,s^{-1}}$. Viotti et al. (1988) found from IUE observations of the star and the nebula that the nebula reflects starlight and thus contains dust. Dust emission from the nebula in the far-IR was detected and discussed by McGregor et al. (1988). (See also the review of Mc Gregor in this volume). Spectacular high resolution images of the AG Car nebula have been obtained by Paresce (this volume).

AG Car has always been assumed to be at the distance of the Car OB association. New distance determinations based on the radial velocity of the star and the nebula (Humphreys, this volume) and the extinction to the star (Hoekzma and Lamers, this volume) favor a larger distance of up to 7 kpc. This has of course implications for the inferred properties of the nebula, in particular its age and mass. The properties of the nebula (based on a distance of 2.5 kpc) have been reviewed by Stahl (1987). With a revised distance of 7 kpc, the nebula becomes older and more massive.

The object He3-519 has also been discussed by Stahl (1987). The star is not known to be an LBV, but its spectrum is similar to AG Car in minimum, suggesting a relation with the LBV's. It is surrounded by a ring nebula with a diameter of about $1'$. The distance of the object is uncertain, but likely larger than the distance to AG Car, since its reddening is larger and the distance on the sky to AG Car is only $20'$. A detailed study of the object has not yet been done but is certainly worthwhile.

2.3 P Cygni

Since it is well known that P Cygni had an outburst in the 17th century, several people have tried to find a remnant similar to the shells around η Car and AG Car. Indeed Wendker (1982) and Baars and Wendker (1987) found extended radio-emission ($\approx 30''$ radius) around P Cygni which may be due to the outburst. Although initially thought to be of non-thermal origin, the emission is now considered to be thermal. No optical counterpart of the radio emission is known, but there is no report of a serious attempt to find it.

Recently Leitherer and Zickgraf (1987) (hereafter called LZ) reported the detection of a small ($\leq 8''$) circumstellar shell around P Cygni by direct imaging with narrow-band filters around Hα and [NII]λ6583 and a CCD detector. Interestingly, they found that the radial slope of the emission was consistent with the assumption that the emission was due to a continuous ejection of small shells with constant average mass-loss rate This is different from all other known cases of circumstellar shells around LBV's where the emission arises in a detached shell.

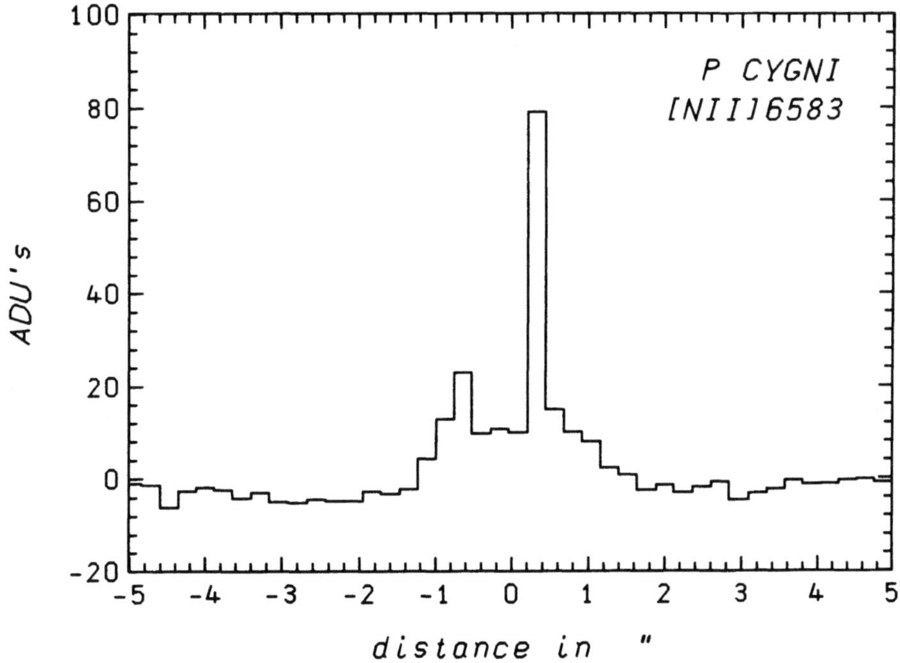

Figure 1: Difference of the line and continuum cross-dispersion profiles of a long-slit spectrogram of P Cygni in the line [NII]λ6583 and in the neighboring continuum. The difference from zero is not considered to be significant. One ADU very roughly corresponds to $\approx 5\,10^{-13}$ erg cm^{-2} s^{-1} arcsec^{-2}.

I observed, together with F.-J. Zickgraf, P Cyg with the coudé spectrograph of the 2.2m telescope of the DSAZ at Calar Alto, Spain in Sep. 1987 using the coudé spectrograph with the f/12 camera which gives a reciprocal linear dispersion of about 2.2 Å/mm in second order. The slit was opened to about $0\rlap{.}''5$.

We compared the cross-dispersion profile around the [NII]λ6583 line with cross-dispersion profiles around this wavelength. The cross-dispersion profile was averaged from 6580.5 to 6586.5 Å to give the spatial profile in the line and from 6574 to 6580 and 6587 to 6593 to give the spatial profile in the adjacent continuum (average of the flux in a 0.1 Å wavelength bin). Assuming that the shell emission is negligible close to the center of the profile, we matched both profiles in the central pixels. The difference *line-continuum* of the spatial profiles is shown in Figure 1.

The difference from zero is typically smaller than 1% of the total flux and not considered significant. The absolute calibration in line fluxes is difficult since it depends on assumptions about slit losses and the line profile from the shell. One ADU very roughly corresponds to $\approx 5\,10^{-13}\,\mathrm{erg\,cm^{-2}\,s^{-1}\,arcsec^{-2}}$. The upper limit is then (at distances larger than $\approx 2''$) $\approx 10^{-12}\,\mathrm{erg\,cm^{-2}\,s^{-1}\,arcsec^{-2}}$. LZ measured $\approx 5\,10^{-12}\,\mathrm{erg\,cm^{-2}\,s^{-1}\,arcsec^{-2}}$ at $2''$, decreasing with the distance to the power of -3. This is significantly larger than the upper limit derived from the data shown in Figure 1.

How can the discrepancy be explained? LZ determine the radial profile averaged over all position angles, where we have only the profile at one position angle. A strongly flattened nebula could thus explain the difference. It is unlikely, however, that LZ would not have noticed such a nebula. A more likely explanation for the effect is that LZ seriously undersampled their images: LZ mention a seeing (FWHM) of $0\rlap{.}''6$ with a pixel size of $0\rlap{.}''46$. In such circumstances the radial profile of the star close to the center is very poorly determined and it strongly depends on the center of the star with respect to the center of the pixels. Since LZ compared the radial profiles of P Cygni with the profiles of a comparison star, a slightly different placement of the stars with respect to the center of the pixels may change the radial profile in the center drastically. We would then expect that the two radial profiles of the logarithm of the brightness are parallel outside the central pixels. This seems indeed to be the case, at least within the error bars given by LZ. Thus there is no convincing evidence that P Cygni is surrounded by an extended emission region. A more thorough search may be worthwhile.

3 Objects in the Magellanic Clouds and beyond

3.1 Spectroscopy

Nebular emission lines of circumstellar origin are very frequently observed in spectra of emission-line stars (see e.g. Stahl and Wolf, 1986). The strongest lines usually are [NII]λλ6548, 6583. In almost all cases, the lines are narrow (typically FWHM ≈ 40 $\mathrm{km\,s^{-1}}$. All the known S Dor-variables in the LMC show these emission lines (see Figure 2). The B[e]-supergiants also frequently have [NII] lines, but in these stars

[OI]$\lambda\lambda$6300, 6363 are the dominant forbidden lines (see also Zickgraf, this volume). Other emission-line stars also frequently have forbidden emission lines. In some cases the electron density of the shell can be determined from diagnostic line ratios. The range of densities is large, from $\leq 10^3\,\mathrm{cm}^{-3}$ to $\approx 10^7\,\mathrm{cm}^{-3}$. In the high density nebulae, [NII]λ5755 is strong.

For most stars of this sample not much is known about the origin of the forbidden lines. In the B[e] supergiants, the forbidden lines are assumed to originate in the outer parts of the circumstellar disk. In very few cases the nebulosities have been spatially resolved.

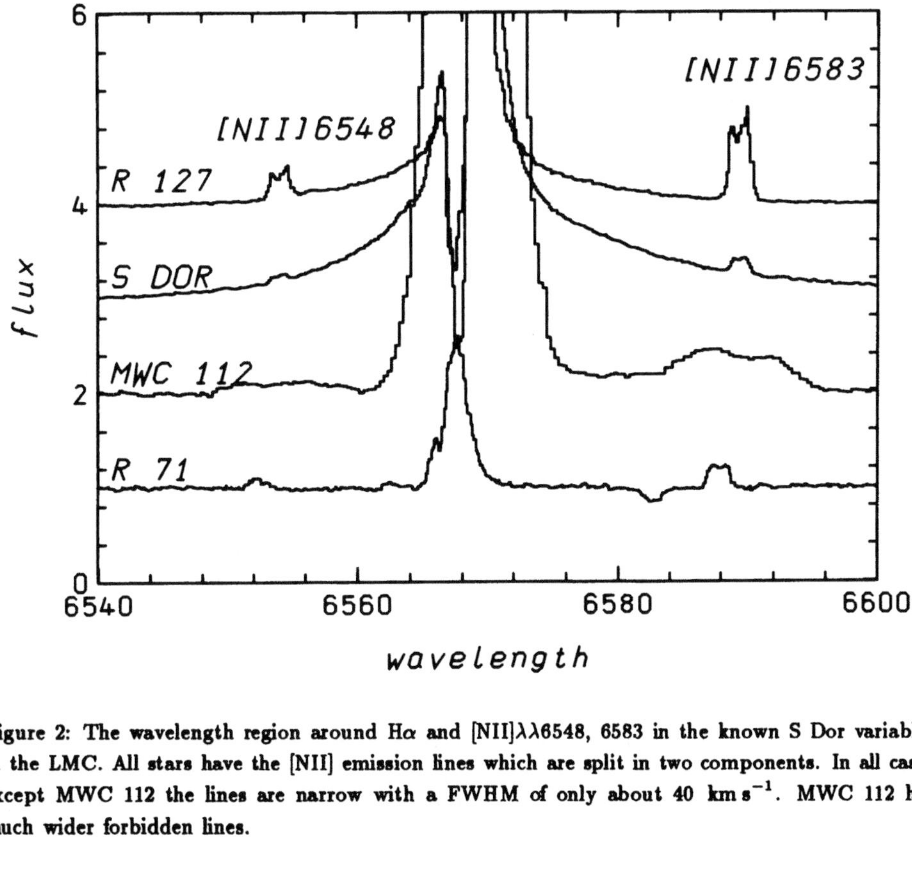

Figure 2: The wavelength region around Hα and [NII]$\lambda\lambda$6548, 6583 in the known S Dor variables in the LMC. All stars have the [NII] emission lines which are split in two components. In all cases except MWC 112 the lines are narrow with a FWHM of only about 40 $\mathrm{km\,s^{-1}}$. MWC 112 has much wider forbidden lines.

3.2 Imaging

Scaling the nebula around AG Car to the distance of the LMC, we expect an angular size of a few arcseconds. Although this should be easy to resolve from the ground, the emission is close to a bright star and so the detection of such shells is difficult due to the small contrast. Only the brightest shells can be expected to be seen. I conducted a survey of direct imaging in narrow-band filters of luminous emission-line stars in the Magellanic Clouds. Part of the results have been published by Stahl (1987). A larger sample has actually been observed. Only 2 out of 24 objects observed could be resolved. These objects are R 127 and S 61 (=Sk-67 266). These two objects have already been found by Walborn (1982) to show strong [NII]$\lambda\lambda$6548, 6583 emission lines. The two resolved objects appear to be objects similar to the AG Car nebula, even more similar, if the larger distance to AG Car is adopted.Some of the unresolved objects in the MC's may also have extended emission, which is just too faint to be seen against the background of the bright star.

Only R 127 is a known LBV, but S 61 is probably spectroscopically almost a twin to R 127 in minimum (Walborn, 1982). The failure to resolve the other objects may be partly due to the limited sensitivity of the observations. However, some nebulae are probably very small.

3.3 High-density nebulae

This shall be demonstrated with the star MWC 112, which is surrounded by a high-density nebula. In addition to the [NII] lines around Hα, MWC 112 also has rather strong [FeIII] lines. The following analysis is due to Kris Davidson and will be published in more detail elsewhere. From the intensity ratio of the [FeIII] lines a likely nebular electron density of $\approx 10^7$ cm^{-3} can be derived. The mass of the gas can be found from the [NII]λ6583 line ($\approx 10^{-3.5}$ M$_\odot$), assuming abundances appropriate for CNO processed matter. From the density and the mass a volume can be inferred and thus a minimum radius ($\approx 10^{15.2}$ cm) for the shell. For this radius, the travel time is of the order of only a few years. Thus, the nebula is clearly different from the large shells observed around AG Car and R 127. The most likely origin of the nebular lines is in this case a shock where the present high-velocity wind shocks the matter expelled in an outburst a few years ago.

3.4 The Hubble-Sandage variables in M31 and M33

Nebular emission lines have also been found in the spectra of the Hubble-Sandage variables in M31 and M33 (see Humphreys et al., this volume), demonstrating again that these objects are physically very similar to the S Dor variables. AE And is the most extreme case among the HS-variables, showing extremely strong nebular emission lines.

4 Chemical abundances and relation to other objects

Our knowledge about the chemical abundances of the ejecta is rather limited. The outer condensations of the η Car nebula have been studied by Davidson et al. (1982). They found a very high abundance of N with respect to O and C, consistent with

CNO-processed matter. S 61 has been analysed by Walborn (1982) and was also found to be N-rich and O-poor, if compared with LMC HII regions. The AG Car nebula has recently been analyzed by Dufour and Mitra (1987). They also give an O/H underabundance. So, from this limited number of observations, there is evidence for CNO-processed matter in the nebulae.

Since the LBV's may evolve to WR stars, one may suspect that the circumstellar shells around LBV's evolve to the well-known ring nebulae surrounding some WR stars. However, these nebulae are in most cases much more massive than the LBV nebulae and may consist mainly of swept-up interstellar matter. The bipolar nebula around the Of?p star HD 148937 may also be related to the nebulae around the LBV's. It is also N-rich (Leitherer and Chavarria, 1987; Dufour et al., 1988).

The bipolar structure of the nebula around HD 148937 may also be related to the double-peak structure seen in the spectra of the nebular lines of the LBV's in the LMC. In fact, there is ample evidence for bipolar ejection of the nebulae. The cause of this bipolarity is unclear. It is also not clear, why the nebular lines are so narrow in almost all cases.

References

Appenzeller, I., Wolf, B., Stahl, O.: 1987, in *Instabilities in Luminous Early-type stars*, eds. C. de Loore, A.J. Willis, Reidel, Dordrecht, pg. 241

Baars, J.W.M., Wendker, H.J.: 1987, *Astron. Astrophys.* **181**, 210

Davidson, K., Walborn, N.R., Gull, T.R.: 1982, *Astrophys. J. Lett.* **254**, L47

Dufour, R.J., Mitra, P.: 1987, *Bull. A.A.S* **19**, 1090

Dufour, R.J., Parker, R.A.R., Henize, K.G.: 1988, *Astrophys. J.* **327**, 859

Johnson, H.M.: 1976, *Astrophys. J.* **206**, 469

Leitherer, C., Zickgraf, F.-J.: 1987, *Astron. Astrophys.* **174**, 103

Leitherer, C., Chavarria-K., C.: 1987, *Astron. Astrophys.* **175**, 208

McGregor, P.J., Finlayson, K., Hyland, A.R., Joy, M., Harvey, P.M., Lester, D.F.: 1988, *Astrophys. J.* **329**, 874

Milne, D.K., Aller, L.H.: 1975, *Astron. Astrophys.* **38**, 183

Stahl, O.: 1986, *Astron. Astrophys.* **164**, 321

Stahl, O.: 1987, *Astron. Astrophys.* **182**, 229

Stahl, O., Wolf, B.: 1986, *Astron. Astrophys.* **158**, 371

Stahl, O., Wolf, B.: 1987, *Astron. Astrophys.* **181**, 293

Thackeray, A.D.: 1950, *Monthly Notices Roy. Astron. Soc.* **110**, 524

Thackeray, A.D.: 1977, *Monthly Notices Roy. Astron. Soc.* **180**, 95

Viotti, R., Cassatella, A., Ponz, D., Thé, P.S.: 1988, *Astron. Astrophys.* **190**, 333

Walborn, N.R.: 1982, *Astrophys. J.* **256**, 452

Wendker, H.J.: 1982, *Astron. Astrophys.* **116**, L1

Zickgraf, F.-J., Wolf, B., Stahl, O., Leitherer, C., Appenzeller, I.: 1986, *Astron. Astrophys.* **163**, 119

DISCUSSION

Schulte-Ladbeck: Franz-Josef Zickgraf and I have measured linear polarization as a function of wavelength for a number of Galactic B[e] stars, and the results are indicative of the presence of circumstellar disks (paper submitted to *Astr.&Astrophys.*).

Lamers: Recent observations of the radio arc around P Cygni indicate that the arc is due to thermal radiation (Baars & Wendker 1987: *Astr.&Astrophys.* **181**, 210).

Heap: What is the evidence **against** AG Carinae being the central star of a planetary nebula?

Stahl: I think everybody agrees that the distance to AG Carinae, though uncertain, is at least 2.5 kpc. Therefore the bolometric luminosity is brighter than -8, significantly more luminous than central stars of planetary nebulae.

Davidson: Some of your objects have [N II] profiles that are flat-topped at first glance, double-peaked on closer examination, with low velocities in most cases but faster in the case of MWC 112. As you have remarked sometimes, these profiles hint at bipolar or axial structure. But it's worth mentioning that the precise form of this axial symmetry is not obvious. In particular, the [N II] does not necessarily have to be in an equatorial disk; similar line profiles might originate in a symmetrically incomplete expanding shell. I say this because hypothetical disks sometimes have embarrassingly small volumes for emitting forbidden lines, especially if orbit-velocity assumptions require them to be only a few a.u. across.

Humphreys: You said very little about R71. Is there evidence for circumstellar ejecta around this star? It is the lowest-luminosity LBV and the one most likely to have been through a red supergiant stage. It does not have a ring nebula or shell, so far as I know. R71 does have a cool dust shell, possibly a remnant of a red supergiant stage.

Stahl: R 71 has [N II] λ6583 emission, not very strong. We could not spatially resolve this shell; maybe it is just slightly too small to resolve from the ground.

Walborn: (1) The two LMC stars with resolved shells, R 127 (= HDE 269858) and S 61 (= Sk -67° 266), are also the two originally found to have the most similar Ofpe spectra and the strongest, velocity-doubled [N II] lines. (2) If the B[e] stars have strong circumstellar oxygen emission but not nitrogen, they may be less evolved than LBV's. η Car shows strong [N I] and [N II] but no [O I] or [O II] and this immediately suggests anomalous abundances. (3) MWC 112 (= HDE 269582) shows strong spectral variations, but I am not familiar with its light curve. What is its LBV pedigree?

Stahl: (1) That is true. (2) The B[e] stars are indeed probably not as nitrogen-rich as η Car. They show strong [O I] and no [N II]. I do not know about [N I], so the difference might be a difference in ionization. (3) Not much is known about the light curve of MWC 112. Hoffleit remarked in the *Harvard Annals*, 1933, that it is an irregular variable with an amplitude of one magnitude. We have begun to monitor it photometrically.

Hillier: Two comments. (1) HD 316285 shows P Cygni profiles and is even more extreme than P Cyg. It has broad [N II] and some [Fe II] lines, these being formed in the stellar wind. Broad [N II] λ6583 is present but is badly blended with the electron scattering wings of Hα. (2) In MWC 112, the [N II] lines are probably formed in the stellar wind. Since λ5755 and λ6583 have different critical densities for collisional de-excitation, you need to take their different formation regions into account.

Appenzeller: The [O I] and [N II] lines have highly different critical densities. Hence their ratio may reflect mainly different physical conditions in the B[e] shells, rather than different abundances. To learn something about the abundances, lines with similar critical densities should be compared.

Walborn: I agree; η Car shows strong [N I] and [N II] but no [O I] or [O II]. If a circumstellar shell shows [O I] but not [N I], it is probably not processed material.

Davidson: About these critical densities -- [N I] is a $2p^3$ configuration like [O II], so I suppose it has a slow transition probability and is collisionally de-excited at lower densities than [O I] which is a $2p^4$ configuration.

{Order-of-magnitude critical electron densities for collisional de-excitation are roughly $log(n_e^{crit}/cm^{-3}) \approx 3.5, 4.9, 6.3,$ and 3.9 for [N I] λ5200, [N II] λ6583, [O I] λ6300, and [O II] λ3727 respectively.}

Hillier: The presence of oxygen is not necessarily inconsistent with enhanced nitrogen. In Maeder's models, you can get nitrogen enhancement through carbon depletion. The depletion of oxygen occurs on a longer time scale.

Maeder: It is worth remembering that before ejecting the CNO-processed material, the star must first eject its original outer solar-like-composition material.

Werner Schmutz *et al.*

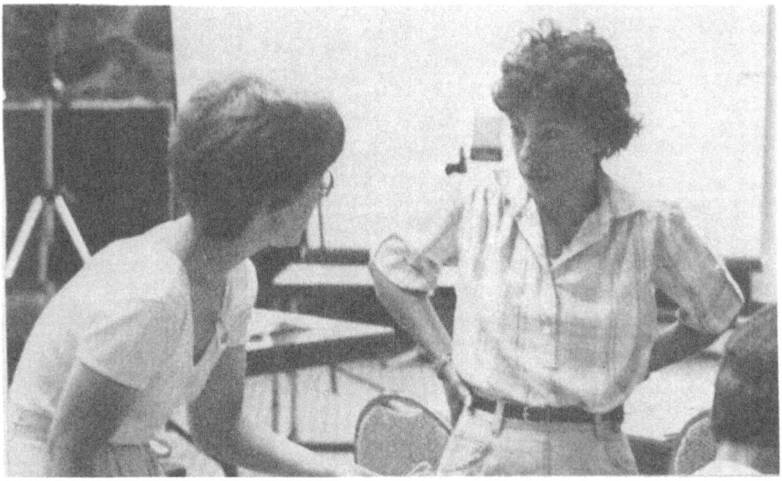

Humphreys, Rossi

HIGH RESOLUTION CORONOGRAPHIC IMAGING OF AG CARINAE[1]

ANTONELLA NOTA[2] and FRANCESCO PARESCE[3]
Space Telescope Science Institute,
3700 *S.Martin Drive,*
Baltimore, MD 21218.

ABSTRACT. We present the first high resolution observations of the ring nebula around the galactic LBV AG Car obtained through broad band B, V, R, I filters with the STScI coronograph on the ESO 2.2 m. telescope. Highly anisotropic continuum emission in all these bands is clearly detected, in the shape of a bipolar structure elongated along the NE-SW direction extending from a few arcsecs from the star out to $\simeq$ 15 arcsecs corresponding to a projected linear distance of $\simeq$ 0.5 parsecs, with a surface brightness in the range 17.9–19.4 mag/arcsec^2. Its essentially neutral color with respect to the star (Δ(V-I) = -0.07) suggests that the observed radiation is starlight scattered by dust grains of diameter $> 1\mu$.

1. Introduction

The shell of line emitting gas surrounding AG Car was first detected by Thackeray in 1950, appearing as an elliptical ring of $39'' \times 30''$ size, with major axis in PA $\simeq 150°$. Its first spectroscopic observations (Johnson, 1976) showed an almost complete absence of [OIII], as might be expected for an evolved object, and the presence of strong Hα to Hδ, [NII] 6548, 6584. Subsequent spectroscopic studies (Thackeray, 1977) derived a remarkable splitting of the [NII] lines (Δv $\simeq$ 119 km/sec) occurring not along the radial expansion direction but parallel to the shell indicating that some portions of the nebula were site of violent motions concentrated close to the star and in a portion of the ring, probably due to a mass loss effect still persisting after the outburst that created the nebula. Recent CCD observations by Stahl (1987) in the light of the nebular emission lines Hα and [NII] have shown the asymmetry of the shell, but without reaching the resolution and the contrast necessary to resolve the structures in the inner part of the ring, due to the brightness of the central star. The problem is even more dramatic in the continuum, where the light of the star ($m_V \simeq 7.7$) dominates the nebula even in a few seconds exposure.

A detailed investigation of the AG Car nebula in the light of the continuum is fundamental, however, to reveal the presence of other emission mechanisms such as

[1]Based on data collected at the European Southern Observatory, La Silla, Chile.
[2]On leave from the Observatory of Padova, Italy.
[3]Affiliated to the Astrophysics Division, Space Science Department, ESA

K. Davidson et al. (eds.), Physics of Luminous Blue Variables, 159–164.

non-thermal or dust-scattered radiation. IR and UV data have recently confirmed that there is a substantial amount of dust in the shell. Observations in the far infrared (McGregor *et al.*, 1988), show evidence of two well resolved peaks at 50 and 100 μ, located at about 9 arcsecs from the star, respectively in direction SW and NE and interpreted by the those authors as being due to a clumpy distribution of cool graphite and silicate dust grains of size $> 1\ \mu$. IUE observations of the nebula (Viotti, 1987) show a nearly constant nebular/stellar continuum ratio between 2000 and 3000 Å suggesting that the main source of UV radiation is the scattering of stellar light by dust particles in the nebula.

2. Observations and Data Reduction

Accurate observations of the two dimensional dust distribution around an object this bright ($m_V \simeq 7.7$) down to a few arcsecs from the star are hampered by the glare from the central star that swamps the faint nebular signal in a broad enough band to detect the scattered continuum. This problem has been recently overcome by using the STScI coronograph mounted on the ESO/MPI 2.2 m. telescope at La Silla.

Briefly, the observing configuration of the coronograph is the following: the light of the central star is focussed onto an occulting mask located in the telescope focal plane (Paresce, Burrows and Horne, 1987). This mask is shaped in the form of a long, thin wedge, that can be moved longitudinally by a micrometer in order to vary its projected width on the sky from 1 to 10 arcsecs depending on the seeing conditions and source brightness. An achromatic doublet reimages the telescope focal plane with a magnification of 5 onto the detector so that the effective focal ratio of the system becomes f/40. A specially designed apodizing mask reduces the stellar light diffracted from the 2.2m telescope pupil. The detector used is a RCA CCD of 320 × 512 30 micron pixels, giving an unrestricted field of view of 22.5 × 36 arcsecs, with a pixel size of 7.03×10^{-2} arcsecs.

AG Car was observed on 1987 December 21 in this configuration, in the broad band pass filters B, V, R, I and in the narrow band interference filters centered on the Hα, at 6563 Å, [NII] at 6583 Å, and [SII] at 6716 Å. The wavelength dependance of the response of these filters closely resembles those of the standard Johnson B, R and the Cousins I_c filters. The complete system has been calibrated absolutely using a standard stellar calibration source (Feige 25) observed in the same broad band filters. The V filter ($\lambda_o = 5591.5$, $\Delta\lambda = 744$ Å) was chosen such that its bandpass is in a region free of emission lines in order to avoid contamination from line emission: comparison calibrations on Feige 25 with the standard Johnson V have shown that our narrow V magnitude is for all purpose identical to the standard V magnitude.

AG Car is first imaged in the unocculted portion of the field and with a neutral density filter of optical density 2.0 and the narrow V filter in order to provide an estimate of its flux at observation time (m_V=7.7, consistent with the AAVSO measurements for that epoch). Then, using the telescope TV autoguiding system, AG Car is placed behind the occulting wedge in a series of short acquisition exposures. The centering of the star behind the mask is achieved by commanding increasingly smaller offsets to the telescope till the spillover light distribution above and below the wedge becomes as symmetrical as possible; at this point, all the neutral density filters are removed and the scientific exposures started. Due to the excellent seeing conditions (FWHM in focus $\simeq$ 0.77 arcsecs) the occulting wedge was set to a width at the center of the stellar profile of $\simeq$ 2 arcsecs, running in the direction EW.

All the images were preprocessed according to standard procedures. Bad columns, blemishes and cosmic rays have been removed from all the frames, then the CCD bias level has been measured in the overscan columns of the frames, averaged and removed. The frames have then been dark subtracted and flat fielded. Results of the processing are shown in Fig.1-a,b. The image on the left corresponds to a 10 min. exposure in the narrow V band and on the right to a 10 min. exposure in the Hα band; north is up, east is to the left, and the cross marks the geometric center of the stellar image. No further image processing has been carried out on these images: diffraction spikes are still visible in the NW-SE, N, NE directions from the star. In all the frames the occulting wedge has been masked out in order to avoid software problems in correspondance to the saturated pixels at its edges.

3. Analysis and Discussion

As we can see from Fig.1a, the images taken in the broad band continua provide a completely different view of the AG Car system, with respect to the well defined ring-shaped nebula observed in the light of the nebular emission lines (Fig.1b). In the continuum there is unambiguous evidence of a *jet-like structure* extending in direction SW from a distance of approximately 5 arcsecs from the center of the star towards the ring at $\simeq$ 15 arcsecs from the center. This feature is extremely well defined in the B, V, I images and seems to consist of two thick helical filaments spiralling towards the external portion of the nebula. It is not visible at all in any of the narrow band images, its shape and size are apparently wavelength independent, and its intensity remains approximately constant with increasing distances from the center. In the opposite direction, towards the NE, at a distance of approximately 9.5 arcsecs from the central star, there is evidence of an extended feature of comparable surface brightness, completely detached from the star and immersed in the weak ring nebulosity. Its sharp inner boundary is reminiscent of the interaction region between a hot stellar wind and the remnants of previous ejecta. The characteristic shell nebula, bright and well defined in the nebular line images, is barely visible, and shows abrupt shock-like edges in the NE and NW inner regions. At the center of the image, the glow due to the scattered light distribution of the central star, not totally suppressed by the coronograph, is still evident, extending to $\simeq$ 4 arcsecs from the center of the star. An investigation of the behaviour of the dust distribution in the closest portion to the star has been attempted by subtracting from the V frame a suitably chosen and scaled reference star, in order to suppress the central glow. Irregularities in the PSF do not allow us to understand if the dust extends closer than 5 arcsecs from the star. However, it is clear that more material is concentrated in the SW direction, and much closer to the star with respect to the NE feature.

The narrow band image in Hα (Fig.1b) shows the nebular structure at unprecedented resolution. The two known areas of highest surface brigthness are well visible along the SW-NE axis. In addition, it seems possible to identify, especially in the NW and SW regions, an inner ring partially detached from the outer, thicker shell, suggestive of two different, subsequent outbursts experienced by the star in its mass loss history. The same details are found in the [NII] image, even if at a lower signal/noise. A comparison with the broad band images indicates that the dusty region of the jet-structure roughly overlaps with the brightest region of the emission nebula; however, there is no evidence of an helical structure in the nebular lines. A comparison with the data published by McGregor *et al.*, within the limits allowed by the resolution of the infrared measurements, indicates that the brighter SW peak concides with the brightest area of the jet, at $\simeq$ 9-10 arcsecs from the star, the NE peak is approximately located in the region of the counterjet.

An estimate of the integrated flux of the jet structure gives $m_V = 17.94$ mag/arcsec^2 and $m_I = 17.25$ mag/arcsec^2, in the brightest area. Assuming the photometry published by McGregor *et al.* (1985), [$(V\text{-}I)_{star} = 0.76$] we obtain: $(V\text{-}I)_{jet} = 0.69$, $\Delta(V\text{-}I) = -0.07$, where Δ is the colour difference between the jet and the star. Within the uncertainties, these values seem compatible with a stellar light scattering model due to large dust particles distributed between the star and the nebula and not yet swept up by the stellar wind.

4. Acknowledgements

We wish to thank all the people who have contributed to the realization of this project: C.Burrows, D.Hofstadt and the ESO operations group at La Silla; A.Ferrari, D.Shore, K.Davidson for useful discussions. We acknowledge the AAVSO observational data sent to us by the Director of the AAVSO, Janet A. Mattei. We are particularly grateful to Roberto Viotti for suggesting the observation of AG Car.

5. References

Johnson, H. M. (1976) 'Spectra and Fabry-Perot interferometry of AG Car and the nebula', Astrophys. J., 206, 469.

McGregor, P.J., Finlayson, K., Hyland, A.R., Joy M., Harvey, P. M., Lester, D.F. (1988) 'Far infrared emission from the AG Carinae ring', Astrophys. J. 329, 874.

Paresce, F., Burrows, C., Horne, K. (1988) 'The structure of the inner R Aquarii nebula in the light of Hα and [NII] λ 6584 Å emission', Astrophys. J., 329, 318.

Stahl, O. (1987) 'Direct imagery of circumstellar shells around Ofpe/WN9 stars in the galaxy and in the LMC', Astron. Astrophys., 182, 229.

Thackeray, A.D. (1950) 'Some southern stars involved in nebulosity', Monthly Notices Roy. Astron. Soc., 110, 524.

Thackeray, A.D. (1977) 'Spectra of the low-excitation nebulosities around AG Carinae and HD 138403', Monthly Notices Roy. Astron. Soc., 180, 95.

Viotti, R., Cassatella, A., Ponz,D., The, P.S. (1987) 'AG Carinae. II. IUE observations of the nebula', Astron. Astrophys., 190, 333.

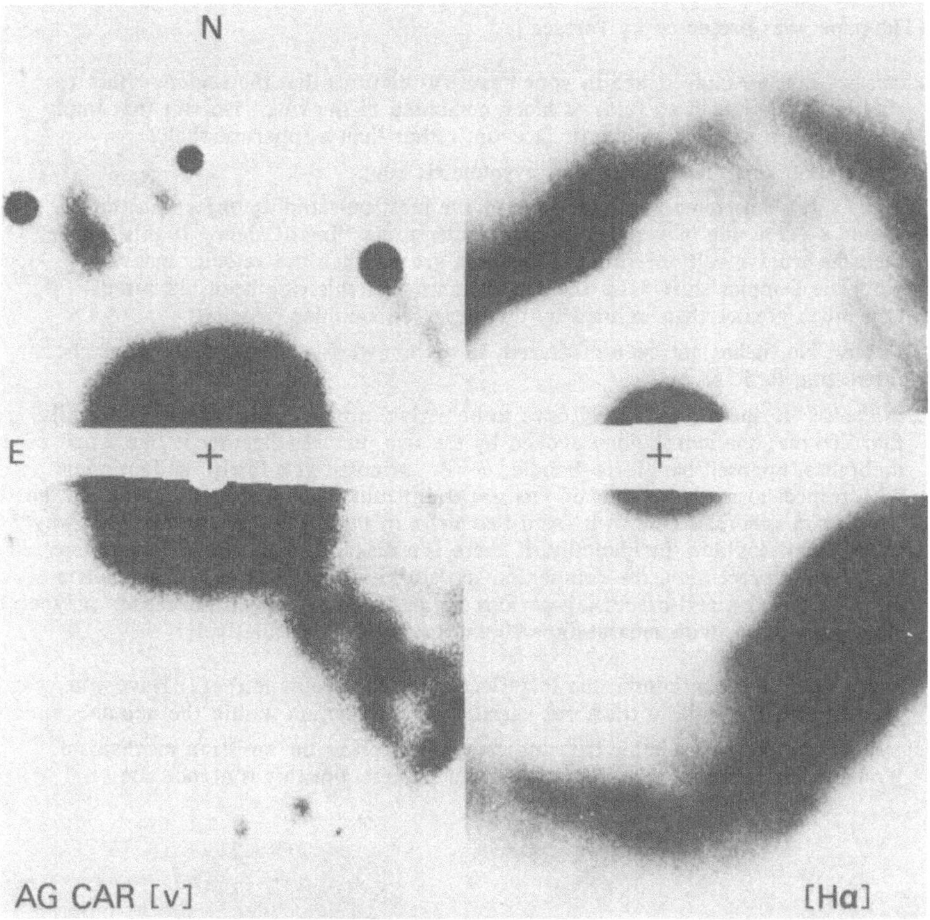

Fig.1a,b - on the left, 10 min. exposure in the V band of the AG Car system. A cross marks the center of the star, whose light is almost completely suppressed by the coronograph. The occulting wedge has a width at the center of $\simeq$ 2 arcsecs. A jet-like structure is well visible in direction SW from a distance of approximately 5 arcsecs from the star towards the outer portion of the ring. On the right, 10 min. image of the characteristic shell nebula in the light of Hα.

DISCUSSION

[The paper was presented by Paresce.]

Lamers: It is amazing to see in your beautiful pictures that the region within the ring in [N II]-light is so faint or black compared to the ring. Doesn't that imply that we are seeing a ring almost face-on, rather than a spherical shell?

Paresce: Perhaps, but if so, a very asymmetric ring.

Owocki: It seems remarkable that one of the most outstanding objects in the sky should eject a ring of material perpendicular to our line of sight. If this is true then the true velocity of ring expansion is greater than the velocity measured from the Doppler shift. Has the proper motion of this ring been measured? Is it much greater than implied by the measured Doppler velocity?

Paresce: No, it has not been measured, to my knowledge. It certainly would be interesting to do so.

Davidson: It doesn't absolutely have to be a ring, not with those jets pointed like that. To me, one morphology evoked by the two pictures together is like a pair of umbrellas, opposed handle-to-handle, (←→), oriented at a fairly random angle with respect to us. Or a pair of jets squirting fluid toward opposite points on the inside of a spherical fishbowl. And that helix in the SW jet is marvelous -- why does it look double? Incidentally, if there is a disk and a jet and a companion star causing precession, the companion star's orbit period will be the geometric mean of the disk and precession periods, in an order-of-magnitude sense: maybe ~ 20 years or so, with separation ~ 20 a.u. or so, very speculatively!

Heap: (1) Is the continuum due to reflection or recombination? (2) Have you used the $H\alpha/H\beta$ ratio to trace out variations of extinction within the nebula?

Paresce: All of the available data indicate reflection as the emission mechanism. We have not taken the $H\alpha/H\beta$ ratio as you suggest, but this is a good idea.

Francesco Paresce

INFRARED OBSERVATIONS OF CIRCUMSTELLAR EJECTA AROUND LUMINOUS BLUE VARIABLES

Peter J. McGregor
Mount Stromlo and Siding Spring Observatories
Institute of Advanced Studies
The Australian National University
GPO Box 4, A.C.T. 2601, Australia

ABSTRACT. Recent infrared spectroscopy and imaging of LBVs lead to the following results. CO overtone emission at 2.3 μm has been found in 13 LBVs in the Galaxy and the LMC. This emission is collisionally excited in warm (3000 - 5000 K), dense ($N_H > 10^{10}$ cm^{-3}) circumstellar material. Circumstellar disks offer favorable conditions for the formation and excitation of CO molecules and are very likely the location of the observed emission. It is proposed that the LBVs showing 2.3 μm CO overtone emission possess the highest density circumstellar disks.

A group of eight LBVs has been identified in the LMC with He I 2.058 μm emission stronger than H I Brγ. This group includes the CO emission star HD 37836. Helium is over-abundant in these stars with *N*(He)/*N*(H) ranging from 0.2 to >0.5. Remarkably five of the helium strong stars belong to the small class of Ofpe/WN9 stars and a further two are probably related to this class.

Slit scans of the galactic LBV AG Car have resolved the far-infrared emission from this star, clearly showing it to originate from cool dust in the circumstellar ring structure. Thermal equilibrium considerations require large grains in the ring in order to match the measured grain temperature and radial distance. Similar slit scans of the galactic B[e] star HD 87643 fail to resolve the far-infrared emission from this star at the 10" level.

1. INTRODUCTION

It has long been recognized that the near-infrared (1 - 5 μm) continua of luminous blue variables (LBVs) are dominated by emission from their circumstellar envelopes in the form of either dust emission or free-free emission from the ionized wind (Allen 1973). Medium resolution spectra of these stars have, however, been difficult to obtain because of instrumental limitations. Technological advances in the last few years have overcome this problem and detailed spectra in the 1 - 5 μm region of LBVs in the Galaxy and the Magellanic Clouds are now available.

Some LBVs have also been detected in the far-infrared by *IRAS*. Cool dust might be expected in the envelopes of LBVs showing strong

K. Davidson et al. (eds.), Physics of Luminous Blue Variables, 165–176.

thermal near-infrared continua, but the detections at far-infrared wavelengths by *IRAS* of galactic and LMC LBVs, such as AG Car (McGregor *et al.* 1988) and R71 (Wolf and Zickgraf 1986), which lack thermal near-infrared emission are surprising results. These recent near-infrared and far-infrared observations of LBVs are reviewed below.

2. NEAR-INFRARED SPECTROSCOPY

2.1. Galactic Stars

Spectra of galactic LBVs in the 1 - 2.5 μm region (Allen, Jones, and Hyland 1985; McGregor, Hyland, and Hillier 1988) show atomic emission lines of H I, He I, Fe II, [Fe II], Mg II, Na I, and O I. While no one object displays all the features listed, a generic envelope structure can be defined, based on the near-infrared spectra of galactic LBVs, which broadly accounts for those features present in each star. The presence of an ionized stellar wind is required to account for the H I and He I emission. The Brackett decrement indicates that Brγ is optically thick in all objects and this is confirmed by calculations based on the Sobolev approximation. A common property of galactic LBVs is that Balmer continuum ionizations from the H I n = 2 level are needed to account for the observed level of hydrogen ionization in the wind. He I $3p^3P^0$-$4d^3D$ 1.700 μm emission is formed by recombination in a He II region. This line is present in AG Car, HD 316285, η Car, BI Cru, and HD 326823. The strength of He I $2s^1S$-$2p^1P$ 2.058 μm relative to H I Brγ suggests that helium is over-abundant in AG Car, HD 316285, HD 326823, and η Car. Allen, Jones, and Hyland (1985) estimate $N(\mathrm{He})/N(\mathrm{H}) > 0.16 \pm 0.03$ in η Car from a consideration of the He I 1.700 μm and 1.197 μm line strengths relative to H I Brγ.

The strongest [Fe II] lines are $a^6D_{9/2}$-$a^4D_{7/2}$ 1.257 μm and $a^4F_{9/2}$-$a^4D_{7/2}$ 1.644 μm, which are prominent in HR Car and η Car. These lines arise from the same upper level, but other weaker lines such as $a^6D_{5/2}$-$a^4D_{5/2}$ 1.294 μm, $a^6D_{1/2}$-$a^4D_{3/2}$ 1.298 μm, and $a^4F_{7/2}$-$a^4D_{3/2}$ 1.599 μm probe densities near ~ 10^5 cm^{-3} where these transitions collisionally de-excite. An electron density of log (N_e) = 4.6 ± 0.3 in the [Fe II] region of η Car is derived from these lines (Allen, Jones, and Hyland 1985).

Na I 2.206/9 μm emission is seen in CPD -57°2874, He3-1359, η Car, BI Cru, and possibly in HD 316285. This emission arises from fluorescent absorption of the 3303 Å continuum. The 2.206/9 μm feature is anomalously strong, based on a calculation of the Sobolev optical depth at 3303 Å (McGregor, Hyland, and Hillier 1988). Strong line blanketing by Fe II and other metals in the ultraviolet may lower the ionization level of sodium within the envelope while maintaining the 3303 Å flux necessary to excite the fluorescence.

CO first-overtone band-head emission at 2.3 μm has been detected in the high luminosity galactic stars CPD -57°2874, CPD -52°9243, GG Car, HR Car, BI Cru, and He3-1359 (McGregor, Hyland, and Hillier 1988). The first three stars are probably related to the B[e] stars, HR Car may be of the S Doradus type, and the nature of BI Cru and that of He3-1359 are less certain. This was the first detection of a molecular component associated with stars of the LBV type. CO molecules cannot survive in the hot photospheres of these stars so this emission must have a circumstellar origin. Vibrational temperatures of ~ 3000 - 5000 K are derived from the approximately equal strengths of the (2-0), (3-1), and (4-2) bands. Similar rotational temperatures are required to populate the high rotational levels near the band-head (J ~ 50). CO emission can be excited in any of three ways; by collisions, by ultraviolet fluorescence in cascades from excited electronic states, or by direct pumping in the fundamental bands at 4.6 μm (Scoville, Krotkov, and Wang 1980; Krotkov, Wang, and Scoville 1980; Scoville *et al.* 1983). Collisional excitation at high densities ($N_H > 10^{10}$ cm^{-3}) is probably responsible for the CO emission in the LBVs because the band strength ratios are inconsistent with ultraviolet fluorescence, and a collisional mechanism is required to excite the high J levels seen. Direct infrared pumping may contribute in some stars but HR Car, which has on occasions shown weak CO emission, lacks significant thermal near-infrared emission and so cannot be pumped in this way.

The mass of emitting CO gas generally lies in the range 10^{-9} - 10^{-8} $M_\odot$ which for solar carbon and oxygen abundances relative to hydrogen corresponds to gas masses in the range 10^{-7} - 10^{-6} $M_\odot$. These masses are comparable to the ionized wind mass out to the densities of ~ 10^9 cm^{-3} required for collisional excitation, so the CO emitting material is a significant component of the circumstellar envelope. Nevertheless, the low excitation CO emitting region must be separate and shielded from the ionized hydrogen zone as hydrogen ionizing photons can dissociate CO. This fact requires that the CO emitting material forms an additional component in the circumstellar envelope, separate from the ionized stellar wind and confirms that both high and low excitation envelope regions coexist in at least some galactic LBVs (Zickgraf *et al.* 1986). Recently discovered ultraviolet predissociation bands in the CO molecule are largely responsible for its radiative dissociation under these conditions (Letzelter *et al.* 1987; Viala *et al.* 1988). These transitions will quickly become optically thick and allow efficient self-shielding of the CO gas to occur in this region, as long as some CO can form initially.

High resolution, higher signal-to-noise ratio spectra of the CO emission bands in the galactic stars (McGregor, Hillier, and Hyland 1989; Figure 1) confirm the approximately equal strengths of the (2-0), (3-1), and (4-2) bands and show what may be weak CO emission in the galactic B[e] star HD 87643. The higher resolution spectra also reveal the presence of ^{13}CO emission in CPD -52°9243 and possibly HD 268835. If the ^{12}CO emission in these stars is optically thin, the ^{12}C/^{13}C ratio is ~ 3. However, the optical depths in the ^{12}CO first-overtone bands are difficult to determine and remain a major uncertainty in the analysis of the CO emission spectra. Nevertheless, values of the ^{12}C/^{13}C ratio as low as 7 have been found in red supergiant atmospheres (Hinkle, Lambert, and Snell 1976), and even lower values may arise in evolved massive blue supergiants as material processed in the CNO cycle is exposed. Prantzos *et al.* (1986) predict ^{12}C/^{13}C ~ 3 during the Of stage of the evolution of a 60 $M_\odot$ star. Similar results were obtained by Maeder (1983, 1987). Thus if the low ^{12}C/^{13}C ratio really applies in the envelope of CPD -52°9243, this envelope material could either have been expelled by the present blue supergiant or while the star was a red supergiant.

Hot dust emission is seen in the B[e] stars and η Car, but is absent in the S Doradus star AG Car and in HD 316285, a star similar to P Cygni. The blackbody hot dust radii are generally near 10^{15} cm and correspond to the inner edge of the circumstellar dust shell as the grain temperatures inferred are near the evaporation temperatures of silicate and graphite grains. Hot dust masses in the range 10^{-8}–10^{-7} $M_{\odot}$ are derived. For a canonical gas-to-dust mass ratio of ~ 100 this corresponds to gas masses in the range 10^{-6}–10^{-5} $M_{\odot}$, significantly larger than the mass of emitting CO gas. The dust envelope is therefore far more extensive than the CO emitting region of the envelope.

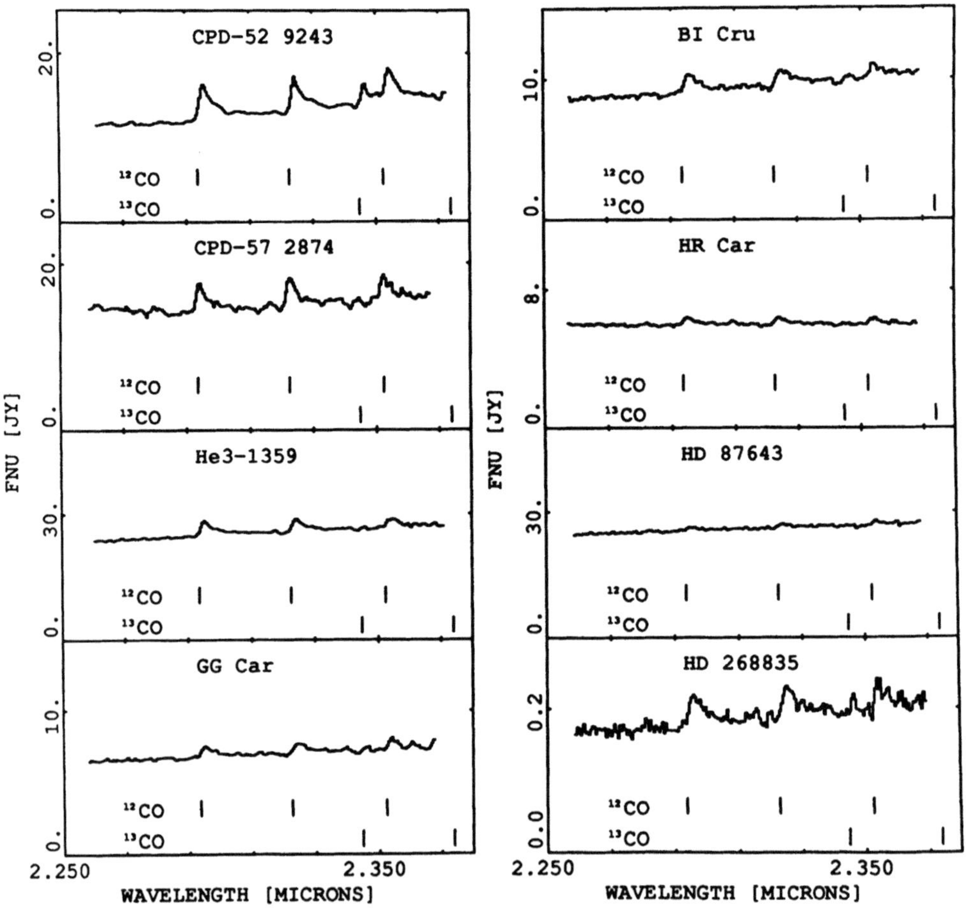

Figure 1. Spectra of the CO first-overtone emission bands in galactic supergiants and the LMC star HD 268835. The spectra were measured with a resolving power $\lambda/\Delta\lambda$ ~ 2000. The (2-0) band head of ^{13}CO is clearly seen in CPD-52°9243 and possibly HD 268835. Weak CO emission may be present in HD 87643.

Several LBVs have been detected in the far-infrared by *IRAS* (McGregor, Hyland, and Hillier 1988). This emission is thermal from cool dust with temperatures near 100 K and masses in the range 10^{-3}-10^{-2} $M_{\odot}$. Again, for a canonical gas-to-dust mass ratio this corresponds to a gas mass range of 0.1-1.0 $M_{\odot}$. At a steady mass loss rate of 10^{-5} $M_{\odot}$ yr^{-1} the star takes 10^4-10^5 yr to lose this material. Considerable evolutionary changes can occur in high mass stars on these timescales so the cool dust emission may contain a fossil record of the mass loss histories of these stars over a significant fraction of their evolution.

2.2. Magellanic Cloud LBVs

2.2.1. CO Emission

A clearer picture of trends in the near-infrared spectra of LBVs is obtained from the larger numbers of LBVs in the Large Magellanic Cloud. Spectra in the 2.0 - 2.4 μm region of 63 luminous supergiants in the LMC have been obtained by McGregor, Hillier, and Hyland (1988). CO emission at 2.3 μm was detected in six stars (Figure 2); HD 37836 is a peculiar O supergiant, HD 268835, Hen S12, and HD 38489 are B[e] stars, and HD 269953 and HD 269723 are G supergiants. HD 37836 in particular is a most unusual object in that it displays high excitation O star wind features in the ultraviolet (Stahl and Wolf 1987) and yet molecules exist in its circumstellar envelope. The LMC CO emission stars show a range of other properties which are also seen in program stars lacking CO emission. Clearly stellar effective temperature does not strongly influence the existence of CO emission since both O and G supergiants show the emission. The lack of reported photometric variability among the LMC CO emission stars suggests that these are relatively stable objects and that the CO emission is probably not transient. Circumstellar disks are believed to be associated with all B[e] stars (Zickgraf *et al.* 1986) so such disks are expected around the three B[e] stars showing CO emission. Although different from the B[e] stars, HD 37836 also possesses a circumstellar disk (Stahl and Wolf 1987). Circumstellar disks are expected to provide conditions that are conducive to CO formation, they can reach the high densities required to collisionally excite CO overtone emission and shield the CO gas from dissociating stellar radiation, and they are likely to be relatively stable envelope structures. It is therefore most probable that the CO overtone emission seen in LMC LBVs arises in the circumstellar disks associated with these stars. Since CO overtone emission is an excellent tracer of high density ($N_H > 10^9$ cm^{-3}) material, *the LBVs showing this emission will be those with the highest density disks.* Similar stars lacking detectable CO overtone emission, such as the other B[e] stars, may possess circumstellar disks but their densities must be below that required to collisionally excite this emission. Other LBVs, such as the S Doradus variables, appear to lack circumstellar disks.

The origin of the circumstellar disks associated with high luminosity evolved stars is of fundamental importance. Near-infrared spectra indicate that high density envelope material is present around

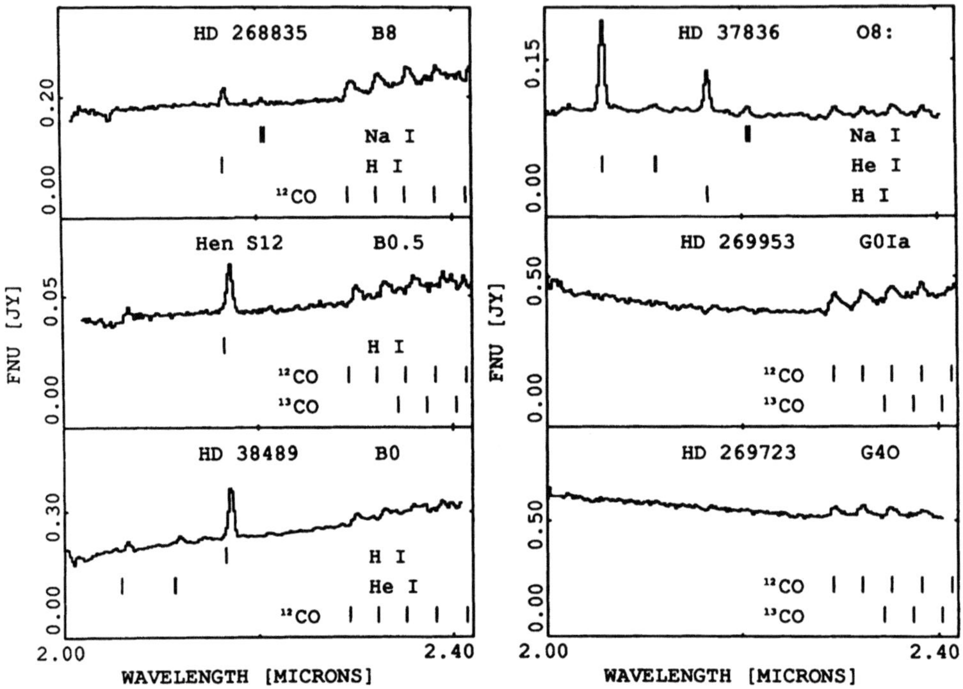

Figure 2. Spectra of LMC CO emission stars. HD 268835, Hen S12, and HD 38489 are B[e] stars, HD 37836 is a peculiar O star possibly related to the Ofpe/WN9 stars, and HD 269953 and HD 269723 are G supergiants.

supergiants with a wide range of temperatures. This may suggest that the circumstellar disk structure seen in B[e] stars forms earlier in their evolution, possible as red supergiants. The CO emitting G supergiants HD 269953 and HD 269723 are then viewed as post-red supergiants and may be the precursors of B[e] stars. Indeed, HD 269953 possesses a cool dust envelope which has been detected at 3.6 and 10 μm (Glass 1974; Ney 1972; Glass 1984), and there is abundant evidence for the production of dense circumstellar dust envelopes by high luminosity red supergiants such as the supergiant OH/IR stars VY CMa and VX Sgr. However, HD 37836, for example, possesses a dense circumstellar disk but is believed to be of too high a luminosity to have evolved through a red supergiant phase. This fact and the presence of similar high luminosity stars with circumstellar disks make a red supergiant origin for all such disks difficult to maintain. An intrinsic, but as yet poorly understood, property of the star appears responsible. CO emission has been seen in the galactic F supergiants ρ Cas and BS 8752 (Lambert, Hinkle, and Hall 1981) where high resolution spectra showed a dense shell being expelled and subsequently falling back onto the star. If unrelated to the B[e] star phenomenon, the CO emission in HD 269953 and HD 269723 may have a similar origin to the galactic star emission.

2.2.2. Helium Strong Stars

Nine LMC stars have been identified with He I $2s^1S-2p^1P$ 2.058 μm stronger than H I Brγ (McGregor, Hillier, and Hyland 1988) and include the O8 CO emission star HD 37836 (Figures 2 and 3). Five of these stars belong to the group of Ofpe/WN9 stars [HD 269445, Hen S61 (Walborn 1977, 1982); HD 269582 (Stahl 1986); Hen S9 (Walborn 1986); HD 269687 (Bohannan and Walborn 1988)] and a further two are probably related objects [HD 37836 and Hen S131 (Stahl and Wolf 1987)]. The remaining two helium strong stars are HD 268840 and Hen S101, the latter of which being a border-line member of the group. The Ofpe/WN9 class is distinguished by the presence of N III and He II emission in comparable strength with N II and He I emission (Walborn 1977).

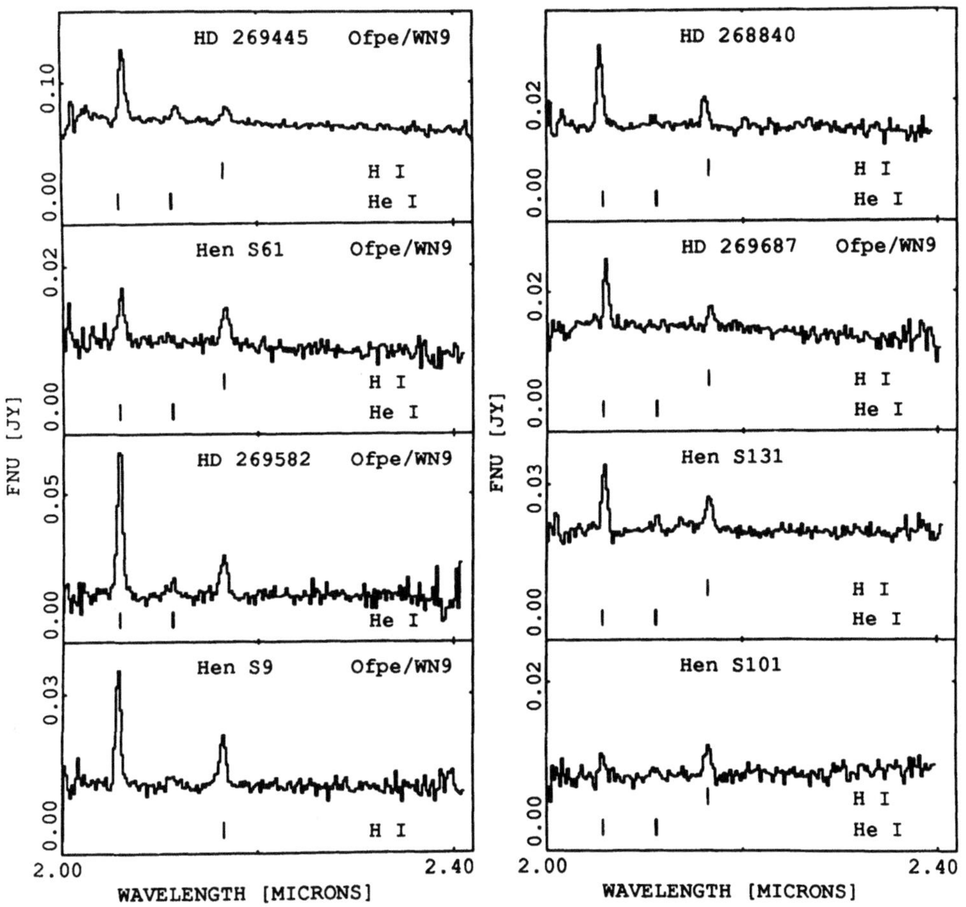

Figure 3. Spectra of LMC supergiants showing He I 2.058 μm emission stronger than H I Brγ 2.167 μm emission. The CO emission star HD 37836 also belongs to this class. Five of these stars are classed as Ofpe/WN9.

Enhanced nitrogen abundances relative to hydrogen and oxygen suggests CNO processing of the envelope material (Walborn 1982). An evolutionary state intermediate between Of and WN stars may be indicated (Stahl 1986).

The strength of the near-infrared He I emission indicates that helium must also be over-abundant relative to hydrogen in the atmospheres of these stars; although difficult to quantify, $N(\mathrm{He})/N(\mathrm{H})$ is estimated to range from ~ 0.2 in Hen S9 and Hen S61 to greater than 0.5 in HD 269445 and HD 269582. Certainly the near-infrared spectrum provides a much clearer indicator of helium over-abundance than is possible from optical spectra. Of the Ofpe/WN9 stars not listed above, HD 269227 is dominated by its M supergiant companion in the near-infrared, BE 381 and HD 269927C were not measured, and the S Doradus star HD 269858f, which was classified as OIafpe in 1977 (Walborn 1977), now lacks near-infrared He I emission.

The 2 μm spectrum of the galactic supergiant HD 326823 is similar to those of the helium rich LMC stars (McGregor, Hyland, and Hillier 1988); this star may therefore also be related to the Ofpe/WN9 stars.

3. FAR-INFRARED MEASUREMENTS

Far-infrared emission from cool dust was detected by *IRAS* in the galactic LBVs η Car, AG Car, HR Car, HD 87643, and CD -42°11721 (McGregor, Hyland, and Hillier 1988), and in the LMC S Doradus star R71 (Wolf and Zickgraf 1986). The far-infrared emission in η Car, HD 87643, and CD -42°11721 is accompanied by near-infrared emission from hot dust, suggesting that the range of thermal emissions is due to temperature gradients in the dust shells. In contrast, AG Car and R71 lack thermal emission from hot dust in the near-infrared making it unlikely that the cool emitting dust is associated with the immediate circumstellar environment. Slit scans of AG Car in a NE-SW direction at 50 and 100 μm obtained from the Kuiper Airborne Observatory clearly resolve the far-infrared emission from this object into two peaks centered on the circumstellar ring structure (Figure 4; McGregor *et al.* 1988). Maximum entropy deconvolution shows the peaks to be at radii of ~ 9" from the star with very little emission coming from between the peaks at the position of AG Car itself. The emitting grains have a temperature of ~ 60 K and the 100 μm optical depth is < 5 x 10^{-5}. Thus the emitting dust absorbs very little of the stellar luminosity. The dust mass is estimated to be between 0.002 and 0.02 $M_{\odot}$ depending on the adopted grain characteristics. Large grains with radii of ~ 1.0 μm are required if they are in thermal equilibrium with the stellar radiation field at the observed distance from the star. However, the lack of mid-infrared emission from the ring indicates that no corresponding population of small grains exists. The presence of dust in the AG Car ring is also inferred from the ultraviolet spectrum of the ring which is due to scattered starlight (Viotti *et al.* 1988). AG Car shows similarities in the optical and ultraviolet to the LMC S Doradus star R71 and to other stars showing spectroscopic evidence for detached ionized shells, the Ofpe/WN9 stars in particular (Walborn 1982; Stahl 1986). Thus circumstellar gas and dust ring structures like the AG Car ring may frequently be associated with evolved massive emission-line stars.

The origin of the AG Car ring bears on questions of the nature of the S Doradus phenomenon and the mass loss histories of massive stars and has been considered by McGregor *et al.* (1988). They conclude that the ring is most likely a "mass loss bubble" formed as the present hot supergiant wind compresses material lost in a cooler supergiant wind (cf., Johnson 1980; Weaver *et al.* 1977). Steady blue supergiant mass loss and a nova-like explosion were rejected. The formation of a mass loss bubble requires that previous mass loss occurred at a slower velocity than the present blue supergiant wind velocity. If wind velocity decreases abruptly in cooler supergiants (as seems to be the case) the requirement may be simply that the high luminosity stars have experienced *a cooler evolutionary phase* rather than that they have necessarily experienced significant red supergiant evolution. Main sequence mass loss is excluded because the present wind velocity is only ~ 290 km s^{-1}. The model proposed is similar to current theories of planetary nebula formation (Kwok, Purton, and FitzGerald 1978), but is here applied to a more massive star. The ring structures may well be precursors of the mass loss bubbles seen on larger scales around Wolf-Rayet stars (Chu 1982; van der Hucht *et al.* 1985).

Similar unpublished slit scans at 50 and 100 μm of HD 87643, a galactic B[e] star, show that this star is unresolved at the 10" level

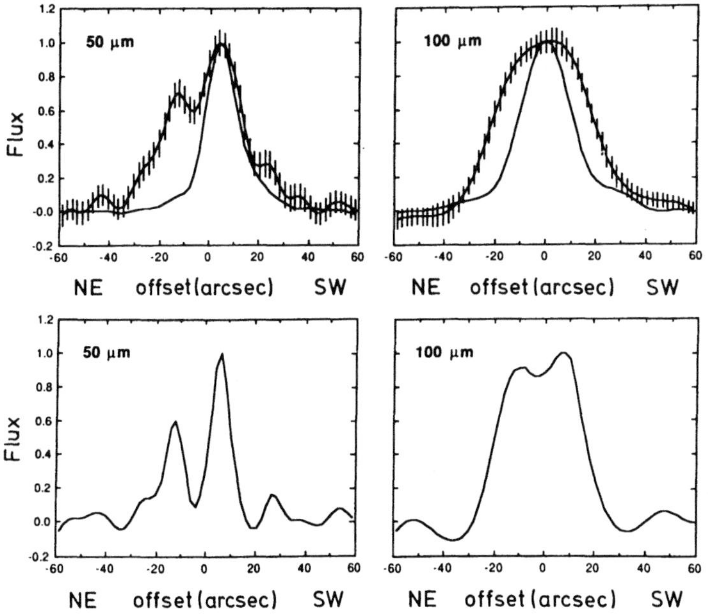

Figure 4. Slit scans of AG Car at 50 and 100 μm compared with similar scans of a point source (*upper frames*) show the far-infrared emission to be resolved. The deconvolved scans (*lower frames*) clearly show the far- infrared emission to come from two peaks at the position of the ring with little emission from the position of AG Car itself.

at these wavelengths, despite the presence of prominent optical reflection nebulosity over a region ~ 90" x 65" around the star. The measured flux densities at 50 and 100 μm are 290 Jy and 248 Jy, respectively. These compare well with the *IRAS* results. The cool dust mass in HD 87643 is ~ 5 times that detected around AG Car. Unlike AG Car, the hotter inner regions of the dust shell in HD 87643 shield the cool dust. This hot dust is not present in AG Car.

REFERENCES

Allen, D. A. 1973, *M.N.R.A.S.*, **161**, 145.
Allen, D. A., Jones, T. J., and Hyland, A. R. 1985, *Ap. J.*, **291**, 280.
Bohannan, B., and Walborn, N. R. 1988, in preparation.
Chu, Y.-H. 1982, *Ap. J.*, **254**, 578.
Glass, I. S. 1974, *M.N.R.A.S.*, **168**, 249.
__________ 1984, *M.N.R.A.S.*, **209**, 759.
Hinkle, K. H., Lambert, D. L., and Snell, R. L. 1976, *Ap. J.*, **210**, 684.
Johnson, H. M. 1980, *Ap. J.*, **235**, 66.
Krotkov, R., Wang, D., and Scoville, N. Z. 1980, *Ap. J.*, **240**, 940.
Kwok, S., Purton, C. R., and FitzGerald, P. 1978, *Ap. J. (Letters)*, **219**, L125.
Lambert, D. L., Hinkle, K. H., and Hall, D. N. B. 1981, *Ap. J.*, **248**, 638.
Letzelter, C., Eidelsberg, M., Rostas, F., Breton, J., Thieblemont, B. 1987, *Chem. Phys.*, **114**, 273.
Maeder, A. 1983, *Astr. Ap.*, **120**, 113.
__________ 1987, *Astr. Ap.*, **173**, 247.
McGregor, P. J., Finlayson, K., Hyland, A. R., Joy, M., Harvey, P. M., and Lester, D. F. 1988, *Ap. J.*, **329**, 874.
McGregor, P. J., Hillier, D. J., and Hyland, A. R. 1988, *Ap. J.*, in press.
__________ 1989, in preparation.
McGregor, P. J., Hyland, A. R., and Hillier, D. J. 1988, *Ap. J.*, **324**, 1071.
Ney, E. P. 1972, *Pub. A.S.P.*, **84**, 613.
Prantzos, N., Doom, C., Arnould, M., and de Loore, C. 1986, *Ap. J.*, **304**, 695.
Scoville, N. Z., Kleinmann, S. G., Hall, D. N. B., and Ridgway, S. T. 1983, *Ap. J.*, **275**, 201.
Scoville, N. Z., Krotkov, R., and Wang, D. 1980, *Ap. J.*, **240**, 929.
Stahl, O. 1986, *Astr. Ap.*, **164**, 321.
Stahl, O., and Wolf, B. 1987, *Astr. Ap.*, **181**, 293.
van der Hucht, K. A., Jurriens, T. A., Olnon, F. M., Thé, P. S., Wesselius, P. R., and Williams, P. M. 1985, *Astr. Ap.*, **145**, L13.
Viala, Y. P., Letzelter, C., Eidelsberg, M., and Rostas, F. 1988, *Astr. Ap.*, **193**, 265.
Viotti, R., Cassatella, A., Ponz, D., and Thé, P. S. 1988, *Astr. Ap.*, **190**, 333.
Walborn, N. R. 1977, *Ap. J.*, **215**, 53.
__________ 1982, *Ap. J.*, **256**, 452.
__________ 1986, in *IAU Symposium 116, Luminous Stars and Associations in Galaxies*, ed. by C. W. H. de Loore, A. J. Willis, and P.

Laskarides, (Dordrecht: Reidel), p. 194.
Weaver, R., McCray, R., Castor, J., Shapiro, P., and Moore, R. 1977, *Ap. J.*, **218**, 377.
Wolf, B., and Zickgraf, F.-J. 1986, *Astr. Ap.*, **164**, 435.
Zickgraf, F.-J., Wolf, B., Stahl, O., Leitherer, C., and Appenzeller, I. 1986, *Astr. Ap.*, **163**, 119.

Peter McGregor

DISCUSSION

Gallagher: In your discussion of ejection processes, you used a nova analogy. Because of the large gravitational binding energy on the surface of a white dwarf, the mechanical energy in a nova is at least comparable to the integrated photon luminosity. This may give you more room to work with in a model in which material is ejected because of a transient energy input deep in the system.

McGregor: The model I propose is similar to that suggested for planetary nebulae; the ring is a shock region between a fast blue supergiant wind and an earlier slow wind.

Zickgraf: (1) For one of the Galactic B[e] stars with CO emission, CPD -52° 9243, Swings found polarimetric evidence for a disk. (2) In our study of the Magellanic Cloud B[e] supergiants we found evidence for a dependence of the TiO emission strength on the density in the proposed disk. This agrees with your finding of CO emission in denser environments. (3) Polarization measurements of Galactic B[e] stars (to appear in *Astr.&Astrophys.*) show that the polarizing particles should not be much smaller than 0.1 μm. This would agree with the large dust particles that you find around AG Car.

McGregor: The TiO emission that you observe may require even higher densities than those needed to produce the CO emission; this should be calculated.

Maeder: You mentioned Na I lines in two of your O and B stars. Is this element overabundant in relation to recent results by Dearborn and others? Also, do you think you can measure the $^{16}O/^{17}O$ ratio? It should differ from the Solar value by about 2 orders of magnitude.

McGregor: Sodium abundances will be difficult to determine from the fluourescent 2.206/9 μm lines, as the sodium ionization depends on blanketing in the UV. Possibly the optical Na I D lines will help here. $^{16}O/^{17}O$ will be difficult to determine in the near-infrared but should be considered in analyzing the TiO emission bands in B[e] stars reported at this meeting.

Humphreys: It has frequently been mentioned that the ejecta of some of these stars may be fossil remnants from the red supergiant stage, but many LBV's are above the luminosities where we observe M supergiants. Do you know of any IR data, say from IRAS, that show potential red supergiants at these luminosities? For example, two IRAS sources in the LMC are considered to be highly obscured M supergiants (Elias, Frogel, & Schwering 1986, *Astrophys. J.* **302**, 675), but their luminosities are not above the RSG limit.

McGregor: I know of no evidence for the existence of red supergiants at luminosities above your limit. In referring to a star evolving through a red supergiant phase, I meant normal red supergiants; this can't happen for the most luminous stars.

Walborn: A very interesting current study by Linda Smith, Max Pettini, *et al.*, shows interacting fast and slow material in the ring nebula RCW 58 which surrounds the WN8 star HD 96548; this suggests a post-red-supergiant interpretation. There may be similarities to the AG Car nebula.

Humphreys: The kinematic distance for AG Car seems to rule out a red supergiant stage. With the luminosity estimated by Henny Lamers in a poster paper at this meeting, $M_{bol} \approx -8.8 - 5\,log(D/2.5\,\mathrm{kpc})$, any distance beyond 3.5 or 4 kpc implies a luminosity too high for RSG's.

THE FORMATION OF SHELLS IN THE WIND FROM P CYGNI

F. D. KAHN
Department of Astronomy
The University
Manchester M13 9PL
England

ABSTRACT. A mechanism is proposed for the formation of shells around P Cygni. These shells seem to arise at approximately two monthly intervals and each appears to undergo a steady acceleration of some 0.6 cm s^{-2} during the six months that it is present. In the process to be described the formation of the shells takes place in a stellar wind that moves outwards ballistically from the photosphere. Small fluctuations in wind speed are introduced there and eventually lead to the condensation of shells at a radial distance of the order of 10^{14} cm. The density in the shells is large enough to account for their acceleration by the Lyman continuum emitted by the star. It is also proposed here that the observed time scale of two months is set by the dynamics of the flow field in interaction with the random fluctuations imposed on it at the photosphere.

1 Introduction

In his invited review given at the Lunteren meeting two years ago Lamers (1987) describes the characteristic variability of LBV stars and also gives an account of the episodic formation of shells around P Cygni. One curious feature of the process is immediately apparent. The currently observed fluctuations in visual magnitude have an amplitude of a few per cent and appear to show no preferred period. The intervals between the appearance of successive shells are also rather irregular, but they do show a preference for a period of about two months. An even more significant difference is that the variations in magnitude are most probably due to relatively small changes in the photospheric radius and can be described as a small perturbation superposed on a star in a steady state; the formation of a shell in a wind is however a grossly non-linear effect. The intention in this paper is to establish the nature of the relation between the variability of the star and the formation of shells in its wind.

The variations in visual magnitude have not always been small. Some four centuries ago P Cygni underwent a giant "eruption", with $\Delta V \sim 3$ mag, and as a result lost, over an interval of only one decade, an amount of energy equal to its stored thermal energy. Such a major change cannot simply be the consequence of an unstable stratification in an outer layer, whose energy content is necessarily small. The most reasonable conclusion must be that the occasional major eruption is caused by an unsteady physical process located deep within the star, and that the stellar wind and the small fluctuations in visual magnitude are driven by the same mechanism.

K. Davidson et al. (eds.), Physics of Luminous Blue Variables, 177–183.

The following are the basic data for the present treatment:

Mass of P Cygni	$M =$	$35 M_\odot$
Radius of photosphere	$r_e =$	5×10^{12}cm
Luminosity	$L =$	$7.5 \times 10^5 \, L_\odot$
Effective temperature (with fluctuations)	$T_e =$	20 000 K
Mass loss rate into the wind	$\dot{\mathcal{M}} \sim$	$10^{-5} \, M_\odot$/year

The wind is fully ionized, certainly until shell formation occurs. The radiative acceleration outwards due to electron scattering is

$$f_r \;=\; \kappa_e L/4\pi c r^2 \;=\; 2.7 \times 10^{27}/r^2$$

at distance r (expressed in CGS units) while the gravitational acceleration inwards is

$$g \;=\; GM/r^2 \;=\; 4.7 \times 10^{27}/r^2 \,,$$

and so the net inward acceleration becomes

$$g_{\rm eff} \;=\; \mu/r^2 \;\equiv\; 2 \times 10^{27}/r^2 \,.$$

The density in a wind with terminal speed u_∞ is

$$\rho \;=\; \frac{\dot{\mathcal{M}}}{4\pi r^2 \, (2\mu/r + u_\infty^2)^{1/2}} \tag{1}$$

at distance r, and the rate of recombinations per unit volume there is

$$b n_e^2 \;=\; \frac{b\rho^2}{m_a^2} \;=\; \frac{b\dot{\mathcal{M}}}{16\pi^2 m_a^2 r^3 (2\mu + u_\infty^2 r)} \tag{2}$$

where m_a is the mass per free electron.

The total rate of recombinations in the wind from the star out to some distance $r_{\rm ext}$ is

$$\begin{aligned} S_{r_{\rm ext}} \;&=\; 4\pi \int_{r_*}^{r_{\rm ext}} b n_e^2 \, r^2 \, dr \\ &=\; \frac{b\dot{\mathcal{M}}^2}{8\pi m_a^2 \mu} \int_{r_*}^{r_{\rm ext}} \frac{dr}{r\,(1 + u_\infty^2 r^2/2\mu)} \,, \end{aligned} \tag{3}$$

where $b (= 2 \times 10^{-13}$cm^{3}s$^{-1})$ is the usual recombination coefficient in the ionized gas and $m_a = 2 \times 10^{-24}$ gm. The integrated term contributes a logarithmic factor, of order unity. The coefficient outside equals 3.6×10^{47}s^{-1}, with the adopted numerical parameters. For comparison P Cygni produces photons in the Lyman continuum at a rate of some 10^{49}s^{-1}, and so maintains full ionization, unless shell formation takes place.

The Ly c photons come from an atmosphere with temperature 20 000K and have an average energy $< \chi > \doteqdot 15.5$ eV$= 2.5 \times 10^{-11}$ erg. They transfer outward momentum to the gas at a rate $b n_e < \chi > /c$ and produce an acceleration

$$f_{\rm Ly} \;=\; b n_e < \chi > / m_a c \;\equiv\; \lambda \rho \tag{4}$$

where

$$\lambda \equiv b < \chi > /m_a^2 c = 4 \times 10^{13}\,\mathrm{cm^4 gm^{-1} s^{-2}}.$$

In a steady wind, with speed u at distance r,

$$f_{\mathrm{Ly}} = \lambda \dot{\mathcal{M}}/4\pi u r^2 \tag{5}$$

and this acceleration is negligible compared with g_{eff} unless

$$\lambda \dot{\mathcal{M}}/4\pi u > \mu$$

$$\text{or } u < \lambda \dot{M}/4\pi\mu \sim 10\,\mathrm{km\,s^{-1}}\,.$$

Such slow winds do not occur in the present model for P Cygni.

2 The Formation of Shells

The wind will be treated as being cold. This approximation is reasonable enough since the wind speed is generally smaller by an order of magnitude than the (isothermal) sound speed c_i. As long as there are no collisions, that is in the time before the first shell forms, each part of the wind moves like an isolated stream in the effective gravitational field. The equation of motion integrates to give

$$\frac{dr}{dt} = \left\{\frac{2\mu}{r} + W(\tau)\right\}^{1/2} \tag{6}$$

where $W(\tau)$ is twice the energy per unit mass of the stream and varies cyclically with the time of emission τ. If W is negative then the stream is gravitationally bound—until other effects take over. The solution of (6) is most conveniently expressed, for the purpose of this treatment, by writing

$$t - \tau = \int_0^r \frac{dr}{\{2\mu/r + W(\tau)\}^{1/2}} \equiv T(r, \tau)\,. \tag{7}$$

The treatment that follows depends in detail on the sign of W for that part of the wind on which the shell first forms, that is on whether it would have had enough energy, if left to itself, to escape from the reduced gravitational field of the star, or not. For the purpose of this calculation it is suitable to deal with the case where W is positive.

Consider then how the function

$$t \equiv \tau + T(r, \tau) \tag{8}$$

depends on τ, for progressively larger values of r. Close enough to the star the time of arrival, t, of a stream must vary monotonically with τ, that is the streams arrive in the order that they were emitted, and $\partial t/\partial\tau$ is always positive. But with increasing distance r the effect of the fluctuations in $W(\tau)$ becomes progressively more pronounced so that at one stage $\partial t/\partial\tau$ vanishes for just one instant in the cycle, and this condition defines the level r_s and the time t_s where the first shell forms; let the corresponding value of τ be τ_s. Evidently $\partial t/\partial\tau$ at level r_s is positive for $\tau < \tau_s$ and $\tau > \tau_s$, and only vanishes at $\tau = \tau_s$.

Consequently $\partial^2 t/\partial\tau^2 = 0$ at $r = r_s$, $\tau = \tau_s$, and so the time of arrival at level r_s of a stream emitted at time $\tau_s + \delta\tau$ is

$$t_s + \delta t = t_s + \frac{1}{6}T_3(\delta\tau)^3 \qquad (9)$$

where

$$T_3 \equiv \partial^3 T/\partial\tau^3 .$$

Let V_s be the speed of the shell at the time when it forms. The stream which left the star at time $\tau_s + \delta\tau$ lags by a distance

$$V_s\,\delta t = \frac{1}{6}V_s T_3(\delta\tau)^3 \qquad (10)$$

at time t_s, but its speed is larger by an amount

$$\Delta V = \frac{(dW/d\tau)\delta\tau}{V_s} . \qquad (11)$$

It will therefore catch up the newly formed shell at time $t_s + \Delta t$, where

$$\Delta t \equiv \frac{V_s\,\delta t}{\Delta V} = \frac{V_s^2 T_3}{6dW/d\tau}(\delta\tau)^2 = \Lambda(\delta\tau)^2 \qquad (12)$$

say.

The mass per unit area that has accumulated on each side of the shell is, at time $t_s + \Delta t$,

$$\Sigma = \frac{\dot{\mathcal{M}}\delta\tau}{4\pi r_s^2} = \frac{\dot{\mathcal{M}}}{4\pi r_s^2}\left(\frac{\Delta t}{\Lambda}\right)^{1/2} \qquad (13)$$

and its rate of arrival, per unit area, is

$$\dot{\Sigma} = \frac{\dot{\mathcal{M}}}{8\pi r_s^2(\Lambda\Delta t)^{1/2}} . \qquad (14)$$

The stream that arrives at time $t_s + \Delta t$ has a speed ΔV relative to the shell (see relation (11)). The pressure on each side of the shell is therefore

$$P = \dot{\Sigma}\Delta V = \frac{3}{4\pi}\frac{\dot{\mathcal{M}}(dW/d\tau)^2}{r_s^2 V_s^3 T_3} . \qquad (15)$$

Since the (isothermal) sound speed in the shell is assumed to be c_i (= 10 km/s) the density of the gas in the shell is

$$\rho = \frac{3}{4\pi c_i^2}\frac{\dot{\mathcal{M}}(dW/d\tau)^2}{r_s^2 V_s^3 T_3} \qquad (16)$$

and the acceleration, due to the pressure of Ly c photons, is $\lambda\rho$.

This formalism has now to be reconciled with the observations of shells in P Cygni, reported by de Groot (1969), Markova and Luud (1983), Markova and Kolka (1984) and

Markova (1986). The period between the creation of successive shells is about two months, the shells have speeds of about 100 km s^{-1} when they first appear, and over a period of seven months or so they accelerate to 220 km s^{-1}. The typical acceleration is therefore about 0.6 cm s^{-2}.

The first step is to assume a value for r_s, the radial distance at which a shell forms. The value to be used for $dW/d\tau$, in equation (16), can then be calculated by noting that

$$V_s^2 = \frac{2\mu}{r_s} + W(\tau_s) \tag{17}$$

and that a necessary condition for shell formation is that

$$\begin{aligned} 0 &= 1 + \frac{\partial T}{\partial \tau} \equiv 1 + \frac{\partial}{\partial \tau} \int_0^{r_s} \frac{dr}{[2\mu/r + W(\tau)]^{1/2}} \\ &= 1 - \frac{dW/d\tau}{2} \int_0^{r_s} \frac{dr}{(2\mu/r + V_s^2 - 2\mu/r_s)^{3/2}} \, . \end{aligned} \tag{18}$$

The value of T_3 cannot be determined unless the form of variation of $W(\tau)$ is known. To make progress, therefore, set

$$W(\tau) = \overline{W}(1 + \epsilon \cos \omega\tau) \, , \tag{19}$$

so that the fluctuation in the specific energy content is sinusoidal, and set $2\pi/\omega$ equal to two months, the interval between the appearance of successive shells. With this simplification it becomes possible, but is rather tedious, to calculate T_3, and hence ρ and $\lambda\rho$ in terms of assumed values of $\dot{\mathcal{M}}$ and r_s. The initial acceleration of the shell is then the resultant of an acceleration $\lambda\rho$ outwards and μ/r_s^2 inwards, towards the star, giving that

$$\lambda\rho = \frac{\mu}{r_s^2} + 0.6 \, . \tag{20}$$

Representative results are given in Table 1 below, for three assumed values of the radial distance r_s.

These results help determine the value of $\dot{\mathcal{M}}$ appropriate for shell formation at each radial distance r_s. The shell has a mass which ultimately grows to $M_s = 2\pi\dot{\mathcal{M}}/\omega$ and, when it is put under pressure P, the number of Ly c photons required, per second, to keep it ionized is

$$S_{\text{req}} = \frac{b\mathcal{M}_s P}{m_a^2 c_i^2} = \frac{2\pi b\dot{\mathcal{M}}P}{\omega m_a^2 c_i^2} \, . \tag{21}$$

r_s	5×10^{13}	10^{14}	1.5×10^{14}	cm
$P/\dot{\mathcal{M}}$	2.64×10^{-22}	1.86×10^{-23}	3.79×10^{-24}	$cm^{-1}s^{-1}$
T_3	2.68×10^{-12}	1.64×10^{-12}	1.25×10^{-12}	s^{-2}
$f = \mu/r_s^2 + 0.6$	1.40	0.80	0.69	cm s^{-2}
$\rho = f/\lambda$	3.5×10^{-14}	2×10^{-14}	1.73×10^{-14}	gm cm^{-3}
$P = \rho c_i^2$	3.5×10^{-2}	2×10^{-2}	1.73×10^{-2}	dyne cm^{-2}

Table 1:

r_s	5×10^{13}	10^{14}	1.5×10^{14}	cm
$\dot{\mathcal{M}}$	1.33×10^{20}	1.08×10^{21}	4.56×10^{21}	gm s^{-1}
S_{req}	1.20×10^{48}	5.65×10^{48}	2.07×10^{49}	s^{-1}
$1.2 \times 10^7 \dot{\mathcal{M}}$	1.60×10^{27}	1.30×10^{28}	5.47×10^{28}	gm cm s^{-2}
$S_* < \chi > /c$	←	8.33×10^{27}	→	gm cm s^{-2}

Table 2:

Finally the momentum required per unit time to speed up all the shells by 120 km/s is $1.2 \times 10^7 \dot{\mathcal{M}}$: the momentum per unit time carried by the Lyman continuum flux is $S_* < \chi > /c$. Table 2 shows the relation between these quantities.

If r_s were as small as 5×10^{13} cm then the number of Lyman continuum photons needed to keep the innermost shell ionized would be about one tenth of the number produced by P Cygni. This would conflict with the observation that only three or four shells are visible at any one time. On the other hand P Cygni does not produce enough Ly c photons to keep even one shell ionized, for the case of $r_s = 1.5 \times 10^{14}$ cm. When $r_s = 10^{14}$ cm the inner shell soaks up about half the available ionizing radiation, which seems about right, but the Lyman continuum flux does not provide quite enough radiation pressure to accelerate the flow. Interpolation of the data in the table indicates that momentum balance is achieved for $\dot{\mathcal{M}} = 6.8 \times 10^{20}$gm s^{-1} and $r_s = 8.6 \times 10^{13}$ cm, and that then $S_{req} = 4.2 \times^{48}$ s^{-1} so that the innermost shell soaks up about two-fifths of the available Lyman continuum photons.

3 Discussion

It is interesting that the stream on which the shell forms takes about $2\frac{1}{2}$ months to rise to radial distance r_s. In fact it is only to be expected that the dynamics of the wind should determine the interval between episodes of shell formation since P Cygni itself does not possess any likely modes of oscillation that have a period which is even of the right order of magnitude. The observed photometric variations of luminous blue variables look rather like a manifestation of random noise. The real question is therefore why and how the flow selects the frequencies at which shells appear from the broad band disturbances introduced at the star. A paper to follow (Kahn 1989) will address this problem and also deal more thoroughly with the physical nature of the flow.

References

de Groot, M.: 1969, Bull. Astr. Inst. Netherlands, **20**, 235.

Lamers, H. J. G. L.M.: 1987 in *Instabilities in Luminous Early Type Stars*, eds. Lamers, H. G. J. L. M. and de Loore, C. W. H., Reidel, Dordrecht, p.99.

Markova, N.: 1986, Astron. Astrophys., **162**, L3.

Markova, N. and Kolka, I.: 1984, Astrofiz., **20**, 250.

Markova, N. and Luud, L.: 1983, Publ. Tartu Astrophys. Obs., **32**, 55.

DISCUSSION

Davidson: Lyman alpha?

Kahn: Yes, $L\alpha$ becomes important when the shell reaches a place where the Lc runs out. The $L\alpha$ photons press on the neutral H-atoms there.

Appenzeller: Are the shells that you described subject to Rayleigh-Taylor instabilities and could such instabilities modify or destroy the shells?

Kahn: Rayleigh-Taylor instabilities very likely become important in the outer, less ionized layers that I just mentioned, where $L\alpha$ radiation begins to push hard on the neutral hydrogen.

Sreenivasan: You said that the first shell forms way out in the supersonic region of the flow. What happens to the bits of the shell that fall back? Shouldn't they be lifted upward by the supersonic wind that is coming off the star? Is there a Rayleigh-Taylor instability as a result?

Kahn: None of the material in the shell ought to fall back, because at the high density prevalent in the shell the pressure in the Lyman continuum is too strong to allow it to do so.

Kudritzki: I am wondering how your treatment would have to be modified in order to include the effects of radiation pressure in metal lines.

Kahn: In principle the process of shell formation should be the same, provided that there are no important effects dependent on the density of the gas or on the velocity gradient. If there are such effects, then a drastic re-think is needed.

De Groot: The light curve of P Cygni shows maxima and minima on a timescale of 40 days, the same as a value you mentioned in your talk. Maxima and minima occur at various brightness levels, indicative of the clumpiness or anisotropy of your shell formation process.

Kahn: It seems to me that random fluctuations are superposed on a steady wind in the case of P Cygni. The dynamics of the wind rising from the star acts like a filter that picks out the effects of fluctuations in the right frequency range.

Lamers: The calculations described by Kahn may be too simple for the instabilities in the winds of O-stars (as modeled, *e.g.* by Owocki and colleagues), but they may be well suited for the LBV winds, in particular for P Cygni. IUE observations (Lamers *et al.* 1986) show that the ionization of the shells is lower than in the quiescent wind. Kudritzki has shown that the effective radiation pressure increases drastically when the ionization drops from Fe III to Fe II, *etc.* This is similar to Kahn's assumptions.

Conti: Why is there only **one** P Cygni? If it is so easy for an LBV to eject multiple shells, why don't more stars show this?

Lamers: P Cygni is not the only LBV showing these features. All LBV's have multiple absorption components in their line profiles (visual and UV wavelengths) if you look carefully, but they have been followed in time only for P Cygni. So I suspect that most LBV's have variable shells like P Cygni. I have the impression that these variable components occur mainly during minimum or quiescent states.

Shore, Cassinelli, Hillier

CLOSE BINARY MODELS FOR LUMINOUS BLUE VARIABLE STARS

J. S. GALLAGHER
Lowell Observatory
1400 West Mars Hill Road
Flagstaff, Arizona 86001
U.S.A.

ABSTRACT. The evolution of massive close binary stars inevitably involves mass exchange between the two stellar components as well as mass loss from the system. A combination of these two processes could produce the stellar wind-modulated behavior seen in LBVs. The possibility that LBVs are powered by accretion is examined, and does not appear to be a satisfactory general model. Instead, identification of LBVs with close binaries in high mass-loss rate or common envelope evolutionary phases shows promise.

1. Introduction

This conference has raised a variety of unanswered questions concerning luminous blue variable stars (hereafter LBVs). What physical class of star are they? Why does an unstable phase occur for only a fraction of evolved very luminous stars? How does the instability process operate and for how long? In this review I briefly consider one set of answers; i.e., that some LBVs are produced by the evolution of massive binary stars.

Invocation of binary star evolution is justified by prior experience when dealing with unusual stellar characteristics that apply only to a subset of stars in a given physical class. Examples of successful applications of binary models to peculiar classes of stars include classical, recurrent, and dwarf novae; symbiotic stars: barium stars; and luminous x-ray stars. But there is a price to be paid when including binaries. The evolutionary function E of a single massive star can be represented to zero order as $E(M_0, t, z, \dot{M}_w(t), \vec{L}(t), \bar{B}(t))$, where M_0 is the initial mass, z the mean metallicity, $\dot{M}_w(t)$ the wind mass-loss rate, $\vec{L}$ the angular momentum, and $\bar{B}$ the average magnetic field strength.

This is bad enough! However, the evolutionary function for binaries is even more complex as all of the above terms now apply to both members of the binary, and additional terms relating to orbital period, mass ratio time, and accretion rates (all functions of time) must be added (e.g., Eggleton 1985). The daunting complexity of binary evolution has caused some people to decide that this problem is simply a cleverly disguised opportunity for theorists (and enterprising observers) to ascend into free-parameter heaven. This view, whatever its merits, ignores the fact that many massive stars *are* members of binaries where interactions *must* occur during the normal course of evolution.

Several arguments suggest that binary evolution is relevant to the existence of LBVs: (1) At least 40% of massive stars are born in binary star systems with some potential for

K. Davidson et al. (eds.), Physics of Luminous Blue Variables, 185–194.

binary interaction during their lifetimes (cf. Garmany *et al.* 1980). (2) There is a proven tendency for interacting binary stars to experience episodic outbursts with mass ejection from the system. (3) The lowered gravitational potentials at the surfaces of near Roche–lobe filling members of binary systems should lead to enhanced mass–loss rates even during quiescent phases. (4) The evolution of binary stars takes place over longer time scales than those of single stars of similar mass. These features imply that processes due to the presence of massive binaries will affect only part of the massive star stellar population, are likely to be associated with episodic mass ejections and high mass–loss rates, and may occur in older environments than those where massive single stars are found. Binary stars must be included in possible models for LBVs, as LBVs share all of the above symptoms of ongoing binary star evolution.

2. Accretion–Powered Binary Model

Theoretical models and observations both support the view that mass ejection events and bolometric luminosity variations can be produced by accretion within binary star systems. This aspect of binary systems is perhaps most clearly illustrated by outbursts in symbiotic stars, where disk accretion is often the physical cause of eruptive behavior (cf. Kenyon 1986).

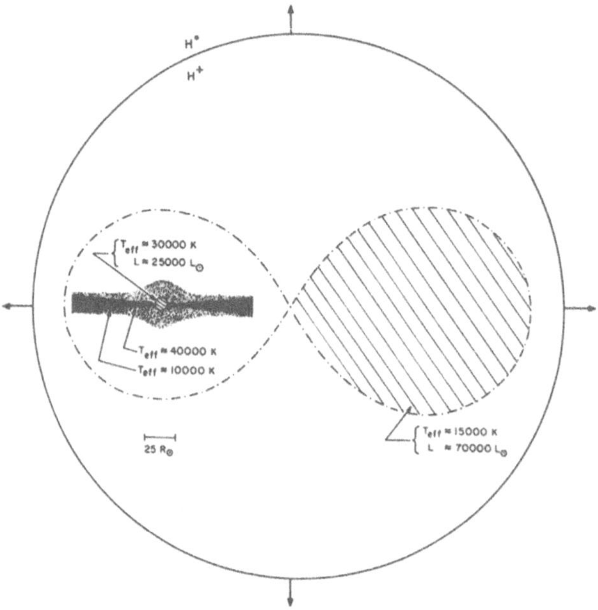

Figure 1. Schematic model of a massive binary system with disk accretion (Kenyon and Gallagher 1985).

In a key paper, Bath (1979; see also Webbink 1979) suggested that the accretion model be extended to the LBVs. The disk accretion model is schematically illustrated in Figure 1, which is reproduced from Kenyon and Gallagher (1985). Bath assumed an α

disk to estimate temperatures of the accreting system and used simple potential energy arguments to derive an accretion luminosity L_a,

$$L_a \approx \frac{GM\dot{M}_a}{R} \approx 10^6\left(\frac{v_{esc}}{1000}\right)^2 \frac{\dot{M}_a}{(6\times10^{-3})}\ L_\odot.$$

Here v_{esc} is in km s^{-1}, and $\dot{M}_a$ is in units of $M_\odot$ yr^{-1}. The last term in this expression is cause for concern. Unless a compact object is allowed such that $v_{esc} \gg 600$ km s^{-1}, then very high accretion rates are required to achieve luminosites in the LBV range.

A compact accretor, however, is unlikely to power most in LBVs. The minimum mass of the accreting object can be estimated from the Eddington luminosity $L_E = 3 \times 10^4(M/M_\odot)\ L_\odot$, where M is the mass of the accreting star. For an LBV luminosity of $L_{\rm bol} \geq 3\times10^5\ L_\odot$, the accreting star should have a mass of $M \geq 10M_\odot$, which exceeds the masses of all but the most extreme stellar black holes.

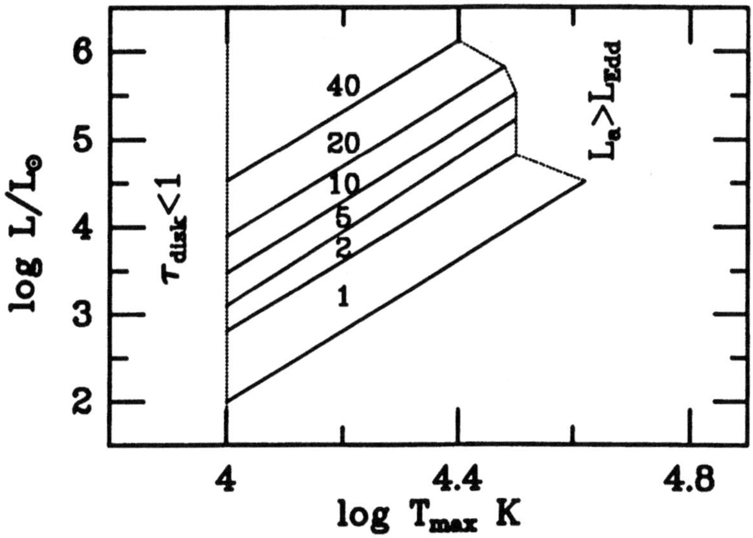

Figure 2. Luminosity–temperature curves for Bath (1979) accretion models. The tracks are labeled by the mass of the accreting star. A lower bound is set when the disk becomes optically thin and the upper bound by $L_a > L_E$.

A second difficulty is illustrated in Figure 2, which is based on Bath (1979). For a system in which the accretion *disk* is a major luminosity source, temperature will *increase* with $\dot{M}_a$; e.g., in Bath's model $T_{\rm max} \propto \dot{M}_a^\gamma$ with $\gamma \approx 0.25$. The behavior of this model is inconsistent with the observations showing that LBVs reach maximum visual luminosity at minimum temperature and the evidence for roughly constant $L_{\rm bol}$ in LBVs. These properties are most readily understood in terms of variations in $\dot{M}_w$ which effect the location of the pseudophotosphere and thus $L_{\rm vis}$, behavior which is well–documented in the early post–maximum development of classical novae (Gallagher and Code 1974; Ruggles and Bath 1979).

Disk accretion models do not seem to fully explain LBVs. Thus Figure 1 might be representative of a quiescent phase LBV where stellar photospheres and the disk all contribute to the visual luminosity, while the disk is also responsible for extra line emission. Detailed comparisons with observations are necessary to see if even this scaled–down disk accretion model is acceptable.

An alternate version of this model is possible if LBVs are experiencing saturated accretion; i.e., if $L_a \geq L_E$. We then can ask whether the wind would carry the excess L_a, as suggested by Żytkow's (1973) and Kato's (1985) models. The Eddington limit provides a natural reason for near constant L_{bol} from the system, and fluctuations in $\dot{M}_w$ are likley to occur and produce the desired visual light variations. We could consider these types of objects as "stealth" binaries, since the binary nature of the underlying system is largely but not completely optically obscured by a pseudophotosphere associated with the accreting star.

We require very high accretion rates of $\dot{M}_a \geq 6 \times 10^{-3} M_\odot$ yr^{-1} to produce a hidden binary from a system where the accretor is a main–sequence star. It is fair to ask if such high rates can be expected in nature. Webbink (1985) has emphasized that there are three natural time scales associated with the mass–losing star which can drive accretion rates in binaries. As a rough estimate of allowable rates for each time scale τ_i from a star of mass M_0 with available mass fraction f for mass loss, we have $\dot{M}_a(i) \approx f M_0 \tau_i^{-1}$. Even for a massive star, nuclear time scales are $\geq 10^6$ yr, so we expect $\dot{M}_a(nuc) \leq 10^{-4}$ $M_\odot$ yr^{-1}. Thermal time scales are similar, so we must rely on the dynamical time scale to produce Eddington–critical accretion rates. Accretion on dynamical time scales occurs when a star cannot readjust to the changes in Roche lobe size in an interacting binary, and in principle can produce the required high values of $\dot{M}_a$.

A detailed study of this problem is in progress as M. S. Hjellming's Ph.D. thesis at the University of Illinois (see Hjellming and Webbink 1987). However, stars with convective envelopes are known to be unstable against dynamical mass exchange, and thus a cool supergiant donor is a good candidate for the model. Possibly stars near the Eddington limit are also sufficiently unstable to make this mechanism work (cf. McClusky and Kondo 1976). A problem that remains to be resolved is the response of the accreting star to rapid accretion. If the accreting star responds by expanding into contact or near contact, then accretion will presumably cease and the model fails (see Kippenhahn and Meyer–Hoffmeister 1977; Webbink 1985).

A best model for an accretion–powered, binary LBV has the following properties: (1) The total initial mass of the system must be $\geq 20 M_\odot$. (2) The orbital period must be long enough to allow an evolved primary with a convective envelope to be present ($P \geq 100$ days). (3) An accreting main–sequence B star is normally optically obscured by a massive wind powered by the accretion process. This picture is consistent with an R81 type of object (Stahl *et al.* 1987), and possibly the related B[e] stars (cf. Shore *et al.* 1987; Zickgraf *et al.* 1988).

However, the accretion–powered binary model has major flaws and is unlikely to provide a general explanation of LBVs: (1) At super–Eddington $\dot{M}_a$ levels, the estimated lifetimes of $\approx 10^3$ yr are too short. (2) Fine tuning is needed to achieve near constant L_{bol}. (3) The response of the accreting star may further limit the duration of this evolutionary phase.

3. Binary Stars and Enhanced Mass–Loss Rates

Binary stars may produce LBVs due to their ability to support high mass–loss rates rather than as a result of high accretion luminosities. This type of model is particularly attractive when one is considering whether a close binary might mimic a very massive single star. For high–mass stars, the bolometric luminosity scales almost linearly with initial mass, $L_{nuc}(M_1 + M_2) \approx L_{nuc}(M_1) + L_{nuc}(M_2)$. A binary star with initial component masses M_1 and M_2 therefore will be nearly indistinguishable from a single star with initial mass $M_1 + M_2$ in terms of nuclear luminosity, and can be even more luminous than the single star if accretion and related processes contribute radiative luminosity. Binaries with initial total masses exceeding about 60 $M_\odot$ will have the nuclear luminosities of LBVs for mass ratios near unity, and such systems are known to exist (e.g., Garmany *et al.* 1980; Hilditch and Bell 1987). Indeed, as A. V. Tutukov (private communication and see Tutukov and Yungel'son 1980) has emphasized, binary systems with initial *total* masses of $\geq 60 M_\odot$ are likely to occur at least as frequently as single stars of similar mass in the Galactic OB stellar population.

3.1 Mass Loss from Main–Sequence Close Binaries

The primary star in a close binary will inevitably expand in radius as a result of nuclear evolution and may fill its Roche lobe during (Case A) or soon after (Case B) the main–sequence core hydrogen–burning evolutionary phase (cf. Vanbeveren *et al.* 1979; Vanbeveren 1982; Nakamura and Nakamura 1984). The models suggest that the ensuing mass transfer to the secondary is likely to lead to a contact binary and extensive mass loss from the system before mass exchange is complete (see also Ziółkowski 1979). This process may be further enhanced in high–mass binaries by the effects of radiation pressure (McCluskey and Kondo 1976). The precise period ranges and other details of the process are not fully established (e.g., the lower period bound for Case A in a given system depends on stellar mass–loss rates and degree of convective core overshooting; Doom 1984; Sybesma 1985, 1986a,b), and quantitative predictions about frequencies of occurrence are thus not available.

The qualitative aspects of this process, however, are interesting in the present context. As discussed above, the necessary luminosity is supplied by nuclear burning in the two binary members for $M(1+2) \geq 60 M_\odot$. Mass exchange rates during Roche lobe overflow can be very high, $> 10^{-4} M_\odot$ yr^{-1} in massive binaries. Thus if the mass lost from the system is even a modest fraction of the mass exchange rate, $\dot{M}_w \approx 10^{-4} M_\odot$ yr^{-1} can be achieved. This level of $\dot{M}_w$ will produce a pseudophotosphere that is larger than the dimensions of a short–period massive binary for LBVs near maximum light (see Vanbeveren *et al.* 1979; Gallagher *et al.* 1981). Mass loss from the primary should also lead to the evidence for CNO–cycle processing that is found in some LBVs (Tutukov and Yungel'son 1983). A further bonus has been noted by Tutukov and Yungel'son (1980) in that this process will yield objects which appear as the most massive "single" stars due to the mass gained by the original secondary component.

The existence of semidetached and contact O star binaries with both components on or near the main–sequence (e.g., Hilditch and Bell 1987; Leung 1988) shows that this sequence of events has the potential to occur in nature. Additional work is needed to see if the specifics of the mass–loss process and other aspects of this model quantitatively fit observed properties of the LBVs.

3.2 Common Envelope Binaries

In massive binary systems with compact components, the less evolved member can readily supply sufficient $\dot{M}_a$ such that $L_a > L_E$. Under these circumstances a common envelope will surround both components of the binary. The envelope is driven by dynamical effects and can carry off large amounts of matter and angular momentum from the system (e.g., Ostriker 1975; Paczynski 1976). The existence of short-period binaries containing compact objects is direct evidence of the importance of the common envelope phase in the evolution of massive binary star systems (e.g., Tutukov and Yungel'son 1979).

The resulting buried binary will appear as an object with a pseudophotosphere whose radius is controlled by $\dot{M}_w$ and is therefore a potential LBV candidate. The luminosities of common envelope binaries can lie in the observed range for LBVs during the initial phase of the interaction between components, which may last for a period of $\approx 10^3 - 10^4$ yr (e.g., Taam *et al.* 1978). An additional signature of this and other binary ejection processes is expected to be an asymmetrical envelope which lies preferentially in the equatorial plane of the binary system (Bodenheimer and Taam 1984; Livio and Soker 1988). The ejection velocities for this process may also be in the appropriate range for LBVs.

3.3 Musings on η Carinae

Since η Carinae is in an area rich in young, massive stars, it stands out as an excellent single star LBV candidate resulting from an instability in an extremely massive star. However, the possible presence of bipolar structure in the ejecta reviewed by K. Davidson (this volume) suggests that an effort should be made to consider a binary model for this most extraordinary LBV.

In this spirit, we can ask what type of event in a binary could lead to an outburst of the magnitude seen in the 19th century in η Carinae? One interesting model that might merit further exploration is a merger between components in a young binary (or multiple star?) system. The thermal energy generated in this type of event should be sufficient to account for both the radiated power and ejection of several $M_\odot$ of material with bipolar symmetry. Is it then possible that η Carinae is an extreme example of the Tutukov mass exchange mechanism operating in a high-mass binary system yielding a single superstar?

4. Conclusions

1. Accretion-powered massive binary models for LBVs as a class require tricky fine-tuning and are unlikely to provide a general explanation for LBVs.

2. Enhanced mass loss can occur in high-mass binary systems during mass exchange and common envelope evolutionary phases. This will produce a superwind which is the key physical characteristic of LBVs. Radiative luminosities of $10^5 - 10^6$ $L_\odot$ can be provided by a combination of nuclear and dynamical sources within known examples of such classes of binary stars. In addition to naturally producing most of the observable characteristics of LBVs, this class of model also provides a natural distinction between LBVs and hypergiants of similar luminosity which do not show the LBV phenomenon.

3. The atmospheric characteristics of LBVs resemble those seen in superwinds from known binary systems (e.g., symbiotic stars). Thus the qualitative argument can be made that binary LBV model will have the desirable propertites of high $\dot{M}_w$, low v_∞, bipolar

ejection, complex spectral line profiles, interesting emission lines (e.g., lots of [Fe II]), and optical variability.

4. We should consider the possibility that the η Carinae outburst was associated with an unusual event in a young binary or multiple star system.

Acknowledgments

I would like to thank my colleagues Otto Franz, Icke Iben, Scott Kenyon, Sasha Tutukov, and Ron Webbink for providing me with an appreciation for the richness of binary evolution. This paper has been supported in part by the Lowell Observatory Research Fund. Travel support from the IAU is also gratefully acknowledged.

References

Bath, G. T. (1979). *Nature* **282**, 274.
Bodenheimer, P. and Taam, R. E. (1984). *Ap. J.* **280**, 771.
Doom, C. (1984). *Astron. Ap.* **138**, 101.
Eggleton, P. H. (1985). In *Interacting Binary Stars*, J. E. Pringle and R. A. Wade (eds.), Cambridge University, Cambridge, p. 21.
Gallagher, J. S., and Code, A. D. (1974). *Ap. J.* **189**, 303.
Gallagher, J. S., Kenyon, S. J., and Hege, E. K. (1981). *Ap. J.* **249**, 83.
Garmany, C. D., Conti, P. S., and Massey, P. (1980). *Ap. J.* **242**, 1063.
Hilditch, R. W., and Bell, S. A. (1987). *Astron. Ap.* **176**, 59.
Hjellming, M. S., and Webbink, R. F. (1987). *Ap. J.* **318**, 794.
Kato, M. (1985). *Pub. Astr. Soc. Japan* **37**, 19.
Kenyon, S. J. (1986). *The Symbiotic Stars*, Cambridge University, Cambridge.
Kenyon, S. J., and Gallagher, J. S. (1985). *Ap. J.* **290**, 542.
Kippenhahn, R., and Meyer-Hoffmeister, E. (1977). *Astron. Ap.* **54**, 539.
Leung, K.-C. (1988). In *Critical Observations vs. Physical Models for Close Binary Systems*, in press.
Livio, M., and Soker, N. (1988). *Ap. J.* **329**, 764.
McClusky, G. E., and Kondo, Y. (1976). *Ap. J.* **208**, 760.
Nakamura, M., and Nakamura, Y. (1984). *Ap. Sp. Sci.* **104**, 367.
Ostriker, J. P. (1975). Paper presented at "The Structure and Evolution of Close Binary Systems," IAU Symposium No. 73.
Paczynski, B. (1976). In *The Structure and Evolution of Close Binary Systems*, IAU Symposium No. 73, P. Eggleton, S. Mitton, and J. Whelan (eds.), Reidel Publishing Co., Dordrecht, p. 75.
Ruggles, C. N., and Bath, G. T. (1979). *Astron. Ap.* **80**, 97.
Shore, S. M., Sanduleak, M., and Allen, D. A. (1987). *Astron. Ap.* **176**, 59.
Stahl, O., Wolf, B., and Zickgraf, F.-J. (1987). *Astron. Ap.* **184**, 193.
Sybesma, C. H. B. (1985). *Astron. Ap.* **142**, 171.
__________. (1986a). *Astron. Ap.* **159**, 108.
__________. (1986b). *Astron. Ap.* **168**, 147.
Taam, R. E., Bodenheimer, P., and Ostriker, J. P. (1978). *Ap. J.* **222**, 269.
Tutukov, A., and Yungel'son, L. (1979). In *Mass Loss and the Evolution of O-Type Stars*, IAU Symposium No. 83, P. S. Conti and C. W. H. de Loore (eds.), Reidel Publishing Co., Dordrecht, p. 401.

_________. (1980). *Pis'ma Astron. Zh.* **6**, 491 (*Sov. Astron. Lett.* **6**, 271, 1981).

_________. (1983). *Pis'ma Astron. Zh.* **9**, 230 (*Sov. Astron. Lett.* **9**, 124).

Vanbeveren, D. (1982). *Astron. Ap.* **105**, 260.

Vanbeveren, D., Grève, J. P., van Dessel, E. L., and de Loore, C. (1979). *Astron. Ap.* **73**, 19.

Webbink, R. F. (1979). In *Changing Trends in Variable Star Research*, IAU Colloquium No. 46, F. M. Bateson, J. Smak, and I. H. Ulrich (eds.), University Waikato, Hamilton, New Zealand, p. 102.

_________. (1985). In *Interacting Binary Stars*, J. E. Pringle and R. A. Wade (eds.), Cambridge University, Cambridge, p. 39.

Zickgraf, F.-J., Wolf, B., Stahl, O., and Humphreys, R. M. (1988). Preprint.

Ziółkowski, J. (1979). In *Mass Loss and Evolution of O-Type Stars*, IAU Symposium No. 83, P. S. Conti and C. W. H. De Loore (eds.), Reidel Publishing Co., Dordrecht, p. 385.

Żytkow, A. (1973). *Acta Astr.* **23**, 121.

Jay Gallagher

DISCUSSION

Wolf: The best way to study the physical nature of LBV's is perhaps to look at them at minimum. R 71 at quiescence was modeled by the Munich group; they could fit the observed continuum fairly well with the continuum of an early B-supergiant model. Is this finding consistent with any of your binary scenarios?

Gallagher: In a qualitative sense, yes. Since a binary can be embedded in a common envelope and/or in an optically thick wind, it is hard to distinguish the nature of the underlying energy source. Perhaps the binary model would be characterized by a non-spherical atmosphere or by effects associated with lower gravity? Maybe Dr. Kudritzki could comment on this issue.

Kudritzki: I don't know. One problem I encountered with my hydrostatic NLTE energy-distribution-fitting of R 71 was that the NLTE model with the correct gravity ($log\ g \approx 1.6$) shows a Balmer jump that is not observed. An extended pseudo-photosphere might be better. But without a calculation it is hard to predict.

De Groot: Since Vanbeveren's remark about a spiralling-in companion star yesterday, I have been wondering about binaries and LBV's. On each day of this meeting we have heard about difficulties with ejected shells and their expansion velocities, [N II] profiles that do not look like outburst shells or red supergiant ejecta ... Binaries have the potential of solving these problems. The cartoon on your last viewgraph resembles Paresce's picture of AG Car.

Friedjung: One must be careful not to mix up classical novae, symbiotic stars, and LBV's. There are good reasons for believing the existence of optically thick winds for classical novae. For symbiotic stars, however, there is not so much evidence for this. Some are thought to contain a red giant and an accreting main sequence star. According to the present orthodoxy (which might change), sometimes there is an accreting white dwarf with surface nuclear reactions. Such a star would expand and move in the H-R diagram along a track like that of a planetary-nebula-nucleus but in the opposite direction. A wind from such an expanded white dwarf would resemble that from a normal hot star.

Gallagher: Certainly one has to be careful about mixed metaphors when comparing LBV's with novae and symbiotic stars. Still, in most of these eruptive variables the optical properties are largely controlled by stellar winds, *i.e.*, supersonic mass flows through a pseudo-photosphere, and in this respect they are similar to LBV's.

Zickgraf: The merger model for B[e] stars is very interesting. But isn't there a problem with the statistics? Presently we know 7 B[e] supergiants in the LMC. Would you expect to have so many mergers?

Gallagher: A merger would be a short-lived phase that is accessible to only a fraction of massive binaries. To make an estimate, we need better statistics on massive binaries in the LMC.

Conti: Almost the only person doing such work on the LMC is Virpi Niemela. She has found a handful of short-period O-type binaries. This work requires considerable observing time and much encouragement.

Appenzeller: Concerning the suggestion that an LBV is a mass-exchange binary containing (as a donor star) a red supergiant with an extended outer convection zone: According to the classical investigations by Paczynski and his group and by Plavec, even with hydrodynamic effects taken into account, in such systems the mass exchange takes place on a time-scale not much larger than the dynamical time-scale. Even with optimistic assumptions, the lifetime of such a state should not exceed a few years. This is much shorter than the empirical lifetime of an LBV as derived, *e.g.*, from kinematic ages of circumstellar matter.

Gallagher: I agree with your comment. The estimated lifetime of 1000 years that I mentioned may too optimistic, and even that is too short to completely explain LBV's.

Davidson: At least some LBV's appear to have bipolar morphologies; the easiest ways to get suitable axes are either by rotation or by orbiting; and yet many of us at this meeting seem uncomfortable with either rotation or binaries (because we'd like to minimize the number of parameters, I suppose). However, if we choose to do so, a fairly close binary companion can be employed in a way that is less ambitious than Gallagher's and Vanbeveren's scenarios: we could use the companion star merely as a device to enforce co-rotation. Its orbital speed, after all, is unaffected by the radiation pressure that reduces effective gravity in the LBV star's atmosphere. Therefore the effective pseudo-Roche limiting surface is modified and a crisis may occur for a smaller (*stellar radius*)/(*companion distance*) ratio. Perhaps the result would mimic rapid rotation by a single isolated star.

Maeder: I see a number of difficulties in binary models for explaining the LBV phenomenon: (1) It would seem difficult to keep L = constant during the accretion process, among binaries with different orbital parameters and mass ratios.
(2) If they are all binaries, why don't we see a clock in any of them?
(3) In view of the diverse binary parameters, why would binary models be confined to the left of the H-D limit? Binaries of all masses exist but the LBV phenomenon is restricted to a limited part of the H-R diagram.
(4) How can the well-known short-period pulsations survive in such distorted stars?
(5) I do not see tidal mixing, which could seriously change the overall picture, included in your models.
(6) Finally, even if some LBV's are binaries, we do not **need** the binary hypothesis; the upper-right part of the H-R diagram is forbidden to static models anyway.

Gallagher: These are all interesting points. Let me respond briefly to each point in order. (1) The accretion luminosity is minor in the preferred models where binary processes supply main the enhanced mass loss. Thus, while variations in accretion rates will be needed to modulate the mass-loss rate, these should not lead to perceptible luminosity variations. (2) When strong interactions occur, it is often hard to see the clocks even in known binaries (*e.g.*, dwarf novae and symbiotic stars). (3) In this conceptual model, the combination of near-Eddington luminosities and destabilization by binary interactions produces an LBV. Quantitative boundaries need to be defined for these processes. (4) Perhaps the pulsations take place in the common envelope, as seems to occur in post-maximum classical novae. (5) I agree that the con-conceptual picture presented here should be confirmed *via* quantitative models. (6) The existence of short-period, high-mass binaries containing compact stars could be viewed as evidence that binary interactions do play a role in determining the properties of massive stellar populations. The question in my mind is to find the relative importance of binary versus single-star evolutionary paths to the LBV's.

THE ROLE OF RADIATION PRESSURE IN LBV ATMOSPHERES

I. APPENZELLER
Landessternwarte Königstuhl
D-6900 Heidelberg 1
Federal Republic of Germany

ABSTRACT. As LBVs have luminosities close to their Eddington limits, their structure is profoundly influenced by radiation pressure. Radiation pressure effects probably cause the highly extended atmospheres and the extreme mass loss observed during the maximum states of the S Dor variables. An opacity-related instability of the radiative acceleration combined with a delayed thermal readjustement of the subatmospheric layers possibly explains the large-amplitude radius variations of these objects.

1 Introduction

The number of mechanisms which have been suggested in order to explain the variability of LBVs probably exceeds the number of reliably identified members of this exciting class of stars. Nevertheless, in view of the observed very high luminosities, everybody seems to agree that radiation pressure effects have to be taken into account and play a major role in realistic models of these stars. As noted by Humphreys and Davidson (1979), Lamers (1986a), Davidson (1987), Lamers and Fitzpatrick (1988), and others, radiation pressure certainly contributes to the exceptionally strong mass loss of the LBVs. Furthermore, instabilities of the radiative acceleration in the expanded, nonstatic atmospheres of the LBVs have been suggested to cause the observed strong variability (Appenzeller 1986, Lamers 1986b). However, in the absence of detailed time dependent model atmosphere computations, other variability mechanisms (see e.g. Stothers and Chin 1983, de Jager 1980, 1984, Maeder 1983, and various contributions to this conference) cannot be ruled out. Hence, this review will be separated into two parts. I shall start with a brief summary of the relatively well established radiation pressure effects in equilibrium atmospheres. This will be followed by a more speculative discussion of the opacity-related radiation pressure instabilities and of a relaxation oscillation scenario of the the observed variability.

K. Davidson et al. (eds.), Physics of Luminous Blue Variables, 195–204.

2 Equilibrium States and Eddington Limit

The atmospheric equilibrium of a nonrotating, nonmagnetic, nonturbulent luminous star can be described by the equation of motion

$$\frac{dv}{dt} + v\frac{dv}{dr} + \frac{1}{\rho}\frac{dP_G}{dr} = -g_{eff} = -g(1 - \frac{g_R}{g}) \tag{1}$$

where v $= dr/dt$ is the flow velocity, t the time , ρ the density, P_G the gas pressure , and g and g_R denote the gravitational and radiative accelerations, given by

$$g = \frac{GM}{R^2} \tag{2}$$

$$g_R = \frac{4\pi}{c}\int \kappa_\nu H_\nu d\nu \tag{3}$$

(M = stellar mass, R = stellar radius, κ_ν = mass absorption coefficient, H_ν = Eddington flux).

Among the various types of solutions of Equ. (1) are the following well known special cases: For $v \equiv 0$ Equ. (1) degenerates to the hydrostatic equation which describes "normal" static stellar atmospheres. As shown by Equ. (1), for $g_R \geq g$ no such static solutions are possible. $dv/dt \equiv 0$ and $vdv/dr > 0$ corresponds to the case of stationary accelerated stellar winds. Finally, we note that for $g_{eff} = g(1 - g_R/g) \rightarrow 0$ the pressure (and density) scale height of hydrostatic models

$$h = \frac{\Re T}{\mu g_{eff}} \tag{4}$$

($\Re$ = gas constant, μ = mean molecular weight) approaches infinity. Hence, with increasing radiation presure effects atmospheres become progressively more extended.

In the idealized case of a gray atmosphere, i.e.

$$\kappa_\nu = const. \tag{5}$$

we have

$$g_R = \frac{4\pi\kappa_\nu}{c}\int H_\nu d\nu = \frac{\kappa_\nu L}{4\pi c R^2} \tag{6}$$

and thus

$$\frac{g_R}{g} = \frac{\kappa_\nu L R^2}{4\pi c R^2 GM} = \frac{L}{L_{EDD}} \tag{7}$$

where L is the stellar luminosity and

$$L_{EDD} = 4\pi c \kappa_\nu^{-1} GM \tag{8}$$

is called the "Eddington luminosity" or "Eddington limit". For very hot stars, where the opacity is dominated by Thomson scattering we, have for a Population I chemical composition

$$L_{EDD} = 3.9\,10^4 \frac{M}{M_\odot} L_\odot \tag{9}$$

Using the luminosity values presented at this conference in the reviews of R. Humphreys and B. Wolf and using stellar mass values derived by matching these luminosities to the evolutionary tracks published by Maeder and Meynet (1987), we obtain for the known S Dor LBVs electron-scattering L/L_{EDD} ratios between 0.5 (R71) and 0.8 (R127). In view of the fact, that pure electron scattering is expected to underestimate the opacities of real LBV atmospheres, these values are surprisingly high and confirm the importance of radiation pressure for these objects.

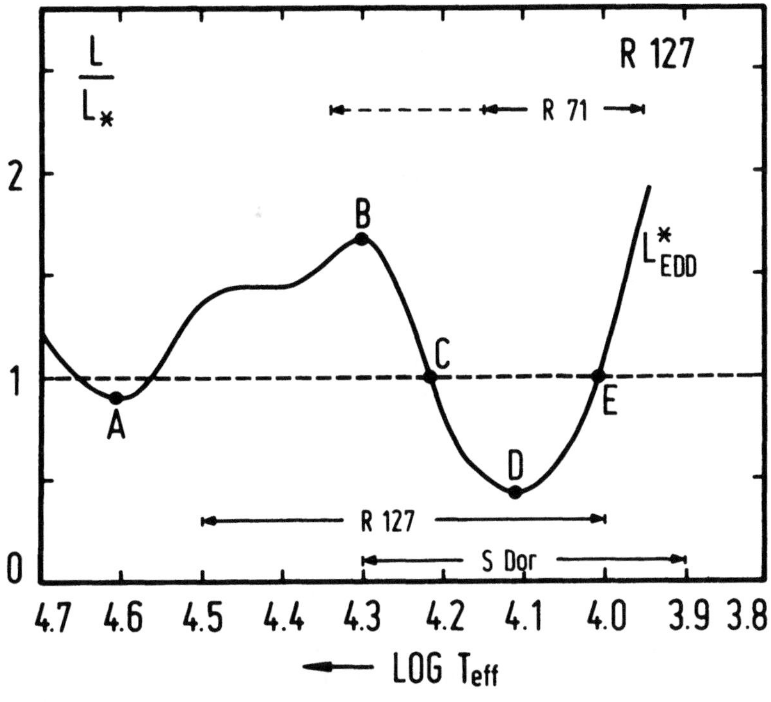

Figure 1: Approximate temperature dependence of the photospheric modified Eddington luminosity of a bright LBV. Also shown are the T_{eff} ranges of three LMC S Dor variables.

In a real LBV atmosphere L_{EDD} has no direct physical meaning as the absorption coefficient depends on the frequency. However, locally the radiation pressure effects can still be characterized by the ratio g_R/g or by the "modified Eddington luminosity" L^*_{EDD}, defined (analogous to Equ.(7)) by

$$L^*_{EDD}/L = g/g_R \tag{10}$$

In contrast to the conventional Eddington luminosity L_{EDD}, which is constant within an atmosphere, L^*_{EDD} normally is a function of the optical and geometrical depths. But, if the atmospheric structure of a star is known, for each depth L^*_{EDD} can be calculated using Equs. (3) and (10).

3 Instabilities

In realistic atmospheres $\kappa_\nu H_\nu$ and thus g_R and L^*_{EDD} depend on the effective temperature. A very approximate plot of this dependence for the photosphere of a bright LBV, such as R127, is given in Figure 1. In the absence of more suitable model grids, the plotted relation $L^*_{EDD}(T_{eff})$ has been based on the functions $g_R(T_{eff})$ computed by Kurucz and Schild (1976) on the basis of hydrostatic LTE model atmospheres. As LTE atmospheres are expected to introduce in the case of LBVs a qualitatively incorrect dependence of the line opacity on the gravitational acceleration, the g-dependence of κ_ν was neglected. While the static Kurucz and Schild models are probably reasonable approximations for the slowly expanding atmospheres of S Dor variables, they are unrealistic for high wind velocity O stars. Hence, at high tempreatures Figure 1 becomes a progressively poorer approximation. Using the more recent line acceleration values (based on non-static atmospheres) tabulated by Abbott (1982) gives qualitatively similar results. But, as many weak lines, which are important at the low temperature of the S Dor variables at minimum, were not included in the nonstatic computations, the Kurucz and Schild models are probably the more adequate approximation for the present purpose.

The most important feature of the $L^*_{EDD}(T_{eff})$ relation is a conspicuous minimum near log $T_{eff} = 4.1$, followed by a steep increase for lower temperatures. This minimum of L^*_{EDD} and the corresponding maximum of the radiative acceleration is mainly due to a large number of merging lines of singly ionized iron group elements (Fe II, Cr II,...) occuring near the maximum of the stellar flux distribution (Kurucz and Schild 1976). The great importance of these lines for LBV atmospheres is evident from Figure 2.

For a radiation pressure dominated stellar atmosphere Figure 1 predicts a dynamical instability for $4.1 \leq logT_{eff} \leq 4.3$ (Points B to D). In this range we have $dL^*_{EDD}/dT_{eff} > 0$. Hence, any increase of the photospheric radius (or mass loss) at constant luminosity and the corresponding decrease of the effective temperature results in a increase of g_R/g, a lowering of g_{eff}, and thus in a further expansion and further increased mass loss. Thus, any radius or T_{eff} perturbation results in a runaway. Outside this instability range (except for the region to the left of A where the static LTE models are not applicable) Figure 1 indicates dynamical stability, since perturbations will result in a return to the initial equilibrium state.

A sufficiently massive post-main sequence star, evolving with approximately constant luminosity to the right in the HR diagram, will sooner or later encounter the instability outlined above. At a critical effective temperature (which, because of the relation between opacity, gravity, and atmospheric chemical enrichment, will depend somewhat on the initial mass) the photosphere will lose hydrostatic equilibrium and the star will rapidly move to lower effective temperatures, until the two conditions $dL^*_{EDD}/dT_{eff} < 0$ and $L^*_{EDD} > L$ are both fulfilled again. According to Figure 1, the region where LBVs are expected to regain equilibrium coincides well with the observed minimum-state temperature of the S Dor variables, while the maximum-state temperatures are (at least for S Dor and R 127) compatible with a location on the

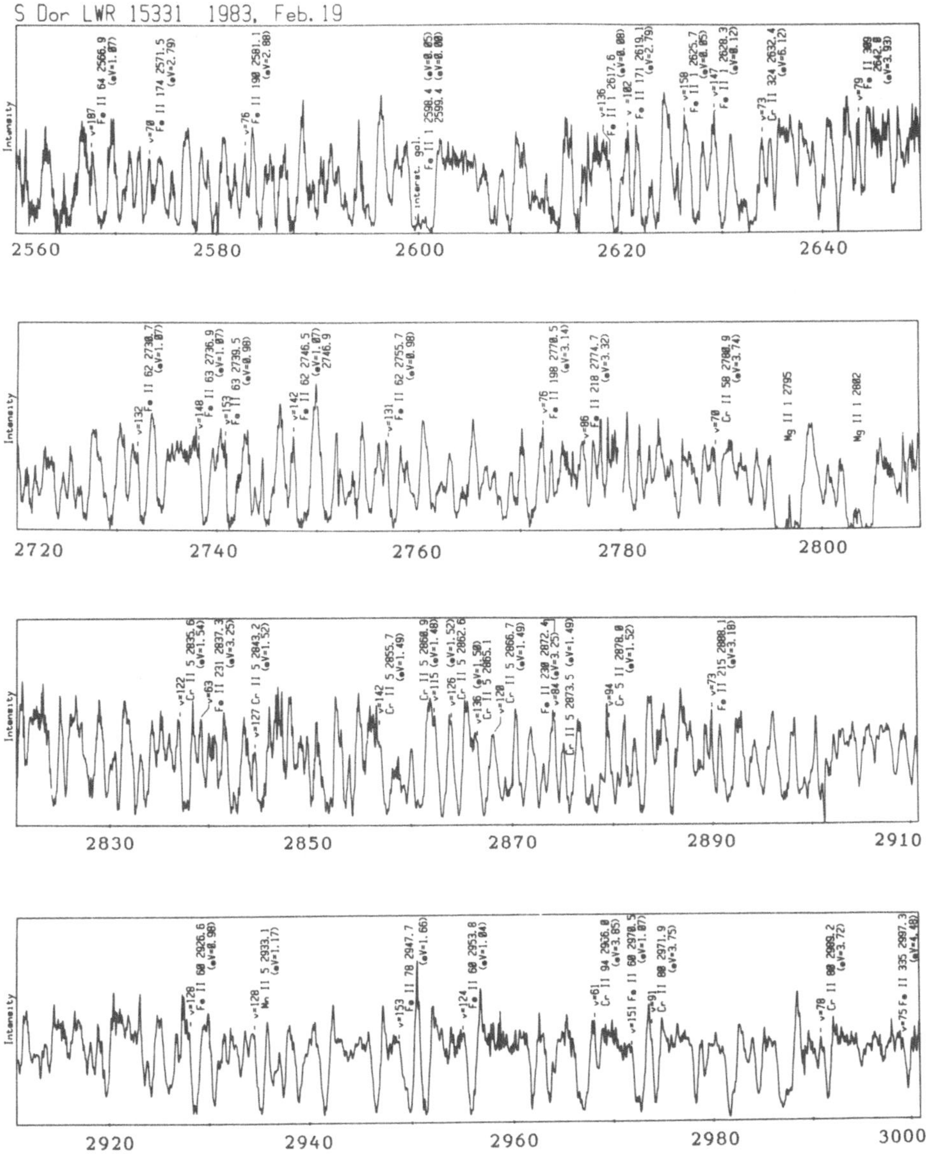

Figure 2: Sections of the 2500 - 3000 Å spectrum of S Dor at maximum. Note the large number of merging absorption features of singly ionised iron group elements. (From Leitherer et al. 1985).

hot stable branch of the $L^*_{EDD}(T_{eff})$ relation. Hence, it seems attractive to identify the extended maximum and minimum states of the S Dor variables with the two different stable configurations predicted (qualitatively) by Figure 1. However, while the above scenario allows to understand the transition from the minimum state to the maximum state and the observed properties of the two states, additional physical processes are needed to cause the observed return (after $\sim$ a few decades) from the stable maximum state to the high T_{eff} minimum state. An example for such a process is discussed in the following section.

4 The Relaxation Oscillation Scenario

The maximum and minimum states of S Dor stars are characterized by highly different effective temperatures and effective ($\tau = 2/3$) radii. On the other hand, the bolometric luminosity remains essentially constant (Appenzeller and Wolf 1981, Stahl et al. 1983, Humphreys et al. 1984, Wolf, 1986, 1987). This indicates that the S Dor variability is mainly an effect of the outer stellar layers. However, as illustrated by Figure 3, for these outer layers the change is dramatic. In the case of S Dor the increase of the effective radius amounts to a factor 7. For R127 the corresponding ratio is about 15.

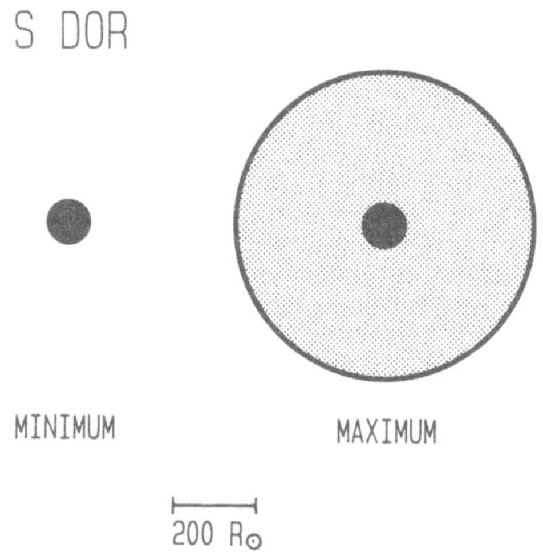

Figure 3: Comparison of the effective ($\tau = 2/3$) radii of S Dor at minimum and maximum.

From the wind models one can estimate that the highly expanded nonstatic layers contain about 10^{-4} of the stellar mass. For the deeper interior the envelope expansion is practically equivalent of (temporarily) removing this mass on a dynamical timescale. Hence the interior will react with a fundamental mode radial expansion.

Because of the central mass concentration of supergiants, the relative amplitude will be small in the deep interior. $\Delta r/r$ will be above 0.1 only for about the outermost ~ 3 percent of the mass.

The question of how a star reacts to the removal of mass from its surface has been investigated thoroughly in the context of mass-exchange binary evolution (Morton 1960, Kippenhahn and Weigert 1967, Plavec 1970, Webbink 1977, Heisler and Alcock 1986). As noted by these authors, a star's reaction depends on the ratio between the mass-loss time scale and the thermal adjustment time

$$\tau_{KH} = \frac{GM^2}{RL} \approx 3.1\,10^7 \frac{(M/M_\odot)^2}{(R/R_\odot)(L/L_\odot)} years \tag{11}$$

which for typical LBVs is of the order $\tau_{KH} \approx 10^3$ years. But, as pointed out by Kippenhahn and Weigert (1967) for stellar surface layers we may have much shorter local thermal adjustment times. A surface layer of mass Δm is expected to regain thermal equilibrium after about

$$\tau_{\Delta m} = \frac{GM\Delta m}{RL} \tag{12}$$

For the outermost 10^{-4} of the stellar mass $\tau_{\Delta m}$ corresponds roughly to its dynamical time scale ($\approx$ a few weeks). Hence, these layers are expected to remain in thermal equilibrium at all times. However, as noted above, the atmospheric expansion will have a significant effect for the subatmospheric layers down to about $M_r/M \approx 0.97$. For this whole volume Equ. (12) indicates a thermal adjustment time of several decades (i.e. of the order of the duration of the minimum and maximum states of S Dor variables). Consequently, these subatmospheric layers expand adiabatically and lose thermal equilibrium. The resulting thermal adjustment requires additional heating of the expanded gas. Hence, a local minimum of the heat flow is expected to occur somewhere in these outer layers. Below the minimum the gas will slowly expand, above the minimum initially a slow contraction will occur. Because of the larger volume of the outer layers, the net effect will probably be an initial net contraction and heating of the subatmospheric layers. Obviously, such a contraction and heating on a thermal time scale could cause the eventual collapse of the extended atmosphere, resulting in a return to the hot minimum state. The collapse of the atmosphere is again expected to occur on the shorter dynamic timescale. Hence, the new minimum state will also be out of thermal equilibrium, resulting in a new thermal adjustment, now probably accompanied by a slow expansion, until the atmosphere again encounters the radiation pressure instability and a new cycle of the relaxation oscillations is started.

In principle, the scenario described above can explain all basic properties of the observed variability of the S Dor stars. However, so far the above suggestions are based on qualitative considerations, very rough numerical estimates, and on the analogy to computed models of mass loosing binary components. To confirm or disprove the above scenario, selfconsistent time-dependent non-static models of LBV outer layers (including the winds, atmospheres and the optically thick subphotospheric layers) are needed. Impressive progress in computing stationary nonstatic LBV models has been reported at this meeting by Kudritzki (1988) and by Leitherer and Abbott (1988). An extension of such computations to time dependent cases will allow a quantitative study of the radiation pressure related instabilities. But, already from the qualitative discussion outlined above it is clear that radiation pressure effects play an important but complex role in the LBV atmospheres.

References

Abbott, D.C.: 1982, *Astrophys.J* **259**, 282

Appenzeller, I.: 1986, *IAU Symp.* **116**, *Luminous Stars and Associations in Galaxies* C. W.H de Loore et al. eds.,p. 139

Appenzeller, I., Wolf B., 1981, in *"The Most Massive Stars"*, S. D'Odorico et al. eds., *ESO Proceedings, Garching 1986*, p. 131

Davidson, K., 1987, *Astrophys. J.* **317**, 760

de Jager, C., 1980, *The Brightest Stars. D. Reidel, Dordrecht* 1980

de Jager, C., 1984, *Astron. Astrophys.* **138**, 246

Heisler, J., Alcock, C., 1986, *Astrophys. J.* **306**, 166

Humphreys, R. M., Davidson, K., 1979, *Astrophys. J.* **232**, 409

Humphreys, R. M., Blaha, C., D'Odorico, S., Gull, T.R., Benvenuti, P., 1984, *Astrophys. J.* **278**, 124

Kippenhahn, R., Weigert, A., 1967, *Zeitschr. Astrophysik* **65**, 251

Kudritzki, R. P., 1988, *elsewhere in this volume*

Kurucz, R. L., Schild, R. E., 1976, *IAU Symp.* **70**, *Be and Shell Stars*, A. Slettebak ed., p. 377

Lamers, H. J. G. L. M., 1986a, *Astron. Astrophys.* **159**, 90

Lamers, H. J. G. L. M., 1986b, *IAU Symp.* **116**, *Luminous Stars and Associations in Galaxies*, C.W.H. de Loore et al. eds., p. 157

Lamers, H. J. G. L. M., Fitzpatrick, E.L., 1988, *Astrophys. J.***324**, 279

Leitherer, C., Appenzeller, I., Klare, G., Lamers, H. J. G. L. M., Stahl, O., Waters, L. B. F. M., Wolf, B., 1985,*Astron. Astrophys.***153**, 168

Maeder, A., 1983, *Astron. Astrophys.* **120**, 113

Maeder, A., Meynet, G., 1987, *Astron. Astrophys.* **182**, 243

Morton, D. C., 1960, *Astrophys. J.* **132**, 146

Plavec, M., 1970, *Publ. Astron. Soc. Pacific* **82**, 957

Stahl, O., Wolf, B., Klare, G., Cassatella, A., Krautter, J., Persi, P., Ferrari-Toniolo, M., 1983, *Astron. Astrophys.* **127**, 49

Stothers, R., Chin, C., 1983, *Astrophys. J.* **264**, 583

Webbink, R. F., 1977, *Astrophys. J.* **211**, 486

Wolf, B., 1986, *IAU Symp.* **116**, *Luminous Stars and Associations in Galaxies*, C.W.H. de Loore et al. eds., p. 151

Wolf, B., 1987, *IAU Symp.* **122**, *Circumstellar Matter*, I. Appenzeller and C. Jordan eds., p. 409

DISCUSSION

Owocki: If the radiative forces everywhere exceed gravity, then this does not imply a steady-outflow wind, because that would require a subsonic region that is fixed at the base. I think the situation implies instead that there must be a time-dependent, not steady, mass outflow. I believe then that we must begin to address this time dependence explicitly. With some reasonably simplified formulation for the opacity effects that you have described, this is quite feasible to do numerically.

Appenzeller: In the interior layers of LBV's we have $|g_{rad}| < g_{grav}$. But time-dependent model computations are clearly desirable.

Kudritzki: A comment to Stan Owocki's remark. I totally agree: time-dependent calculations for simplified or parametrized wind line opacities are important and can be done. We plan to do such calculations as well. This was why yesterday, on the river, I was so anxious that Stan should find his way back into the raft ...

Langer: LBV's are probably in global thermal **dis**equilibrium: the H-burning shell continuously adjusts as the core contracts toward helium ignition. This may be a continuous cause of envelope expansion on a thermal time-scale $\sim 10^3$ y.

Appenzeller: Yes. But thermal readjustment of just the outer layers following outbursts will have a time-scale $\sim$ 10 y.

Maeder: Is it possible to estimate the total amount of mass ejected in one of the events that you described in the framework of the radiation pressure model?

Appenzeller: An order-of-magnitude estimate is $\dot{M} \times duration \sim 10^{-2} M_{\odot}$. Much higher values (as for η Car) could be achieved only by radiation pressure effects at high optical depths, $\tau >> 1$.

Walborn: Does the bulk of the material producing the A-type maximum spectrum (false photosphere, shell) leave the star or does it fall back?

Appenzeller: About 10^{-2} of the stellar mass is affected by the expansion. $\dot{M}$ is not large enough to remove this much matter during the eruption. So most of the expanded layers are expected to contract again.

Wolf: R 127 starts to get unstable at a temperature higher than expected in your diagram of the modified Eddington limit. In my talk I showed that Ofpe/WN9 stars have enhanced absorption lines, *e.g.* Fe III originating from metastable levels and even Al III lines, normally unexpected in O stars. Can this additional line opacity make these more luminous stars radiatively unstable at higher temperature?

Appenzeller: Although Fe II and other singly-ionized iron-group elements are most important for the minimum of the modified Eddington luminosity, many other ions (including Fe III at higher temperatures) also contribute.

Davidson: A pair of mildly naïve heuristickal remarks --

(1) The **basic idea** of a temperature-dependent modified Eddington limit, with an opacity crisis (= recombination crisis) as a possible cause for LBV eruptions, is one of the oldest items in our modern LBV lore, antedating the term LBV. In a semi-popular-level review article written in 1983 (*Science* **223**, 243), Roberta and I were careful to include a qualitative description of this intuitive idea, which was then a dozen years old for one LBV, at least -- need I say which?! So, while it has been disappointing that sophisticated calculations have taken such a long time to follow naive intuition, it is also gratifying that analyses like those by Kudritzki's group seem

to be vindicating the basic idea so far. Their two-mode results so closely resemble what some of us have been expecting/hoping-for, that it seems to me that nature is playing a joke on us if they are **not** on the right track. (In saying this, of course I am being sloppy about which source of opacity is the main culprit.)

(2) Here's a related empirical remark. Recently I calculated a few old-fashioned, simplified static atmospheres with plausible grey-opacity dependences and with a mass-luminosity relation appropriate for LBV's: pocket-calculator work. Suppose we characterize each of these idealized models by its maximum value of $\Gamma = |g_{rad}/g_{grav}|$, which in the parameter range of interest occurs at some optical depth of order unity -- let's ignore the normal winds for my purposes! Then it turns out that *the observed Upper Limit of stars in the H-R diagram nicely follows the locus of* $\Gamma_{max} \approx 0.8$ (or maybe 0.7 or 0.9, but **not** 1.0, whose locus is distinctly different). Two poster papers only a meter apart upstairs, one by the Munichers and one by Nieuwenhuijzen and de Jager, evoke this same conjectural point when viewed together, if one is in the right frame of mind. Is there a critical value of the idealized Γ_{max} defined this way, marking the onset of instability?

In any case, we need some calculations aimed specifically at isolating the hypothetical instability -- analyses that include *only those processes that are necessary for the instability* (if it exists). Sophisticated NLTE atmosphere-and-wind models like those described by the Munich and JILA groups are valuable in demonstrating that interesting effects do indeed occur, but their industrial-strength levels of detail cause some distraction from the immediate problem. It seems unlikely that details of line formation, departures from LTE, *etc.*, are crucial to the LBV instability. I don't know whether a largely analytic stability analysis or hydro runs on a big computer would be more appropriate, but in either case I suspect that simplified opacity dependences and simplified radiative transfer would suffice, and that the hydrodynamics are more important. Some work along these lines should have been inspired by Appenzeller's excellent discussions at two meetings in 1985--1986.

Immo Appenzeller

THE ROLE OF ROTATION IN LUMINOUS BLUE VARIABLES

S.R. Sreenivasan and W.J.F. Wilson

Department of Physics
The University of Calgary
Calgary, AB, T2N 1N4, Canada

ABSTRACT: We demonstrate that differential rotation plays a significant role in the essential features of LBV's, among other types of objects.

1. INTRODUCTION

It has been known for some time that rotation is a common characteristic of stars in the upper-left-hand quadrant of the HR diagram. Rotational speeds display a bimodal distribution with maxima around 100 km/s and 300 km/s (Conti and Ebbets 1977; Useugi and Fukuda 1982).

However, this aspect has not received as much attention among those studying massive stars, partly because rotational effects are difficult to model and partly because there is still a considerable amount of confusion about their importance for massive stars (Chiosi and Maeder 1986). Classical studies of rotating stars have held that meridional circulation is too slow to cause significant change and that mean molecular weight gradients effectively prevent mixing (see Tassoul 1980 for a review). But it is becoming increasingly clear that there is evidence for mixing of material between the core and envelope of massive stars right across the HR diagram (see recent reviews in Nomoto 1988) including the example of SN1987A. It has also been known for some time that mass-loss rates in some types of massive stars such as WR stars, LBV's, and Be stars are much higher than predicted by radiatively driven wind theories. Be stars display different mass-loss rates between their polar and equatorial regions (see Abbott and Conti 1987, Slettebak and Snow 1987). Finally, pulsations in many early-type stars are believed to be non-radial and the precise driving mechanisms are still being discussed and debated (see Smith 1988 and Baade 1988 for recent reviews). We have advanced the hypothesis that rotation has a significant role to play in the structure and evolution of stars in the upper-left-hand quadrant of the HR diagram, as well as amongst intermediate-mass stars. We have tested this hypothesis in a number of cases and find that it is possible to understand these different features on the basis of a single hypothesis.

2. STRUCTURAL CHANGES DUE TO ROTATION

In studying the effects of stellar rotation, it has been customary to assume that stars either rotate like rigid bodies or that they

K. Davidson et al. (eds.), Physics of Luminous Blue Variables, 205–209.

possess rotation laws that are stable to begin with. When one does that in stellar models, they are found to possess smaller cores, lower effective and central temperatures, lower luminosities, higher central density and pressure and longer main sequence lifetimes (Bodenheimer 1971). There are slight quantitative differences between rigidly rotating models and those for which the specific angular momentum is proportional to the mass enclosed within a cylinder of radius ω. But there is no *a priori* reason why stars should obey either of these assumptions. It is equally reasonable to suppose that stars have differential rotation, as they are gas-masses and as viscosity and turbulence in them are anisotropic. The shear that is generated as a consequence brings the star into a state of dynamic equilibrium with some mean rotation rate for the core. Mixing between the core and envelope will be a natural consequence as is efficient meridional circulation and redistribution of angular momentum.

We have taken the latter stand and found that if the rotation law is specific-angular-momentum-conserving, then independently of whether or not the envelope is differentially rotating the core size is *smaller* if rigid rotation is imposed on the core, and the other consequences are similar to those found by Bodenheimer (1971). The model evolves as one of lower luminosity etc. until helium is completely burnt in the core (Sreenivasan and Narasimha 1986). But, if the core is presumed to be differentially rotating as well, following the same law: $\Omega\omega^2 =$ *constant*, one gets a larger core and the resulting mixing of material between the core and envelope makes the models bluer. This effect causes a widening of the main sequence, explains the observed blue/red supergiant ratios, and results in the appearance of CNO-processed products at the surface when mass-loss rates are significant. The change is brought about because the radiative gradient for differentially rotating models is *larger* and the adiabatic gradient is *smaller* than those of non-rotating models which are otherwise similar in properties. We wish to emphasize that this effect does *not* require very high initial rotational velocities. The higher the rotational speed, the more pronounced is the effect (Sreenivasan and Wilson 1988).

3. EVOLUTIONARY CHANGES DUE TO ROTATION

As a star evolves, the core shrinks and spins up while the surface spins down due to mass loss and the accompanying loss of surface angular momentum. This results in angular momentum transfer from the core to the envelope, increased shear turbulence, and more efficient meridional circulation. The mechanical energy flux generated by shear turbulence drives a non-thermal wind *in addition to* the radiatively driven wind. Thus, not only does differential rotation increase the mass-loss rate rate in early-type stars, but this increase increases with evolution. One can therefore understand the higher mass-loss rates of WR stars as well as LBV's. One can also understand the origin of circumstellar shells around these objects, because a larger rate of mass loss increasing with evolution produces a sweeping-up process as new wind catches up with material from an earlier wind, reminiscent of theories describing planetary nebula formation (Kahn 1983, Kwok 1983, Sreenivasan and Wilson 1986). Whether stars turn around in the HR diagram or not is strongly controlled by the mean molecular weight, the opacity in the outer layers, and the mass fraction of the helium core. Some stars do and some do not (Chiosi *et al.* 1978, Sreenivasan and Wilson 1988a,b). But the foregoing mechanism of the LBV stage is independent of the mass-loss effect. Hence one can understand LBV's which are *not blue*! Lately, we have found that differentially rotating models of massive stars are overstable for non-radial prograde sectorial g-modes. There are many such modes that are excited with frequencies close to each other. This leads to mode-coupling and the result is a long time modulation of a higher frequency oscillation with periods ranging from hours to fractions of a day. We believe that this affords a qualitative explanation of the pulsation properties of many B-type stars including LBV's. The pulsations are driven by differential rotation and they could explain the line-profile variability observed in 53-Persei-type objects (Smith 1988). In addition, variability in the winds of O stars has been inferred from P Cygni resonance line variations in the spectra (Prinja 1988) which correlate with their rotational velocities. Also, Moffat and his collaborators (1988) have observed variabilities in the winds of WR stars. Both these variabilities could be the results of interaction and instabilities in the winds caused by radiation pressure on the one hand and the non-thermal wind driven by mechanical energy flux due to shear-generated turbulence on the other. Our pulsation models are combined with evolutionary models of massive stars losing mass and angular momentum (Narasimha and Sreenivasan 1987) and the results are in qualitative agreement with those of Lee and Saio (1987,1988).

Since rotation is latitude-dependent, our models lead to the natural conclusion that mass-loss rates at equatorial regions are much higher than those in the polar regions. There is virtually no enhancement of the mass-loss rates due to radiatively driven winds in the polar regions and the enhancement is confined to an equatorial belt. This agrees with what is seen in Be-star winds. We have advanced a similar model for the pulsation of β Cephei stars, which are slow rotators, whereas Be stars are rapid rotators. We believe on the basis of our models that β Cephei stars are slowed down by mass loss and surface angular momentum loss, whereas the Be stars are those objects which have already generated circumstellar disks and hence are revealing interior layers exhibiting more rapid rotation assisted by angular momentum transfer from the spun-up shrunken cores.

Finally, we point out that our model leads to the conclusion that different initial rotation speeds on the zero-age main sequence result in differing rotational histories of massive ($>15\,M_\odot$) and intermediate-mass stars. We can thus understand the evolutionary features of massive stars displaying different mass-loss rates with other properties remaining much the same, or with higher effective temperatures that are now suspected for WR stars ($1.2\,M_\odot < M_* < 9\,M_\odot$); the abundance anomalies of peculiar red giants amongst intermediate-mass stars as well as OB stars, yellow and red supergiants; the enigmatic evolutionary features and location of high-latitude F supergiants and the reasons for large-scale mixing in the progenitor of SN 1987A. We do not need to invoke problematic convective overshooting or excessive rotational speeds to cause the mixing (Sreenivasan and Wilson 1982, 1985abc, 1986, 1987). They simply become the result of including rotational effects.

ACKNOWLEDGEMENT
Our work is supported by an NSERC research grant of Canada.

REFERENCES

Abbot, D. and Conti, P.S. 1987, *Ann.Rev.Astr.Ap.* **25**, 113.
Baade, D. 1988, in *O and WR Stars* (ed. P.S. Conti and A. Underhill; NASA SP-497).
Bodenheimer, P. 1971, *Ap.J.* **167**, 153.
Chiosi, C. and Maeder, A. 1986, *Ann.Rev.Astr.Ap.* **24**, 329.
Chiosi, C., Nasi, E., and Sreenivasan, S.R. 1978, *Astr. Astrophys.* **63**, 103.
Conti, P.S. and Ebbets, D. 1977, *Ap.J.* **213**, 438.
Kahn, F.D. 1983, in *Planetary Nebulae* (ed. D.R. Flower; Reidel, Dordrecht), p. 305.
Kwok, S. 1983, in *Planetary Nebulae* (ed. D.R. Flower; Reidel, Dordrecht), p. 293.
Lee, U. 1988, *M.N.R.A.S.* **232**, 711.
Lee, U. and Saio, H. 1987, *M.N.R.A.S.* **224**, 513 and **225**, 64.
Moffat, A. 1988, these proceedings.
Narasimha, D. and Sreenivasan, S.R. 1987, in *Stellar Pulsation* (ed. A. Cox; Springer, New York), p. 106.
Nomoto, K. (ed.) 1988, *Atmospheric Diagnostics of Stellar Evolution* (Springer, New York).
Prinja, R. 1988, *M.N.R.A.S.* **231**, 21P.
Slettebak, I. and Snow, T.P. (eds.) 1987, *Be Stars* (Cambridge University Press).
Smith, M.A. 1988, in *Pulsation and Mass-loss in Stars* (ed. R. Stalio and L.A. Willson; Reidel, Dordrecht), p. 251.
Sreenivasan, S.R. and Narasimha, D. 1986, *Bull.A.A.S.* **17**, 895.
Sreenivasan, S.R. and Wilson, W.J.F. 1978a, *Astron.Astrophys.* **70**, 755.
----- 1978b, *Astrophys.Space Sci.* **53**, 193.
----- 1980a, in *Stellar Pulsation Instabilities* (ed. D. Fischel *et al.*; NASA TM 80625), p. 363.
----- 1980b, in *Evolution of O Stars* (ed. P.S. Conti; Reidel, Dordrecht), p. 367.
----- 1982, *Ap.J.* **254**, 287.
----- 1985a, in *Non-radiative Heating and Momentum in Hot Stars* (ed. A.G. Michalitsianos and A.B. Underhill; NASA CP 2358), p. 177.
----- 1985b, *Ap.J.* **290**, 653.
----- 1985c, *Ap.J.* **292**, 506.
----- 1986a, in *Luminous Stars and Associations in Galaxies* (ed. C.W.H. DeLoore and A.J. Willis; Reidel, Dordrecht), p. 385.
----- 1986b, *Bull.A.A.S.* **17**, 600.
----- 1986c, in *Mass-loss from Red Giants* (ed. M. Morris and B. Zuckerman; Reidel, Dordrecht), p. 261.
----- 1986d, in *Cool Stars, the Stellar System and the Sun* (ed. D. Gibson and M. Zeilik; Springer, New York), p. 426.
----- 1987a, in *Circumstellar Matter* (ed. I. Appenzeller and C. Jordan; Reidel, Dordrecht), p. 425.
----- 1987b, in *Internal Angular Velocity of the Sun* (ed. B. Durney and S. Sofia; Reidel, Dordrecht), p. 229.
----- 1988, *Ap.J.* (to be published)
Tassoul, J.L. 1980, *Stellar Rotation* (Princeton University Press).
Useugi, A., and Fukuda, I. 1982, *Stellar Rotational Velocities* (Kyoto University)

DISCUSSION

Langer: If a rotation law leads to appreciable convective core enlargement, this would resemble the effects of large convective overshooting. Don't you then run into the same kind of problem as in stellar models with large overshooting, *i.e.*, that very massive stars avoid the LBV phase after hydrogen burning and instead turn toward the blue and directly become Wolf-Rayet stars?

Sreenivasan: Only a relatively small amount of rotation is required to produce appreciable core growth; e.g., for a $70M_{\odot}$ star, a value of ($v \sin i$) ~ 150 km/s increases the core size from about 50 $M_{\odot}$ to 56 $M_{\odot}$. The evolutionary tracks for such a population I model go to the red before turning blueward. So, what you are worried about does not happen. Only when unreasonably large overshoot is imposed and excessively large mass-loss rates (exceeding what is allowed by radiatively driven stellar winds) are invoked does that happen.

Vanbeveren: Do you think that rotation effects may be important in close binaries with periods, say, larger than ~ 4 days?

Sreenivasan: Yes!

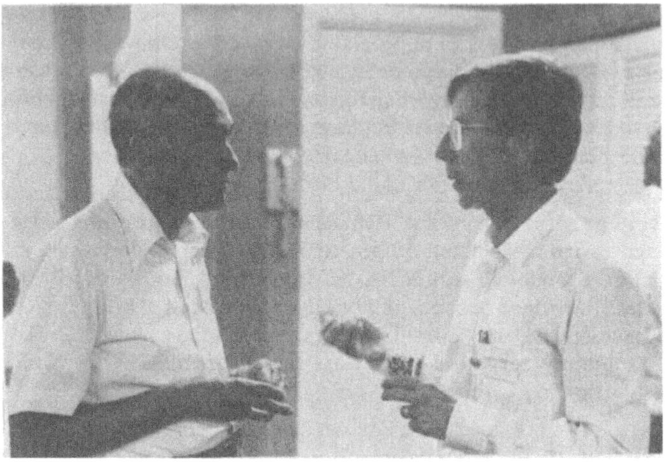

Sreenivasan and Sterken

de Jager
and
Humphreys

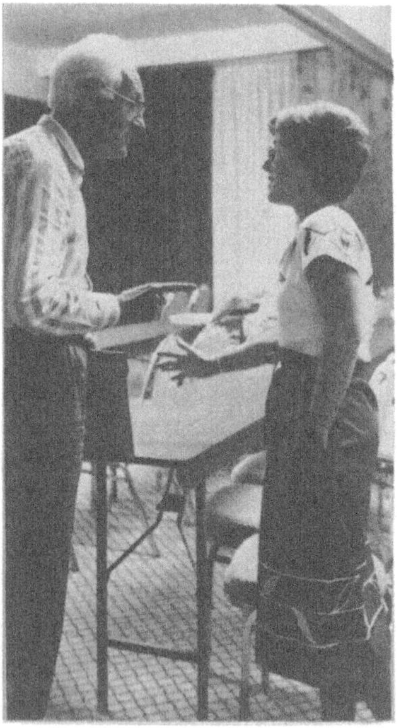

ATMOSPHERIC DYNAMICS OF LUMINOUS STARS

Cornelis de Jager, Joost Carpay, Alex de Koter, Hans Nieuwenhuijzen, Erik Schellekens
Laboratory for Space Research and Astronomical Institute
Beneluxlaan 21, 3527 HS UTRECHT, The Netherlands

ABSTRACT. A review is given of data and theories on the motion fields in super and hypergiants with special reference to LBV's. We show that the radiative momentum flux is incapable of driving the episodical bursts of mass loss of these stars, and that there are several indications that the LBV-phenomenon is hydrodynamically driven. The sum of turbulent and radiative accelerations in the atmospheres of the most luminous stars compensates the gravitational acceleration for stars near the Humphreys-Davidson limit. This explains their atmospheric near-instability. The motion field in the atmosphere of a typical LBV consists mainly of low-order gravity waves, while acoustic waves are rapidly damped. These gravitation waves may be stochastic rather than coherently ordered. These stochastic pulsations are assumed to be responsible for the LBV phenomenon.

1. PHYSICAL CONDITIONS IN THE ATMOSPHERES OF VERY LUMINOUS BLUE STARS

1.1. Momentum of stellar winds at the upper brightness limit of stellar existence

A review of mass loss rates in the upper part of the Hertzsprung-Russell diagram was published recently by De Jager, Nieuwenhuijzen and Van der Hucht (1988). These data allow us to compare the momentum flux of the stellar winds Mv_∞ (where M is the rate of mass loss and v_∞ the asymptotic stellar wind velocity) with that of the stellar radiation flux L/c. We made these calculations for stellar parameters valid at the Humphreys-Davidson limit, for which we know L and T_{eff}. The values of M were read from our mass loss review while for v_∞ a relation was used derived earlier by De Jager (1984). The results, shown in Table 1, demonstrate that along the whole upper limit of stellar existence the mass flux momentum losses are smaller than those of the radiation flux momentum; hence the stellar radiation flux could, at least in principle, account for the quiescent average stellar mass loss momentum flux. But it is clear from the data in Table 1 that this will no longer be the case when the rate of mass loss would increase by a factor > 3 to 5, as happens in LBV mass ejections. Another, presumable hydrodynamic, explanation has then to be sought. This is our <u>first indication</u> that the

K. Davidson et al. (eds.), Physics of Luminous Blue Variables, 211–219.

LBV phenomenon may be hydrodynamically driven.

Table 1. Comparison of radiative versus wind momentum fluxes along the Humphreys-Davidson limit;

$\log T_{eff}$	$\log \frac{L}{L_\odot}$	$\log \dot{M}$ ($M_\odot yr^{-1}$)	v_∞ ($km\ s^{-1}$)	$\dot{M}v_\infty \times 10^{-28}$	$\frac{L}{c} \times 10^{-28}$	$\frac{\dot{M}\ v_\infty c}{L}$
4.4	6.3	- 4.5	520	10.3	25.5	0.39
4.1	6.1	- 4.5	160	3.09	16.1	0.19
3.8	5.9	- 4.0	47	2.96	10.1	0.29
3.5	5.9	- 3.5	14	2.79	10.1	0.27

1.2. Accelerations and stellar wind velocities in the upper blue part of the Hertzsprung-Russell diagram

For stellar parameters valid in the upper part of the Hertzsprung-Russell diagram Nieuwenhuijzen and De Jager (1989) investigated the physical properties and the various accelerations, such as radiative, dynamic wind and turbulent accelerations. We refer to their Figure which gives the ratios g_{out}/g_{grav}, g_{turb}/g_{rad} and v_{wind}/s (where s is the velocity of sound). Here, g_{out} is the sum of all outward accelerations: those due to radiation pressure, turbulence and the stellar wind.
In those figures the relevant quantities are given for an optical depth $\tau_R = 0.1$, which is about the level where most of the Fraunhofer lines of average strength are formed in the atmospheres of hot stars.
Noteworthy, and perhaps important for our understanding of the LBV phenomenon is that the area occupied by the LBV's coincides fairly well with the area where the rate g_{turb}/g_{rad} assumes its largest values, with a maximum of about 0.5. This is our <u>second indication</u> that the LBV phenomenon may be driven by hydrodynamic phenomena.

In addition it appears that the upper boundary of the LBV region coincides fairly well with the line where $v_{wind}/s = 1$ at $\tau_R = 0.1$. This means that only a slight disturbance of (a part of) the atmosphere, such as a local (pulsating) outward motion, would be sufficient to shift the sonic point of the stellar wind into much deeper layers, where the density is larger, which would then cause a large increase of the rate of stellar mass loss in that area of the stellar disc. This may explain the observed semi-regular mass ejections, and this may be our <u>third indication</u> that the LBV phenomenon is a hydrodynamic one.

We give an example applied to a real star. For the L and T_{eff} values at the position in the HR diagram of a typical LBV, P Cyg, we take (De Jager et al.), 1988; star # 145): Log $T_{eff} = 4.3$; $\log(L/L_\odot) = 5.93$, the quiescent rate of mass loss is $\log\dot{M}$ [$M_\odot$/yr] $= 2.10^{-5}$. Its "estimated mean mass" is 32 $M_\odot$. In the photosphere, at $\tau_R = 0.1$: $s = 21$ km s^{-1}. If that outward velocity would occur at the depth level $\tau_R = 2/3$ (just to give an example) the star's mass loss would increase to 2×10^{-4}; an enhancement by a factor 10! If, as seems likely, we are not dealing with strict radial pulsations but rather with localized (non-radial or stochastic) pulsations the increase in the rate of mass loss would be smaller, proportional to the area involved. But for pulsations for which the above mentioned outward velocity would occur at larger depths the rate of mass loss would become larger, proportionally to the increased density at the level considered.

1.3. The influence of turbulent pressure on atmospheric instability

Any directed gas motion excerts a dynamic pressure, which is the transport of momentum ρ**v**.**v**. associated with the gas motion. In a turbulent medium the gas motions are distributed stochastically. Each moving "element" has its own dynamic pressure, and the integrated dynamic pressure of a turbulent gas, the so-called turbulent pressure is the momentum flux density tensor $\rho.v_i.v_j$. For a field of accoustic waves the turbulent pressure is

$$P_t = 0.5\ \rho<\xi_t^2>,$$

where $<\xi_t^2>^{\frac{1}{2}}$ is the mean squared turbulent velocity component. For a field of shock waves the factor is 1/3 (Ulmschneider et al., priv. comm.).

The value of the turbulent acceleration g_t in a stellar atmosphere can be determined on the basis of a determination of the microturbulent velocity component, if that quantity is known as a function of the height in the stellar atmosphere. This has so far been done observationally for six stars; Figure 1 gives $g_{out}/g_{grav} = (g_t + g_{rad})/g_{grav}$ as a function of Δ log L, where the latter quantity is the vertical separation in the Hertzsprung-Russell diagram of the star's luminosity from the luminosity at the Humphreys-Davidson limit (Humphreys and Davidson, 1984).

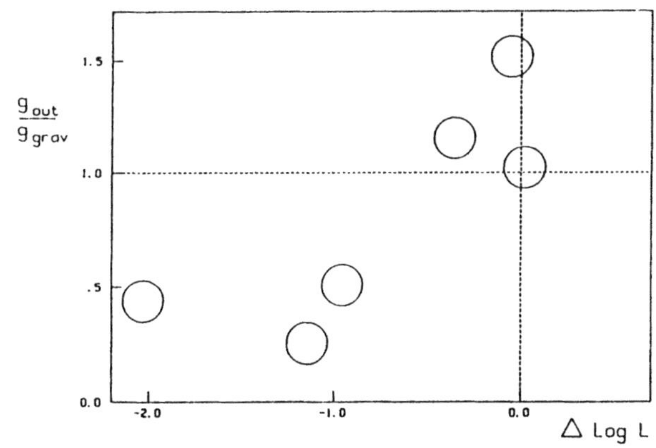

Figure 1: Ratio g_{out}/g_{grav} for six well-studied super and hypergiants. The data are plotted against the vertical distance of the star to the Humphreys-Davidson limit. The absolute value of the ratio approaches unity at the limit.

It appears from the Figure that for $\Delta \log L \to 0$, $g_{out} \to g_{grav}$. This explains the stellar atmospheric instability at and above the Humphreys-Davidson limit.

Thereby a distinction should be made between hot and cold stars. For $T_{eff} \lesssim 10^4$, $g_{out} = g_{turb}$ (the influence of radiation is negligible), but for hotter atmospheres g_{rad} becomes important. So far we have investigated one hot star, HD 80077 (B2 Ia+; Carpay et al., 1989). For this object we found (Table 2) that $g_{turb} < g_{rad}$ but g_{turb} is certainly not negligible, being one-fourth to one half the radiative acceleration. It would be highly important to study more hot stars of various temperatures, along the HD limit.

Table 2. Accelerations in HD 80077 (B2 Ia+) at two optical depths. The g-values are in cm s^{-2}.

τ_{Ross}	g_{grav}	g_{rad}	g_{turb}	g_{eff}
0.24	- 100	+ 80	+ 15	+ 5
0.08	- 100	+ 73	+ 20	+ 7

1.4. The large-scale ("macroturbulent") motion field

This motion component is difficult to determine since it needs high-dispersion spectra, while also knowledge of all other broadening mechanisms is required. For Alpha Cyg (A2 Ia) we found a bimodal distribution of the large-scale vertical motion component, with values of + and - 14 km s^{-1}, (Boer et al., 1988). This velocity is supersonic. This observational result implies the occurrence of large up- and downward elements on the stellar surface. Since the average (r.m.s.) stellar velocity fluctuation is 2.3 km s^{-1}, there must be on the average 30 to 40 elements on the visible surface with, hence, average diameters of 30 x 10^6 km. Also in the less extreme supergiant Alpha Sco (M1.5 Iab) there are indications for a bimodal distribution of the large-scale motions, but the amplitudes are smaller (rather of the order of 5 km s^{-1}) and it is therefore more difficult to determine them (De Koter, 1989).

2. THE PHYSICS OF THE MOTION FIELD

2.1. The dispersion function

Assuming plane waves and adiabatic disturbances the dispersion function gives the relationship $\omega(k)$ between the wave number k and the wave frequency ω. Neglecting magnetic effects two kinds of waves remain for consideration:
the G waves (gravity waves; gravity being the restoring force);
the P waves (pressure or acoustic waves; pressure difference restores).

Instead of giving the $\omega(k)$ relation it adds to clarity by giving the T(L) relation instead, where $L = 2\pi/k$ is the wavelength and $T = 2\pi/\omega$ is the wave period. Figure 2 gives such a diagram, constructed for the L, T_{eff} and M values expected to apply to a star with characteristics similar to those of a typical LBV, P Cyg.

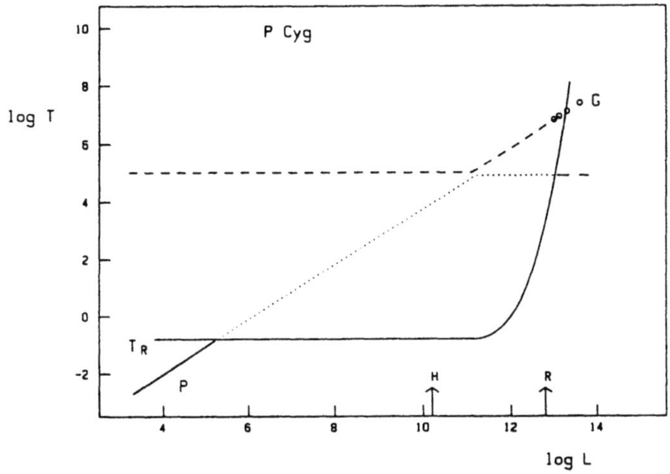

Figure 2: Diagnostic diagram for the motion field in a typical S Dor star (LBV); we chose the parameters of P Cyg.

The diagram is valid for an optical depth $\tau_R = 0.1$, where we derived (Nieuwenhuijzen and De Jager, 1989):

log T = 4.188; log P = 0.760; log ρ = -11.467. The stellar mass is taken as 32 $M_\odot$, and the accelerations are

$g_{grav} = -143$; $g_{rad} = 112$, $g_{wind} = 4$; $g_t = 14$. Hence $g_{eff} = -15$.

For other stars similar graphs can easily be derived.

Important lengths are indicated in the diagram: H is the atmospheric density scale height; R the stellar radius.

2.2. Motions cannot exist at all wavelengths; restriction of the domains

Although, in principle, waves can occur all along the lines marked G and P in Figure 2, the actual domain of their occurrence is restricted because for certain wavelength regions the waves do not develop or are extinguished in a time shorter than one wave period. There are several such damping mechanisms (for more details we refer to De Jager et al., 1989).

a. Radiative damping of waves occurs because of radiative exchange of energy between the "hot" and "cold" parts of the wave. The line labeled T_R gives the e-folding time of radiative damping as a function of wavelength L. Only in those parts of the diagram where the P or G lines are situated below the line T_R, waves can exist virtually undamped during a period longer than one wave period. Above it they are damped in less than a wave period and thus cannot develop. The effect is only valid for non-compressible waves such as gravity waves; for acoustic waves the T_R curve is a lower limit.

b. Atmospheric curvature effects tend to reduce the restrictions due to radiative damping because that effect is working less efficiently when the wavelengths of the surface waves are so long that the curvature of the atmosphere prohibits radiative exchange. For a star like P Cyg this effects starts to work for $L = 0.1\ R_{star}$. The effect is smaller and the consequent radiative damping time longer for longer wavelengths, but the precise course of the line has not yet been calculated. Awaiting further research, the line drawn in Figure 2 serves merely as an illustration.

On the basis of these considerations we conclude that in the atmosphere of a star like P Cyg the P waves only occur for short wavelengths. The G-waves are only possible for fairly long wavelengths, longer than the stellar radius. At the long wavelengths of the gravity waves standing waves can only develop for discrete wavelengths given by $L(m) = 2\pi R/m$, where m is an integer. But since radiative damping still plays a role for these waves they may exist only for a few small m values.

2.3. Coherent non-radial or stochastic pulsations?

Another possible consequence of the situation as outlined in Figure 2 is the fact that it appears difficult for a system of standing non-linear oscillations to develop. Any single wave, once excited, will fairly rapidly damp out. This is the reason why we have suggested in various places (De Jager, 1980; Boer et al., 1988) that the large-scale motion field in super- and/or hypergiants may be a stochastic motion field rather than an ordered system of nonlinear pulsations. To show the evidence we give in Figure 3 the variation of the radial velocities of Alpha Cyg (A2 Ia) as given by Lucy (1976) and as modelled by Schellekens et al. (1989) by a field of stochastic motions.
This is no proof, evidently, that the motions on Alpha Cyg are like that, but it shows the possibility.

We think that detailed analyses of the atmospheric motion fields of S Dor stars may be very rewarding.

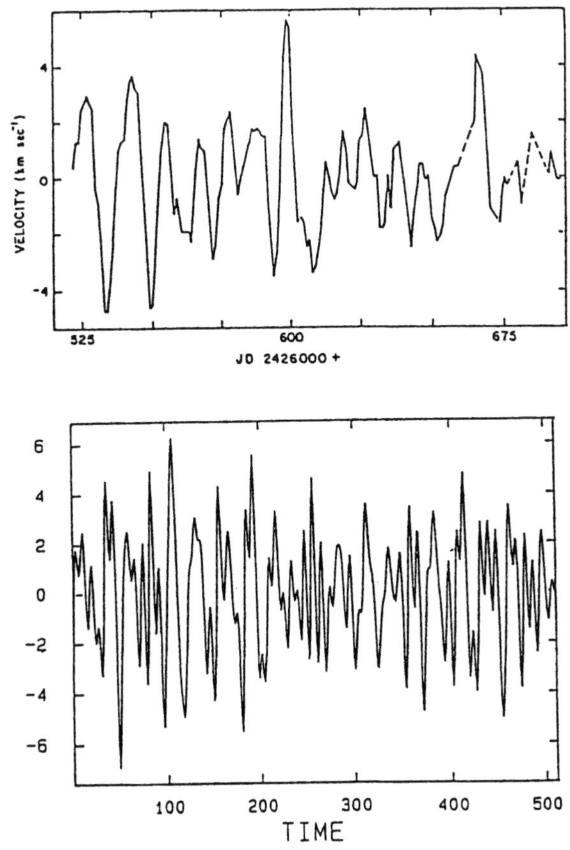

Figure 3: Comparison of observed (above) and modelled radial velocity variations for α Cyg; note the different scales of the abscissae.

3. CONCLUSION ON THE MOTION FIELD IN SUPER AND HYPERGIANTS

We have shown in this review that there are various indications that the LBV phenomenon may have hydrodynamical causes. Radiation pressure and its fluctuations do not seem capable of driving the stellar bursts of episodic mass loss, while pulsations of a part of the stellar surface may well provide a working mechanism. The motion field in a typical LBV (we choose the parameters of P Cyg) may consist mainly of low-mode gravitation waves, and there is evidence that the motions are stochastic and not coherently ordered. By such stochastic pulsations, not necessary all of the same amplitude, episodical bursts of enhanced mass loss may occur.

REFERENCES

Boer, B., De Jager, C. and Nieuwenhuijzen, H.: 1988, Astron. Astrophys. **195**, 218
Carpay, J., De Jager, C. and Nieuwenhuijzen, H.: 1989, Astron. Astrophys., submitted
De Jager, C.: 1980, The Brightest stars, Reidel, Dordrecht
De Jager, C.: 1984, Astron. Astrophys., **138**, 246
De Jager, C., Nieuwenhuijzen, H. and Van der Hucht, K.A.: 1988, Astron. Astrophys., Suppl. Ser. **72**, 259
De Koter, A.: 1989, Astron. Astrophys. in prep.
Humphreys, R.M. and Davidson, K.: 1984, Science, **223** 243
Lucy, L.B.: 1976, Astrophys. J., **206**, 499
Nieuwenhuijzen, H. and De Jager, C.: 1989, These Proceedings
Schellekens, E., De Jager, C. and Nieuwenhuijzen, H.: 1989, Astron. Astrophys. in prep.

C. de Jager

DISCUSSION

Sreenivasan: We have shown, as I mentioned briefly in my talk, that α Cyg has a number of g-modes that are overstable (up to $l = 10$ and $l = |m|$ prograde modes). There are presumably more modes that are overstable. The frequencies are close and mode-coupling is likely. This has the consequence that observations may look like "stochastic" pulsations as you said. That was also the conclusion of Lucy who analyzed Paddock's observations. The strong differential rotation would induce shear turbulence in such stars which could be the source of your turbulent pressure gradient in the atmospheres of these objects.

De Jager: I agree with what you say but wish to stress that it is virtually impossible to decide between stochastic and well-ordered non-linear pulsations on the basis of Paddock's series of observations; three years is still too short!

Maeder: From the relatively long periods of the short-term pulsations of LBV's, you assigned them to blueward tracks. By themselves, these relatively long periods only imply a lower average density, consistent with the fact that LBV's experience high mass-loss rates. They can be on either redward or blueward tracks, or just making horizontal excursions back and forth in the H-R diagram.

De Jager: I see your point and realize that evolved blue supergiants will not return on their tracks early enough if they have not lost enough mass. That is the problem for evolutionists to solve. My point is that there is some evidence that they are far evolved, and hence may start losing mass rapidly only on their blueward track. I realize that this is contrary to what evolutionists want.

Appenzeller: A few blue supergiant stars, such as #12 in the Cygnus OB2 association, are about as luminous as LBV's but show no variability or exceptionally strong winds. Their effective temperatures may be just cool enough to be safe from radiation pressure effects. How important are turbulent-pressure effects in such stars?

De Jager: The star that you mention lies well above the range occupied by LBV's in the H-R diagram. Its absolute magnitude makes it a hypergiant. The apparent faintness of that star makes it difficult to obtain the high-resolution spectra needed for a good analysis of its atmospheric motion field. But I would guess that there is strong atmospheric turbulence.

Humphreys: Cyg OB2 (or VI Cyg) #12 is a B8 hypergiant, an extremely luminous star. HD 33579 in the LMC is a similar star, with apparent visual magnitude about 9.5, and would be a good candidate for your analysis.

De Jager: According to its position in the H-R diagram, Cyg OB2 #12 should have an atmosphere in which g_{grav} is reduced by ~ 70% by radiation and turbulence may account for the remaining 30%; it is a pity that high-resolution spectroscopy of this star is so difficult. Thank you for your suggestion about HD 33579; we will apply for observing time!

Owocki: (1) I would like to point out that pulsation-driven mass loss in late-type stars has been investigated quite extensively by Lee Ann Willson and her colleagues. A nice review of this field by John Castor appeared in the volume honoring Prof. de Jager's retirement (*Instabilities in Luminous Early-Type Stars*, ed. by Lamers and de Loore, 1987). (2) Recent work that I have done, in which the effects of pulsations at the base, but including radiative force, are simulated numerically, seems to suggest that the mass-loss rate should be variable. The timescale, however, is about a day, much faster than the large variations of LBV's.

Hillier, Niemela

Davidson, Nieuwenhuijzen, Klapp, Gosset

THEORETICAL RELATIONS BETWEEN LBVs AND OTHER STARS

Norbert Langer
Universitäts - Sternwarte
Geismarlandstr. 11
D-3400 Göttingen
Federal Republic of Germany

1. Introduction

The so called Luminous Blue Variables (LBVs) form a rather heterogeneous class of very active, luminous (and therefore massive) stars (c.f. R.M. Humphreys, this volume). The following two points are the aim of theoretical work concerning these objects: 1. to know their internal structure from the center up to the atmosphere, i.e. to find out the place of these objects in the course of thermonuclear evolution, or to identify the physical processes which are responsible for the different types of mass loss phenomena, and 2. to conclude which were their preceding evolutionary phases, and to predict their future evolution, i.e. to identify LBV progenitors and descendants, derive their ages, and so on. The basic tool of the theorists is a stellar evolution computer code, which allows to perform evolutionary computations (including different assumptions, which are necessary since several physical ingredients cannot yet be treated from first principles), leading stellar tracks through the part of the HR diagram where LBVs are found. But in this context we meet already a basic problem: in a certain small area of the HR diagram (which may correspond e.g. to some kind of error box) stars of several different types may be identified by the observers, each type may be even devided into several subtypes. Therefore, the correlation of computer models and observed stars is often very difficult or impossible. The various types of stars are mainly classified by spectroscopic criteria, while important "theoretical quantities" as stellar masses or the chemical composition are hard to derive from observations.

A correlation of stellar models with observed stars is also difficult, since there are only few phases of internal evolution in contrast to many different types of stars. For the LBVs, specifically, we will argue in the next Section that their internal evolutionary stage is that of shell hydrogen burning, i.e. the contraction phase between central H- and He-burning, but the variety of LBV subtypes is quite large (see again: R.M. Humphreys, this volume). Consequently, in the case of the LBVs we have to conclude that besides the interior evolutionary state one further parameter must be responsible for the diversity of the LBV phenomena. One may of course think about binarity, rotation, magnetic fields, etc., things which may certianly be important in specific cases. In the present paper we will, however, investigate the most simple possibility, which is that the (actual or initial) stellar mass has some effect on the display of a massive star in the LBV phase. In Section 3

K. Davidson et al. (eds.), Physics of Luminous Blue Variables, 221–227.

we will therefore look for structural differences in stellar models of different masses in the H-shell burning phase. The question, whether some LBVs may be post-Red Supergiants (RSGs), which would correspond to the central helium burning evolutionary phase, is explicitly discussed in Sect. 4.

Concerning the comparison of theoretical stellar sequences with LBV observations, a further important point is the following: it exists (to our knowledge) no stellar evolution calculation leading through the LBV regime of the HR diagram which produces dynamically unstable stellar models. This means that the physics which is responsible for the violent outburst phenomena observed in several LBVs (cf. K. Davidson, this volume) is not included in any stellar evolution calculation. There are two possibilities about the origin of these outbursts: either they are "surface phenomena", i.e. the responsible process acts in the stellar atmosphere, or the outbursts are initiated in the deep interior of the star. In the first case the problem should be solved by stellar atmosphere physicists, and the solution will be implemented as a boundary condition in the stellar structure code. The second case means, however, that the stellar models include unjustified simplifications. Since it seems more likely that the ejection of $\sim 0.1\,M_{\odot}$ (as in the case of η Car) is triggered at the bottom of the shell rather than at its outer surface, it is necessary to look for processes in the deep stellar interior, which may be responsible for such events. A possible candidate is presented in Section 5.

Finally, in Section 6 we summarize our results and present a coherent picture for the evolution of massive stars including the LBV stage as a key phase.

2. The evolutionary phase of LBVs: H-shell burning

In principle it is possible that massive stars evolve to the blue side of the zero age main sequence (ZAMS) already during core hydrogen burning, e.g. in the case of huge convective core overshooting (Langer & El Eid, 1986; Prantzos et al., 1986) or rotational induced mixing (Maeder, 1987). However, in this case the stars never return to the red side of the ZAMS and can thereby obviously never reach the LBV stage. For this reason we concentrate on the other alternative, i.e. the case where no bluewards evolution occurs during core hydrogen burning, which is obviously realized in nature since LBVs exist, and which is the result of stellar evolution calculations with "standard assumptions" (cf. e.g. Maeder & Meynet, 1987).

In this case, all massive stars (when computed with mass loss during core H-burning) show a strong tendency to move towards the Hayashi line in the HR diagram after central hydrogen exhaustion (cf. Langer & El Eid, 1986, Maeder & Meynet, 1987), i.e. they would become RSGs if not something (preferentially mass loss at huge rates) would prevent them, and ignite He-burning at the RSG branch. Now, the RSG configuration is forbidden for stars more luminous than $\sim 10^{5.8}\,L_{\odot}$ for observational reasons (cf. R.M. Humphreys, this volume). Therefore, something must make the stars stop at the Humphreys-Davidson (HD) limit in the HR diagram, causing thereby the LBV phenomenon. This means that helium burning did not yet ignite in this situation. In other words, LBVs (at least those with log $L/L_{\odot} \gtrsim 5.8$) are in the shell hydrogen burning phase of evolution. This means specifically, that the whole star (and not only the stellar envelope) is completely out of thermal equilibrium. Consequently, the whole internal structure changes very rapidly, on a thermal timescale, i.e. $\sim 10^4\,yr$. Note that this is the same order of magnitude as the LBV lifetime estimated from observations (cf. Lamers, this volume), which may be an indication for the core contraction being the virtual origin of the LBV phenomenon (cf. next Section).

3. How the H-shell burning phase affects the further evolution

After central hydrogen depletion a star performs a short phase of overall contraction, which soon leads to the ignition of a hydrogen burning shell around the hydrogen exhausted core. Up to central helium ignition, this shell is the only nuclear energy source of the star. While the contracting hydrogen exhausted core does not react upon the ignition of the H-shell, this has two effects on the outer layers:

1) The whole envelope, i.e. the layers above the shell, start to expand, which makes the star becoming a supergiant, and which is accompagnied by the formation of a more or less extended convection zone close to the surface. (This convection zone may well be relevant to the LBV phenomenon; see Section 5).

2) Just above the H-burning shell an intermediate convection zone (ICZ) forms in layers, which contain gradients of the chemical composition, left behind by the shrinking convective core during central hydrogen burning.

While effect 1) will be discussed in Section 5, we will focus here on the second effect: The formation of the ICZ leads to changes of the chemical profiles. Though part of this zone, especially in stars with $M_{ZAMS} \lesssim 40\, M_\odot$, may be semiconvective rather than convective (cf. Langer, 1985, for a discussion), we will assume here for simplicity, that the whole superadiabatic region becomes rapidly chemically homogeneous. When we designate the mass of the ICZ as ΔM_{ICZ}, and the hydrogen mass fraction which is established by the mixing processes in that zone as X_{ICZ}, we notice in stellar evolution computations, that both quantities are monotonous functions of the stellar mass (or ZAMS mass): ΔM_{ICZ} is increasing and X_{ICZ} is decreasing with increasing stellar mass. We found e.g. for $M_{ZAMS} = 100\, M_\odot$: $\Delta M_{ICZ} = 20\, M_\odot$, $X_{ICZ} = 0.23$, and for $M_{ZAMS} = 30\, M_\odot$: $\Delta M_{ICZ} = 7\, M_\odot$, $X_{ICZ} = 0.50$.

Now it is important to recall, that the redwards evolution is halted at the HD-limit in the LBV stage due to enormous mass loss. Evolutionary calculations show, that the redwards trend itself is stopped, when sufficient mass has been removed in order to drop the surface hydrogen mass fraction below some critical value X_{crit}. In this situation we get a bifurcation of the evolution (cf. Langer, 1987):

In the case $X_{ICZ} > X_{crit}$, the LBV mass loss has to last until all the envelope material up to the bottom of the former ICZ is lost, which means that the He-core itself almost appears at the stellar surface at the end of the LBV phase. This scenario, which is a possibility for stars with $M_{ZAMS} < M_{crit}$ (M_{crit} being defined by the condition $X_{ICZ}(M_{crit}) = X_{crit}$), leads directly to low mass WNE stars as LBV descendants.

In the second case, i.e. $X_{ICZ} < X_{crit}$, the redwards trend is stopped and the LBV phase finished when the top of the H-plateau established by the former of the ICZ reaches the stellar surface. This scenario, which may apply for $M_{ZAMS} > M_{crit}$, results in massive WNL stars as LBV descendants.

Both cases lead to quite different average LBV masses. The first case ($X_{ICZ} > X_{crit}$) applies to the less massive stars considered here, but implies a larger total amount of mass lost in the LBV stage, since the whole former ICZ has to be ejected, while the second case applies to the most massive stars and does not require too much mass to be ejected in the LBV stage, since the former ICZ is kept, forming the envelope of the WNL star lateron.

In summary, we have two evolutionary possibilities for massive stars which enter the LBV phase:

$M_{ZAMS} < M_{crit}$: O-star → low mass LBV → WNE

$M_{ZAMS} > M_{crit}$: O-star → massive LBV → WNL

The exact numbers for M_{crit} and X_{crit} depend greatly on the input physics used in stellar evolution calculations. For the two examples mentioned above we find: $M_{ZAMS} = 30\,M_\odot$, $X_{ICZ} = 0.50 > X_{crit} \Rightarrow \quad M_{LBV,min} = 12\,M_\odot$, and $M_{ZAMS} = 100\,M_\odot$, $X_{ICZ} = 0.23 < X_{crit} \Rightarrow \quad M_{LBV,min} = 60\,M_\odot$, indicating that $30\,M_\odot < M_{crit} < 100\,M_\odot$, and $0.23 < X_{crit} < 0.50$ for standard assumptions.

4. Do post-RSG LBVs exist?

According to R.M. Humphreys (this volume) and others, some LBVs exist with luminosities below the upper luminosity limit for RSGs. For these objects, both evolutionary connections, LBV → RSG and RSG → LBV, are therefore not prohibited by observations.

As for the second possibility, since the star enters in this case the LBV region of the HR diagram the second time, one might wonder why the LBV phenomenon did not occur already on the way towards the RSG branch. However, we note that several physical quantities of a star returning from the RSG branch may be different compared to its state on the way becoming a RSG, though the HR diagram location may be almost identical. First of all it is important, that all stars, when they reach the Hayashi-line, increase their luminosity considerably. This means that a star is more luminous on its way back to the blue side of the HR diagram than it was on its way towards the red side. The relative luminosity increase is larger for less massive stars and may be of the order of 30% for stars with luminosities close to the observed upper luminosity limit for RSGs (cf. Maeder & Meynet, 1987), the numbers depending, however, greatly on the considered internal mixing processes (Langer et al., 1985). It may therefore be possible that a star in its pre-RSG state is not sufficiently luminous in order to encounter the LBV instability, but as a post-RSG its luminosity is high enough.

Further changes that occur at the RSG branch are a modification of the envelope chemical composition due to convective dredge-up, and a considerable decrease of the envelope mass due to the large mass loss of RSGs.

In summary, for a narrow mass range the evolutionary path O-star→RSG→LBV seems possible, and also the alternative O-star→LBV→RSG cannot be excluded for those stars. One may speculate whether the relatively low luminous cool LBVs mentioned by C. de Jager (this volume) are connected with such scenario.

As mentioned above, stellar evolution calculations do not result in dynamically unstable stellar models in neither of the two possibilities (see, however, A. Maeder, this volume). In one sequence we found, however, a thermally unstable situation for post-RSG models at about the LBV position in the HR diagram (cf. Langer & El Eid, 1986), which was related to an adjustment of the H-burning shell, and a similar small loop at $\log T_{eff} \simeq 4.3$ is visible in the $60\,M_\odot$-track of Maeder (1981).

Whether this is a hint towards the existence of post-RSG LBVs or not cannot yet be answered.

Finally, we want to specify a few points, by which post-RSG LBVs may be distinguished observationally from "usual" LBVs. First of all, the slow and dense RSG wind might be swept up by the faster LBV wind, giving rise to the formation of some kind of ring nebula (reminding us to AG Car). And secondly, post-RSG LBVs should be relatively old as compared to normal ones (cf. Lortet and Testor, 1988, for discussion), since they originate from lower ZAMS masses and are additionally more evolved: they are within the central helium burning phase, while "normal" LBVs are in the H-shell burning phase (cf. Section 2).

5. A possible mechanism for violent shell ejections

In Section 3 we have already mentioned, that one consequence of H-shell burning is the expansion of the stellar envelope. When the surface temperature drops to values of the order of $15\,000\,K$, the opacity in part of the envelope has increased so much, that a convection zone forms, which is located close to the stellar surface. In context with the LBVs it is now interesting, that the convective velocity v_{conv} — i.e. the average velocity of the convective eddies — becomes very large, i.e. very close to the velocity of sound v_{sound}. If the equations of the standard mixing length theory (MLT; cf. e.g. Kippenhahn et al., 1967) are applied, even supersonic convection velocities are the result. Note, however, that the MLT is in principle not applicable to such a situation, since dissipative terms are not included. In a real star, the sound velocity may not be exceeded by convection.

However, it is clear that the ratio $r = v_{conv}/v_{sound}$ becomes close to unity in the considered convection zone. The resulting turbulent pressure P_{turb}, which can be estimated according to $P_{turb}/P = \frac{\pi}{2}\Gamma_1 r^2$, will therefore be of the order of the gas- plus radiation pressure P, since $\Gamma_1 = (\partial \ln P/\partial \ln \rho)_{ad}$ is also of order unity. In a study of a $100\,M_\odot$-ZAMS sequence, computed with standard assumptions analogue to that of Langer & El Eid (1986), Kiriakidis (1987) finds the turbulent pressure to have local maxima in the recombination regions of hydrogen and helium. Furthermore, the maximum turbulent pressure as a function of the stellar effective temperature is largets at $T_{eff} \simeq 10\,000\,K$, which coincides well with the location of the LBVs. The acceleration due to turbulent pressure, which is proportional to dP_{turb}/dr, is found to have a large maximum at the hydrogen recombination front, which lies $\sim 0.06\,M_\odot$ deep below the surface within the star in the model with $T_{eff} = 10\,000\,K$; an ammount of mass which might correspond to the shells ejected by several LBVs. Note, however, that the computations mentioned above are not selfconsistent in the sense that the turbulent pressure has not been taken into account in the models. Selfconsistent computations would be highly interesting and are under investigation in Göttingen.

6. Summary: LBVs as a stage of massive stars evolution

According to the observed luminosities of LBVs it is clear, that a lower ZAMS mass limit for LBV formation exists, which we want to designate as M_{LBV}. The possibility of an upper ZAMS mass limit for LBV formation — let us call it M_{Conti} — cannot be excluded, but since η Car is regarded as an LBV it should be $M_{Conti} > M_{ZAMS}(\eta\,\mathrm{Car})$. If M_{Conti} exists, stars with ZAMS masses above that limit evolve into WNL stars already during central hydrogen burnig (cf. Section 1).

In Section 3 we have shown the possibility of the existence of a critical ZAMS mass M_{crit} such that stars above that limit form massive LBVs which finally turn into WNL stars, while stars below it form low mass LBVs, which evolve directly into WNE stars. From the definitions above and the results of Section 3 we conclude that $M_{LBV} < M_{crit} < M_{Conti}$.

Finally, we explored the possibility of post-RSG LBVs in Section 4, and found that stars with ZAMS masses close to but above M_{LBV} might follow this scenario. (Let us call the upper mass limit for RSG formation M_{RSG}.)

Altogether we see that evolutionary connections in the upper HR diagram may be quite complex, as much more in the case where the different Wolf-Rayet subtypes are considered (cf. the "phase diagram" in Langer, 1987). As function of increasing ZAMS mass they may be listed as follows:

$8\,M_\odot < M_{ZAMS} < M_{LBV}$: O,B-star → supergiant → SN
$M_{LBV} < M_{ZAMS} < M_{RSG}$: O-star → RSG → low mass LBV → WNE (→ WC) → SN
$M_{RSG} < M_{ZAMS} < M_{crit}$: O-star → low mass LBV → WNE (→ WC) → SN
$M_{crit} < M_{ZAMS} < M_{Conti}$: O-star → high mass LBV → WNL (→ WC) → SN
$M_{Conti} < M_{ZAMS}$: O-star → WNL (→ WC) → SN

Note, however, that still much theoretical and observational work is required in order to derive values for the critical masses and to discuss other possible evolutionary scenarios.

Acknowledgement: This work has been supported in part by the Deutsche Forschungsgemeinschaft (DFG) through grants La 578/1 and Fr 325/28. The author is grateful to the organizers of the meeting for a travel grant which enabled him to participate at this colloquium.

References

Kippenhahn, R., Weigert, A., Hofmeister, E.: 1967, Meth. Comp, Phys. **7**, 129
Kiriakidis, M.: 1987, Diploma thesis, Göttingen University
Langer, N.: 1987, Astron. Astrophys. *Letters* **171**, L1
Langer, N., El Eid, M.F., Fricke, K.J.: 1985, Astron. Astrophys. **145**, 179
Langer, N., El Eid, M.F.: 1986, Astron. Astrophys. **167**, 265
Lortet, M.-C., Testor, G.: 1988, Astron. Astrophys. **194**, 11
Maeder, A.: 1981, Astron. Astrophys. **99**, 97
Maeder, A.: 1987, Astron. Astrophys. **178**, 159
Maeder, A., Meynet, G.: 1987, Astron. Astrophys. **182**, 243
Prantzos, N., Doom, C., Arnould, M., de Loore, C.: 1986, Astrophys. J. **304**, 695

DISCUSSION

Sreenivasan: I am glad to see your tracks for a 100 $M_\odot$ model. It does not seem to have been generally appreciated that small changes in the mass-loss rates (factors of 2 or 3) can change evolutionary tracks significantly. One can turn tracks around near $T_{eff} \sim 10^4$ K only with large core sizes and high mass-loss rates. Also, by increasing mass-loss rates abruptly, one can reverse the direction of a track. Precisely what rates did you use?

Langer: I used the mass-loss formula of Lamers (1981, *Ap.J.* **245**, 593), and further details of this track can be found in *Astron. Astrophys.* **167**, 265. But I would not characterize a change in mass-loss rate by a factor of 2 or 3 as small.

Maeder: You stated that a very massive star experiences nothing remarkable along its redward track. I have just the opposite opinion. During redward evolution, the models show that the luminosity becomes supercritical in the layers where the opacity peak is located; this is the sign that static models are no longer valid. I think that this is a major event, possibly related to the essence of the LBV phenomenon.

Langer: In the stellar evolution sequences that I have computed, the Eddington luminosity is never actually exceeded within the star while moving toward the Hayashi line. The density inversion in the envelope that you referred to may or may not occur in models, depending on the way convection is treated. In a real star, a density inversion may be prevented by the Rayleigh-Taylor instability. Of course something has to happen when the star crosses the H-D limit redwards, but what I meant to say is that the relevant physics is not yet included in any stellar evolution calculation.

De Groot: I cannot judge whether these oscillations along the blueward track are real or not, but I want to note that P Cygni's basic parameters place it exactly in this part of the H-R diagram for the appropriate mass. I wonder if this is true also for other LBV's; if so then this could be evidence for the reality of these oscillations in the evolutionary track.

Langer: It is of course remarkable that the thermal oscillations occur at the "right place" in the H-R diagram. However, one has to try to understand the physical processes in order to assess evidence that this is related to the LBV phenomen.

Sreenivasan: I agree with Maeder that something remarkable happens to the right of the sloping Humphreys-Davidson line, but I think that we are remarkably ignorant of precisely what that is. We just do not know the physics of what happens there. The effect of opacities on the Eddington luminosity seems clear, but how that relates to the stellar wind of supergiants in that temperature range is not clear. One cannot simply scale upwards the solar wind type of mass loss.

Langer: I agree completely; but, as you say, a proper theoretical treatment has not yet been applied in any calculation.

Moffat and Langer,
demonstrating the contrast
between observational
and theoretical approaches

OBSERVATIONAL CONNECTIONS BETWEEN LBV'S AND OTHER STARS, WITH EMPHASIS ON WOLF-RAYET STARS

A. F. J. MOFFAT, L. DRISSEN and C. ROBERT
Département de Physique, Université de Montréal
C.P. 6128, Succ. "A"
Montréal, Québec H3C 3J7
Canada

ABSTRACT. *We suggest that the LBV mechanism is an essential step to "force" massive stars ($M(ZAMS) \gtrsim 40 M_\odot$) to finally enter the Wolf-Rayet (W-R) domain in the Hertzsprung-Russel diagram (HRD). Just as massive supergiants show* <u>*increasing*</u> *variability as they* <u>*approach*</u> *the Humphreys-Davidson (H-D)instability limit (horizontally in the HRD diagram), so the W-R stars show* <u>*decreasing*</u> *variability as they* <u>*recede from*</u> *the H-D limit (at first horizontally into the WNL domain, then, with their high mass loss rates, plunging irreversably downwards as ever hotter, smaller and fainter, strong-line W-R stars). Among the W-R stars, the luminous WNL subtypes (especially WN8) are the most variable, probably as a consequence of blob ejection in the wind. The underlying mechanism which triggers this ejection is possibly related to wind instabilities and may thus be quite different from the source of variability in luminous supergiants or LBV's in quiescence, where photospheric effects dominate.*

1. Relatives of LBV's

The most obvious immediate pre- and post-LBV objects must both be located below and to the left of the H-D limit, by default. Pre-LBV stars must be situated between the H-burning zero-age main-sequence (ZAMS) and the H-D limit; they obviously include the more luminous supergiants (SG) that have not yet shown any obvious LBV activity beyond quiescent microvariability (see below). They evolve essentially horizontally in the HRD, both before and after encountering the H-D limit (cf. Humphreys 1989).

The exact nature of post-LBV stars is more controversial. However, it does seem likely that after several major LBV eruptions, one should look among luminous stars with enhanced CNO-cycle abundances. The W-R stars are the most obvious candidates, in particular the most luminous ones, of type WNL (late-type W-R stars of subtypes WN6-9). After the WNL stage, W-R evolution progresses downwards in the HRD, as the stars lose more mass, shrink in size and head for the He-ZAMS (see below).

This brief review will be restricted to the blue relatives of LBV's, with emphasis on the W-R stars. Red SG's and the exotic OH/IR sources of high luminosity will not be considered here. The observational point of view is stressed, with minimal reference to theory.

K. Davidson et al. (eds.), Physics of Luminous Blue Variables, 229–240.

2. Variability of Pre-LBV Supergiants

A great deal of precise photometric data for SG's in the upper HRD has been accumulated and summarized by Maeder (1980) and later by Baade (1988). Figure 1 shows a résumé of Maeder's tabulated variability amplitudes, A, converted to standard deviation, σ, by assuming A$\simeq 3\sigma$. Note that the level of variability increases towards higher luminosity for all spectral classes. (Note also that evolution of SG's occurs at $\sim$ constant bolometric luminosity, so that both the absolute visual magnitude M_v and the spectral type change during this evolution.) The amplitudes for the most luminous SG's resemble the microvariations observed in LBV's during quiescence (cf. Lamers 1987). Some of these SG's may have already exhibited LBV-like eruptions.

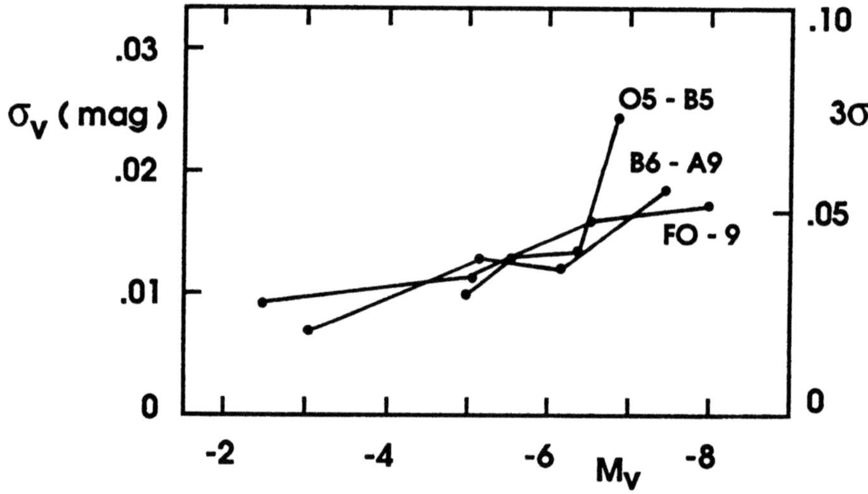

Figure 1. Variability of supergiants as a function of absolute magnitude (extracted from Maeder 1980)

3. Variability of Post-LBV, Wolf-Rayet Stars

W-R stars, with their intense emission lines and hot winds, occur in two main sequences: that of nitrogen and that of carbon (with an extension to oxygen for the latter). As for ordinary stars, they are binned into subclasses, in which higher values mean later (i.e. cooler) types. Calibration of W-R stars in Galactic open clusters and in the Large Magellanic Cloud, yields a *very tight correlation between M_v and subclass* (Lundström and Stenholm 1984, updated by van der Hucht et al. 1988; Breysacher 1986). With mean bolometric correction $M_{bol} - M_v \simeq -4.5 \pm 0.2$ for most W-R subclasses (Smith and Maeder 1988), this implies that the total luminosity also increases systematically from earlier to later WN or WC subtypes (e.g. $M_{bol} \simeq -11$ for WN7 to -8 for WN4 or $M_{bol} \simeq -9$ for WC8 to -8 for WC5).

W-R stars also show remarkable similarity in variability behaviour compared to

SG's. Figure 2 (adapted from Lamontagne and Moffat 1987) shows that both WN and WC stars exhibit greater variability for higher luminosity by an amount which is similar to the SG's at the same M_v. Variability of W-R stars in polarization shows similar behaviour (i.e. greater amplitude at higher luminosity) and is likely intimately related to the photometric behaviour (cf. Robert et al 1989a,b).

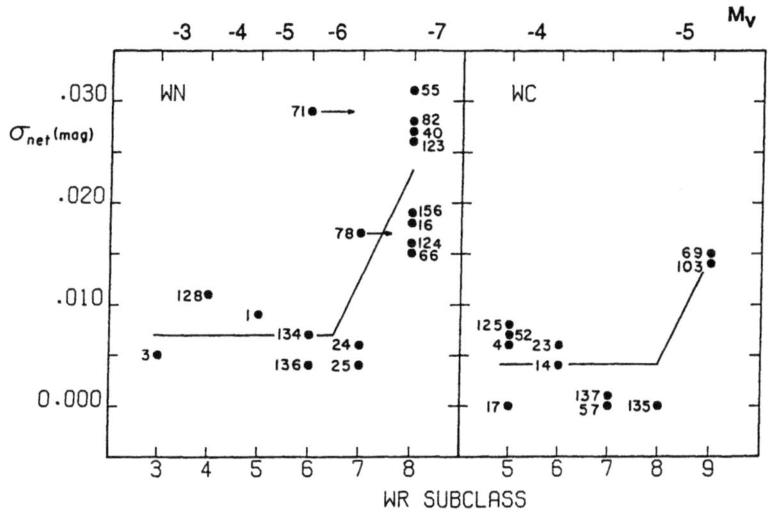

Figure 2. Variability of W-R stars as a function of subtype and absolute magnitude (from Lamontagne and Moffat 1987)

The mechanism that is most likely to cause the variability will be discussed in section 4. First, we look at the possibility of evolution among W-R stars of different subclass.

4. The W-R Evolutionary Scenario

Does each W-R star exist in its own right, independent of other W-R subtypes, or is there a pattern in the evolution, e.g. from one subtype to another? The answer to this question is still rather controversial but is fundamental and very relevant in establishing a connection or not with LBV's (and in the other direction, with supernovae!).

The approach we adopt here is mainly empirical, guided by a simple principle: we look for continuity of various parameters as a function of the most easily obtained quantity, the spectral subclass. We consider the following parameters:

(1) Mass: Although binary-estimated masses do scatter considerably even for a given subtype (the source for this may be intrinsic or spuriously related to the difficulty in eliminating the factor sin^3i), there does exist a tight relation between mass ratio, q=M(WR)/M(O), and subclass. Cooler W-R subtypes tend to have progressively higher ratios (Moffat 1981, 1982; Moffat et al. 1986; Moffat,

Niemela and Marraco 1989=MNM). In particular, the WN6-7 stars have values of q close to unity, while all other observed subtypes (WC or WNE) generally have q $\lesssim$ 0.5, with q$\sim$ 0.2 for the hottest W-R stars. Since binary mass transfer is probably not important in massive systems because of the fast winds (cf. MNM; Sybesma 1980), the mass ratio in W-R + O systems essentially gives a normalized measure of the amount of mass shed by the W-R star, regardless of the original mass. Note that the original mass ratio must have been $\gtrsim$1 (values differing greatly from unity are excluded by observations of O+O binaries: Garmany, Conti and Massey 1980). The fact that the spectral types of the O-companions in W-R binaries show no correlation with W-R spectral subclass implies that all W-R stars can evolve from any massive star (normally $\gtrsim 40M_\odot$; cf. Humphreys, Nichols and Massey 1985).

(2) Luminosity: We already saw in section 2 that the later the W-R subtype, the brighter the star. Especially the WNL stars are the brightest.

(3) Spatial distribution: While WNL stars generally occur in any part of the Galactic disk where active star formation has taken place, other W-R subtypes show a fairly clear frequency dependence on galactocentric distance (van der Hucht et al. 1988).

(4) Line profiles: While WC stars show a clear trend of very broad emission lines in the hottest subtypes to narrow lines in the coolest, the correlation exists too but is noisier for the WN stars. One should probably distinguish broad- from narrow-line WN stars of the same subtype (cf. Hiltner and Schild 1966; Walborn 1974).

(5) Ages: On the basis of the morphology, sizes and kinematics of the ring nebulae surrounding a number of W-R stars, Chu, Treffers and Kwitter (1983) demonstrated that WC stars are generally older than WN stars and WNE are older than WNL.

(6) Abundances: Among the WN stars, it has been known for some time that the WNL stars often contain hydrogen in their winds, while WNE stars are H-poor (e.g. Smith 1973). Among the WC stars, recent evidence also yields a progression of higher C/He from cool to hot WC (Smith and Hummer 1988; but cf. Torres 1988 and de Freitas Pacheco and Machado 1988).

(7) Variability: As noted in section 2, there is a clear trend in continuum light and polarization variability such that cooler (especially WN8) stars reveal larger amplitudes.

(8) Core radii: Although not directly visible in most W-R stars, the stellar radii (where the radial wind velocity is near zero) have been extracted from eclipsing binaries in a few cases. The trend is that cooler subtypes tend to be larger (e.g. $11R_\odot$ for the WN7 star CQ Cep (Leung, Moffat and Seggewiss 1983) and $\sim 3R_\odot$ for the WN5 star in V444 Cyg (Cherepashchuk, Eaton and Khaliullin 1984)). It is assumed that the remaining subtypes follow a continuous trend in radius, as does the luminosity or mass ratio.

All of the above trends, together with the high mass loss rates, favour subtype evolution from cooler to hotter subtypes within each sequence. Our interpretation is as follows: all W-R stars start as WNL, the most massive, luminous and least evolved of all subtypes. As the strong wind peels off the outer layers, the surface abundance gets more exotic with time and the core radius shrinks. Transfer from

WN to WC, basically a surface phenomenon, will occur at a given WN-subtype which depends on the initial metallicity Z (Moffat 1981, 1982). In short, extreme mass loss forces W-R stars to evolve downwards (fainter) and to the left (hotter) in the HRD, more or less toward the He-ZAMS. This is no surprise, since W-R stars are He-burning, with H-rich surface impurities that diminish with time. With typical mass loss rates $\dot{M} \simeq 4 \times 10^{-5} M_{\odot} yr^{-1}$ (e.g. Schmutz and Hamann 1986), the total mass lost in the W-R phase (mean lifetime t $\sim 5 \times 10^5$ yr: Maeder and Meynet 1987) is $\Delta M = \dot{M}t \sim 20 M_{\odot}$ (variations of $\pm$ a factor 2 are possible), which is entirely compatible with the above interpretation. For example, a W-R star in a binary, going from q=1.0 to 0.2, will lose $\Delta q \times M_o \approx 24 M_{\odot}$ for an average $M_o = 30 M_{\odot}$ O-type companion.

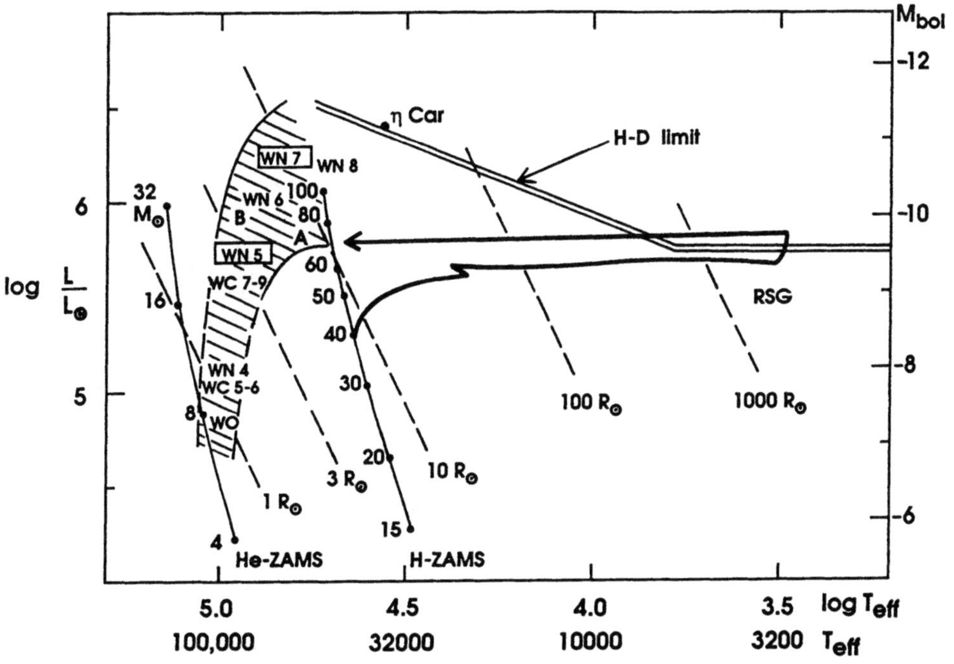

Figure 3. Average positions of W-R stars of different subclass in the HR diagram. The positions of the framed subtypes are constrained by direct eclipse observations of stars of the corresponding type. The hatched, funnel-shaped region allows for the observed spread in M_{bol}. The evolutionary track of a $40 M_{\odot}$ ZAMS star from Maeder and Meynet (1987) is shown for comparison.

Based on the above arguments (especially that hotter W-R stars tend to be smaller in core radii and less luminous), Figure 3 shows our best empirical interpretation of

W-R evolution in the HRD. Note the rough distinction between narrow-line (A) and broad-line(B) WN stars. Both are supposed to have similar L (thus also M), but the WN-B stars have higher v_∞. This makes their radii smaller, assuming the correlation between v_∞ and escape velocity of Abbott (1982) for hot winds: $v_{esc} \simeq v_\infty/3 = \sqrt{2GM(1-\Gamma)/R}$, also applies to W-R stars (Γ is the ratio of radiation to gravitation force). With smaller radii but equal L (and Γ?), the WN-B stars will thus be hotter than WN-A stars. Judging from their approximate, deduced position in Figure 3, WN-B stars probably evolved from more massive progenitors than did WN-A stars.

Figure 3 shows that all W-R stars are extracted ,initially as WNL, out of the region of the HRD where most LBV's are found (M(ZAMS) $\gtrsim 40M_\odot$; cf. Humphreys 1989). Independant information reveals that the progenitor masses of W-R stars must also have been $\gtrsim 40M_\odot$ (cf. Humphreys, Nichols and Massey 1985). After WNL, they evolve down a progressively narrower, funnel-shaped path toward the hotter He-ZAMS. Of course, this scenario is rough and needs some fine-tuning, especially at the faint end. Note that the positions of the W-R subtypes in Figure 3 refer to the cores and not necessarily to $\tau \sim 1$, which often occurs in the winds.

Figure 3 also implies that, in order for a massive star to reach the W-R domain, it first has to encounter the H-D instability barrier (as an LBV), causing it effectively to "bounce" back eventually to the left of the HRD. Once the W-R domain is reached, the star will not return to the H-D limit. Stars well below $\sim 40M_\odot$ do not generally encounter this limit and probably do not become W-R stars either, in general. Stars just below $40M_\odot$ may become "red LBV's" (e.g. OH/IR variable red SG's) and then W-R. Note that there exist several apparently genuine LBV's of relatively low luminosity (e.g. R66, AE And, M31/var 15: Lamers 1987) with deduced masses in the range $\sim 20-30M_\odot$; if the data for these stars are reliable, then their evolutionary future is not clear.

5. Mechanism of W-R Star Variability: Relevance to LBV's

Among W-R stars, those of subtype WN8 (and WN9?) appear to be the closest to LBV's with respect to their high degree of variability, coupled with their relatively narrow emission lines and their high luminosity (although not higher than WN7 stars, which, curiously, are more stable). WN8 stars are also closer to LBV's in the HRD (cf. Figure 3). However, WN8-9 stars show some unusual characteristics. For example, none of 9 Galactic and LMC WN8-9 stars monitored for RV variations show any sign of being a W-R + O binary (Moffat 1989). This contrasts starkly with the 26 monitored WN6-7 stars which are 58% W-R + O. Also, WN8-9 stars tend to avoid clusters, unlike WN6-7 stars, and some appear to be runaways. Possibly WN8-9 stars are slightly lower mass WNL stars associated with lower mass LBV's (cf. Lortet and Testor 1988) that occur more towards the red part of the HRD. On the other hand, it is probably no coïncidence that the massive LBV η Car occurs in a young cluster together with (more massive) WN7 stars.

In order to probe the nature of the variations in W-R stars more deeply, it is essential to carry out high resolution, high S/N, time-resolved ($\lesssim$ 30min) spectroscopy. This allows the additional advantage over continuum studies of being able to explore the velocity domain (hence also the spatial domain if both domains are correlated, say, by a global wind velocity law). Intense monitoring of a few bright W-R stars has been carried out (Moffat et al 1988; Robert, Drissen and Moffat 1989a; McCandliss and Bohannan 1989; Koenigsberger and Massey 1989), but only for the subtypes WN5-6, WC7-8 and for a limited selection of spectral lines. In particular, the first two references reveal strong evidence for random, radial acceleration of blobs which are carried along with the general wind. The study of a symmetric, unblended emission line in this case allows one to see optically thin, emitting blobs being ejected in all directions.

What triggers the blobs? There are two main contenders: (1) non-radial pulsations (Vreux 1985) or (2) stochastic radiative wind instabilities (Owocki, Castor and Rybicki 1989). The former mechanism has been shown to be unlikely to occur in W-R stars (Matthews and Beech 1987; Cox and Cahn 1988; however, cf. Noels and Scuflaire 1987). Direct observational confirmation will be difficult, however, due to the filtering action of the wind. The latter mechanism seems very attractive but has one problem: the theory predicts that fast winds should be less stable, while the observations of W-R stars seem to indicate the opposite, namely that inhomogeneities appear to form more readily and/or have more time to grow in cooler, slower winds.

Beyond the microvariation amplitudes of WN8 stars and LBV's in quiescence, the similarity between W-R stars and LBV's ends. First, the timescales for microvariations are quite different, being ~ 1 d in WN8 stars and ≥ 10 d in LBV's. Also, WN8 stars are not known to show outbursts. Their variability is a wind phenomenon, whose relation to the underlying star, if any, is unknown. In LBV's, the variability is mainly a photospheric effect at the stellar surface. The same difference applies between W-R stars and the SG progenitors of LBV's.

6. Conclusions

In a sense, LBV's represent the peak of a mountain. Stars climbing up to the top as SG's become more and more variable. Stars descending from the peak into the W-R domain become less and less variable. Once the star enters the W-R zone, it has reached the point of no return.

The overall scenario for stars of initial mass above $\sim 40 M_\odot$ may be as follows:

→ Supergiant (looping horizontally in the HRD)
→ LBV (several times, depending on the mass loss and not just in the blue part of the HRD-cf. Humphreys 1989)
→ WNL
↗ (Z low) WNE - - → } WCE - - → WO
↘ (Z high) WCL ⟶ }
→ Supernova of type Ib, probably as WNE, WCE or WO subtypes (cf. Ensmann and Woosley 1988, whose models of SN Ib light curves show a preference for W-R progenitors of mass $\sim 4 - 7M_\odot$, corresponding to the hotter W-R subclasses).

Some stars in the range $30 - 40 M_\odot$ might become LBV-like red SG's and later proceed to the W-R domain. An analogy of the above scenario ($M_i \gtrsim 40 M_\odot$)can be drawn with low mass stars ($M_i \lesssim 8 M_\odot$), which rise far up the asymptotic branch near to the position of the highly variable OH/IR stars, eject planetary nebulae, pass quickly to the left in the HRD and descend down the Harman-Seaton cooling sequence to the white dwarf domain.

References

Abbott, D. C. 1982, *Ap. J.*, **259**, 282.
Baade, D. 1988, in "O stars and Wolf-Rayet stars", eds. P. S. Conti and A. B. Underhill, NASA SP-497, p. 137.
Breysacher, J. 1986, *Astron. Ap.*, **160**, 185.
Cherepashchuk, A. M., Eaton, J. A. and Khaliullin, Kh. F. 1984, *Ap. J.*, **281**, 774.
Chu, Y. H., Treffers, R. R. and Kwitter, K. B. 1983, *Ap. J. Suppl.*, **53**, 937.
Cox, A. N. and Cahn, J. H. 1988, *Ap. J.*, **326**, 804.
Ensman, L. M. and Woosley, S. E. 1988, *Ap. J.*, **333**, 754.
de Freitas Pacheco, J. A. and Machado, M. A. 1988, *A. J.*, **96**, 365.
Garmany, C. D., Conti, P. S. and Massey, P. 1980, *Ap. J.*, **242**, 1063.
Hiltner, W. A. and Schild, R. E. 1966, *Ap. J.*, **143**, 770.
van der Hucht, K. A., Hidayat, B., Admiranto, A. G., Supelli, K. R. and Doom, C. 1988, *Astron. Ap.*, **199**, 217.
Humphreys, R. M. 1989, these proceedings.
Humphreys, R. M., Nichols, M. and Massey, P. 1985, *A. J.*, **90**, 101.
Koenigsberger, G. and Massey, P. 1989, these proceedings.
Lamers, H. J. G. L. M. 1987, in "Instabilities in Luminous Early Type Stars", eds. H. J. G. L. M. Lamers and C. W. H. de Loore, (Reidel:Dordrecht), p. 99.
Lamontagne, R. and Moffat, A. F. J. 1987, *A. J.*, **94**, 1008.
Leung, K. C., Moffat, A. F. J. and Seggewiss, W. 1983, *Ap. J.*, **265**, 961.
Lortet, M.-C. and Testor, G. 1988, *Astron. Ap.*, **194**, 11.
Lundström, I. and Stenholm, B. 1984, *Astron. Ap. Suppl. Ser.*, **58**, 163.
Maeder, A. 1980, *Astron. Ap.*, **90**, 311.
Maeder, A. and Meynet, G. 1987, *Astron. Ap.*, **182**, 243.
Matthews, J. M. and Beech, M. 1987, *Ap. J.*, **313**, L25.
Mc Candliss, S. R. and Bohannan, B. 1989, these proceedings.
Moffat, A. F. J. 1981, in "Effects of Mass Loss in Stellar Evolution", IAU Coll. No. 59, eds. C. Chiosi and R. Stalio, (Reidel:Dordrecht), p. 301.
Moffat, A. F. J. 1982, in "Wolf-Rayet Stars: Observations, Physics, Evolution", IAU Symp. No. 99, eds. C. W. H. de Loore and A. J. Willis (Reidel:Dordrecht), p. 515.
Moffat, A. F. J. 1989, in preparation.
Moffat, A. F. J. , Vogt, N., Paquin, G., Lamontagne, R. and Barrera, L. H. 1986, *A. J.*, **91**, 1386.
Moffat, A. F. J. , Drissen, L., Lamontagne, R. and Robert, C. 1988, *Ap. J.*, **334**, 1038.
Moffat, A. F. J. , Niemela, V. S. and Marraco, H. 1989, *Ap. J.*, submitted.
Noels, A. and Scuflaire, R. 1987, in "Instabilities in Luminous Early Type Stars", eds. H. J. G. L. M Lamers and C. W. H. de Loore, (Reidel:Dordrecht), p. 213.
Owocki, S. P., Castor, J. I. and Rybicki, G. B. 1989, these proceedings.
Robert, C., Drissen, L. and Moffat, A. F. J. 1989a, these proceedings.
Robert, C., Moffat, A. F. J. , Bastien, P., Drissen, L. and St.-Louis, N. 1989b, *Ap. J.*, submitted.
Schmutz, W. and Hamann, W.-R. 1986, *Astron. Ap.*, **166**, L11.
Smith, L. F. 1973, in "Wolf-Rayet and High Temperature Stars", IAU Symp. No.

49, eds. M. K. V. Bappu and J. Sahade, (Reidel:Dordrecht), p. 15.
Smith, L. F. and Hummer, D. G. 1988, *M. N. R. A. S.*, **230**, 511.
Smith, L. F. and Maeder, A. 1988, preprint.
Sybesma, C. H. B. 1986, *Astron. Ap.*, **168**, 147.
Torres, A. V. 1988, *Ap. J.*, **325**, 759.
Vreux, J.-M. 1985, *P. A. S. P.*, **97**, 274.
Walborn, N. 1974, *Ap. J.*, **189**, 269.

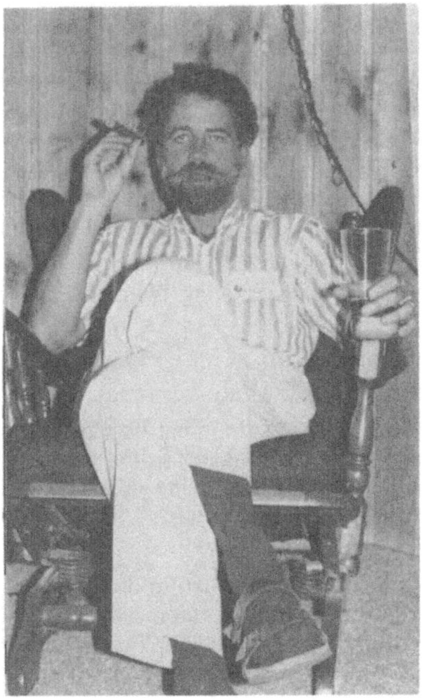

Tony Moffat

DISCUSSION

Koenigsberger: The winds of WNL stars are said to be less optically thick than those of WNE and WC stars. Is it possible that WNE and WC stars are intrinsically just as variable as WNL's and we just don't see it?

Moffat: Yes, this might be so. However, in the context of radiative instabilities à la Rybicki/Owocki, one would expect the instabilities (= blobs?) to grow as they propagate outwards. Since we are seeing the winds farther out for WC/WNE stars than for WNL stars (i.e., the $\tau_{es} \sim 1$ levels), one might expect the amplitude to be even larger in WC/WNE stars, contrary to the observations.

Sreenivasan: I suspect that one cannot make a direct connection between the blobs that you observe and the presence/absence of radial/non-radial pulsations, because the regions that you see are so far out in the wind. Radial pulsations in massive stars may be indicated in the work of Maeder and of Cox and Cahn. The latter authors did not find non-radial pulsations in WR stars, but Kirbiyak and others claimed they did. None of these workers studied the effects of rotation. Non-radial pulsations could be present in WR stars because of differential rotation. Lee and Saoi found non-radial pulsations in 10 $M_{\odot}$ models. The blobs are presumably due to effects such as those considered by Owocki *et al.* in the wind. Non-radial pulsations could affect mass-loss rates but cannot by themselves cause the formation of blobs.

Moffat: Indeed, the connection has yet to be made. However, if rotation or pulsation do play a role in WR stars, this is certainly not obvious judging from the random nature of the observed wind variations.

Friedjung: How close are Wolf-Rayet stars now believed to be to the Eddington limit? Can you have an optically thick wind when such a star is not very close to it?

Moffat: I think that most WR researchers agree that WR stars must be close to their their Eddington limits. The question is what opacity κ to take for the base of the wind to yield $L_{Edd} = 4\pi GcM/\kappa < L$? Certainly electron scattering opacity alone is not sufficient, since the gas is not fully ionized (*cf.* arguments for LBV's given by Lamers & Fitzpatrick 1988, *Astrophys. J.* **324**, 279).

Conti: The "funnel" analogy for the end phases of massive star evolution appears to be a good one, although I am not convinced that a unique WR subtype can be identified at every point "downstream." I would like to take issue with your inferred relationship between variability and absolute magnitude M_V. Clearly there is a correlation with spectral subtype in both WN and WC stars, as you and your coworkers have so ably pointed out. But most of the purported relationship to M_V depends on the relatively bright M_V assigned to WN8 stars. The absence of WN8 stars in H II regions implies that they do not originate from the most massive progenitors, nor from the most luminous stars. These WN8 objects might well come from certain LBV's, whose spatial distributions are similar. To demonstrate a dependence of variability on M_V you need to study stars of similar type but with different M_V's to see if the variability depends on brightness, independent of radius.

Schmutz: The values you used for the basic stellar parameters of the WR stars are more or less guesses. However, reliable parameters are now determined for 30 WR stars (Schmutz *et al.*, *Astron. Astrophys.* in press). The important point of these new results is that for about half of the objects there is no difficulty in determining the temperatures of the stars themselves -- their continuum formation layers are not extended. For another 1/4 of the WR stars the correction between the $\tau = 2/3$ level and the stellar surface is not large. For the remaining 1/4 of the stars there is a large uncertainty; but for the majority of the WR stars there is no longer a large uncertainty in their basic stellar parameters.

Moffat: The stellar parameters that we used are based on observations, *e.g.*, L from WR stars in clusters, masses from binaries, and radii from the few known eclipsing binaries (then $L = 4\pi R^2 T_{eff}$ gives T_{eff}). Where there are gaps I have drawn from arguments of continuity with spectral subclass and Occam's razor. In any case, your theoretical results agree fairly well with the observed values.

Van der Hucht: One of your favorite WR binaries, HD 193793, has an O4 V secondary. Maeder's evolutionary models cannot make a WC7 with the same age as an O4 V star. It is premature to rule out mass transfer: binary component evolution may be different from single-star evolution. This jeopardizes your scenario for subtype evolution.

Moffat: HD 193793 has the longest known period of any known WR + O binary (period 7.9 years!). Even with an eccentricity ~ 0.7, it is difficult to see how even case B mass transfer could have taken place. Even if it did, it should have been highly non-conservative as a result of the extremely hot, dense, fast stellar winds (either O+O, BSG+O, RSG+O or WR+O). Perhaps our understanding of stellar evolution related to spectral types is deficient! More generally, if WR stars in binaries do behave differently from single WR stars, this is not empirically obvious on the basis of spatial and spectral subtype distributions.

Vanbeveren: If observations of WR binaries reveal secondaries with spectral classes as early as O7, O6, O5..., then one needs mass transfer in order to explain this (both components must have the same age).

Moffat: For O-supergiants I would say there is less of a problem. For main-sequence O-stars, I agree that there appears to be a problem with the ages. However, if mass transfer does take place, then it becomes difficult to explain eccentric WR+O binaries. Also, I find it difficult to believe that significant mass transfer took place in the 7.9-year WC7+O4V binary HD 193793, with its large separation and high terminal wind speeds. It would be nice to see some hydrodynamical calculations. Finally, I wonder how good our knowledge is of the spectral-type evolution of the most massive stars.

Gallagher: In considering the role of binary evolution, you also have to include the possible effects of embedded binary cores during the pre-WR evolutionary phases. For example, wouldn't a system like CQ Cep almost certainly have had a common envelope during its supergiant and LBV evolutionary phases? If so, this would have influenced its observable properties, even within your assumption of minimal mass tranfer within the binary.

Moffat: CQ Cep is a poor example (SB1) but nevertheless the only known eclipsing system involving a WN7 star. I assume that **how** it became WN7 is less important in this context than that it yields an estimate of the radius of a WN7 star (11 $R_\odot$ according to Levy et al. 1983). The WN7 phase is only the beginning, when mass loss is relatively minor compared to subsequent WR phases (WNE and/or WC), in which a common envelope in the sense that you mentioned is not very relevant.

Willis: Returning to the variability aspects: Have you any estimate of the mass of an individual blob, and the total mass ejection rate involved in the variability?

Moffat: On the basis of the power contained in the standard deviation profile for pure emission lines, we can say that typically a few percent of the wind is in the form of visible propagating blobs. This fraction appears to vary with subclass (it's higher for cooler stars) and ionization potential (lower-ionization lines are generally more variable). However, it may be that the whole wind consists of discrete blobs, whose size-spectrum we are only sampling at the massive end. At any given time there are only a few of the largest blobs; thus, to find the mass in the blobs, you can take your favorite wind model and convert the observed fraction into mass.

Van der Hucht: Do you think it is realistic to associate blob ejection which occurs at a few R_* with the clumpy structure of RCW 58, whose radius is a few parsecs?

Moffat: Since the central star HD 96548 (WN8) ejected the material that we see in RCW 58 (*cf.* Chu *et al.* 1983, *Astrophys.J.Suppl.* **53**, 937), it is tempting to make some kind of association between the clumpy structure of RCW 58 and one of the intrinsically most variable WR stars in the sky. Possibly the present clumps are related to very large blobs ejected in the past. Note that older WR ring nebulae are generally much smoother after interaction with the interstellar medium.

Langer: I have a question concerning the linear decrease in luminosity for WN stars from late to early spectral types that you reported. Do you think one could conclude that the subluminous LBV's (*e.g.* R 71) must evolve directly into early-WN stars, skipping the WNL phase?

Moffat: As noted by Humphreys (1987, Lunteren meeting, p. 10), most LBV's are brighter than $M_{bol} \sim -10$. The few fainter examples presumably originated as stars with lower initial masses (40-50 $M_{\odot}$). Such stars can still become WR stars first as WNL's, at the lower edge of my funnel entrance. An observational bridge of continuity is lacking for such LBV's to go directly to the very broad-line WNE stars. [Note also that there are discrepancies in the literature and problems with the luminosities: *e.g.*, Lamers in the Lunteren meeting, p. 101, gave $M_{bol} = -10.6$ for R 71. Leitherer (these proceedings) gives a value of -8.8.]

Hillier: Firstly, I believe that many of the conclusions drawn by Tony about WR stars were highly optimistic. Secondly, a lot of high-quality observations concerning variability have appeared at this colloquium. I would, however, like to see more work attempting to correlate the effects of "ejected blobs" on different spectral lines. In particular, the work of Hamman & Schutz, and by myself, indicates that the He II lines should be affected before the He I lines.

Moffat: Sometimes one has to smooth rough corners in order to assess the global scene (forest *vs.* trees). Indeed we (and others) are very aware of the stratification in WR winds, such that higher ionization lines formed generally at smaller radii in the wind should be affected first by blob ejection. But they should also show less variability as they occur earlier in the growth phase, if the Rybicki/Owocki scenario is correct. If telescope allocation committees had been more generous, we might have been able to answer this question observationally by now!

FIRST PANEL DISCUSSION

[On the last day of this meeting two general discussions were held, each beginning with comments by members of a discussion panel. The first panel and discussion concentrated on reviewing the theoretical problem of the LBV eruption mechanism. Members of this panel were I. Appenzeller, C. de Jager, D.G. Hummer, F.D. Kahn, and H.J.G.L. M. Lamers, with K. Davidson acting as moderator or chairman. Remarks by panel members and other participants are given here in condensed form, somewhat reworded for brevity and clarity. It will be noted that the most generally accepted consensus here was that the basic cause of LBV instability is not yet known!]

Appenzeller: Having the privilege of opening this discussion, I shall start with a brief overview and a personal assessment of the different physical processes that have been discussed during this meeting as possible causes of the peculiar properties and dramatic variabilities of the LBV's. To follow some logical order, stellar interior mechanisms will be dealt with first, followed by atmospheric and circumstellar phenomena.

Among the interior mechanisms, we first have the suggestion of multi-mode pulsations induced and powered by interior differential rotation. Sreenivasan has outlined the details of this scenario in his contribution to this volume. They are complex and depend upon the interior structure. Therefore it is difficult to derive a reliable observational signature that could be used to prove the presence of this mechanism.

Another potential deep-interior process involves thermal oscillations as described by Norbert Langer for hydrogen shell burning models. Of course this suggestion depends critically on the preceding evolutionary phases and a star's ability to reach the stage in question.

As Maeder pointed out, a density inversion observed in computed stellar models near the red boundary of the LBV region in the H-R diagram may indicate the onset of violent events at an earlier evolutionary stage. Because of Rayleigh-Taylor instabilities it is very unlikely that density inversions will occur in real stars; but it is not clear how a real star will react to a density gradient trying to change its sign. Strongly enhanced mass loss as observed in LBV eruptions could well result, but a milder reaction (*e.g.*, modified convection properties) also seems possible. Obviously this phenomenon deserves further studies.

From the discussions at this meeting it seems that radiation-pressure-induced mass-loss instabilities are still the most popular concept for explaining LBV eruptions. The new, sophisticated model computations reported by Kudritzki *et al.*, Leitherer *et al.*, and by Owocki seem to provide further support for this type of mechanism. However, only time-dependent computations could actually prove that this mechanism is indeed the cause of the observed variations.

As described in de Jager's review talk, turbulent pressure effects must be present in practically all extremely luminous stars. However, at least in the hotter LBV's (and during minimum phases) radiation pressure probably dominates and turbulent or wave pressure probably can become important only in conjunction with radiation effects.

K. Davidson et al. (eds.), Physics of Luminous Blue Variables, 241–248.

Another mechanism that can explain certain observed properties of the LBV's is rapid mass exchange in massive close binary systems (see Gallagher's excellent review). But, in spite of many high-quality observational results which have been accumulated for the known S Dor stars, in no case has reliable evidence for a close binary nature been detected. If a typical LBV is a binary system, the components must be hidden in common envelopes. But the minimum-state spectra of objects such as R71 are hardly compatible with this assumption.

Summarizing, I personally feel that radiation pressure instabilities are the most likely cause of the LBV variability. However, additional work will be needed to identify the physical mechanism leading to the LBV eruptions without ambiguity.

De Jager: First a few words on very large explosions à la Eta Carinae. That 'explosion' lasted for about ten years and caused a mass ejection of the order of two solar masses. Since such an ejection involves about half the stellar volume, the disturbance leading to it should have been seated somewhere well outside the stellar center but well below the surface. It is remarkable that we have a theory of ordinary novae (atmospheric explosions involving about 0.0001 solar mass) and of supernovae (core explosions involving nearly the whole stellar mass) but not of these slow intermediate-scale events.

With regard to the regular LBV ejection events my ideas are the following. We know that the stars in the upper part of the H-R diagram are pulsating with periods of ~ 4 to ~ 50 days in the LBV region (depending on spectral type and luminosity). The pulsations may be an ordered system of non-radial pulsations or they may have a more stochastic character. The pulsations are gravity waves. An important point is that LBV's differ from other stars in the same part of the H-R diagram in that their pulsations do not show high-frequency components, and have larger amplitudes (judging from the photometric variations). We combine this observation with the fact that the upper limit of the LBV area in the H-R diagram coincides with the line where the stellar wind velocity equals that of sound. Only a relatively small additional (pulsational) outward acceleration of part of the stellar surface would then be sufficient to shift the locus of $(v/\textit{sound speed}) = 1$ inward to fairly dense photospheric layers, thus dramatically increasing the local rate of mass loss. In my talk at this meeting I gave examples. Therefore I think that the stellar non-radial (or stochastic) pulsations are responsible for the mass ejections. If the pulsations are really stochastic, then larger amplitudes may occasionally occur, causing larger ejections.

Hummer: At the beginning of this meeting we heard a quotation from John Herschel relative to the LBV's which included the phrases "fitfully variable to an astonishing extent" and "having no regularity of progression." I would like to speculate on a mechanism that may have a bearing on these phenomena.

Both the Munich and the JILA groups have reported here the operation of what might be called the 'ionization instability' in LBV's. A slight decrease in the degree of ionization of hydrogen in response to some small change in one or more parameters leads to an increase in the opacity in the Lyman continuum, which then causes further recombination and yet further increases the optical thickness, until the hydrogen is essentially neutral (*cf.* the Strömgren sphere). As the Lyman continuum becomes opaque, the dominant lines driving the stellar wind switch from Fe III to Fe II and the mass loss rate increases. In the particular case of P Cyg, the Munich group showed that varying the inner radius of the atmosphere by as little as 1.5% triggered this mechanism.

This instability is an example of a *nonlinear amplifier,* in which the output remains small until the input reaches some threshold value, beyond which the output suddenly increases. If the output above threshold is roughly constant, we speak of a "bistable" device.

What is the 'input' in the case of P Cyg? I suggest that it could be fluctuations of the radius of the base of the photosphere caused by non-radial pulsations. Although theory at present neither predicts nor forbids non-radial pulsations for LBV's, we do know that they flicker on the level of a tenth of a magnitude. Small irregular variations in radius at any point of the stellar surface would have little effect so long as their amplitude is smaller than some threshold value, which is determined by the stellar parameters. If the star is unstable to non-radial pulsations in many modes, then at a given point on the surface the various modes would interfere destructively and nothing would happen. But if occasionally they interfered constructively, the threshold amplitude could be surpassed and a major change in the local atmospheric structure would occur.

Two consequences are implied. First, the characteristic times between outbursts would be much longer than those of the pulsations. Secondly, the size of the ouburst would scale with the area of coherent fluctuations. Thus very large outbursts, covering much or all of the surface, could be rare compared to smaller ones. Localized outbursts would give rise to the formation and ejection of clumps of matter.

Kahn: It has been remarked that a large outburst by a star like P Cyg releases an amount of radiant energy of order 10^{49} ergs (and a similar amount in mechanical energy), and might therefore be regarded as being in some way analogous to a supernova explosion in which the prompt release of radiant energy is rather similar.

Here I want to consider this view a little more deeply, because there are also very obvious differences between supernova explosions, on the one hand, and big LBV eruptions on the other. The total energy release in a Type II SN is of order 10^{53} ergs, including neutrinos, and of order 10^{51} ergs including only baryonic material. Simply making a comparison between the energy released in the optical and UV near the time of the event therefore gives a misleading impression: the major release of energy in a supernova, apart from the neutrinos, is in the form of a violent explosion of material with typical speeds of thousands of km/s. The eruptions of LBV's do not have a comparable counterpart.

There is another big difference between the two phenomena. In the case of a supernova the release of energy takes place essentially in a fraction of a millisec deep inside the star and the effects reach the surface within a few hours. In the case of an LBV the violent eruptions produces a continuing disturbance which lasts for years, and it is therefore much harder to regard it as a single event, since the subsequent response goes on for such a long time.

Nevertheless 10^{49} ergs is a great deal of energy, even for a massive star, almost of the order of its binding energy. In its present state the star radiates this amount in about a century or so. Whatever was the unknown process that led to this event, it must have taken place deep down inside the star. The present manifestations of LBV-dom are of course much milder, and consist of a massive wind driven off at modest speed together with irregular fluctuations in photospheric radius. To me the simplest view is that we are still, nevertheless, witnessing the consequences of some physical state that gave rise to the violent eruption. The star has reached a state in its evolution where no stable static structure is available to it. Consequently, an unstable disturbance has grown to finite amplitude somewhere deep in its interior. The resulting oscillations are rumbling away in an irregular and probably unpredictable fashion and are transmitting their energy in the form of pressure waves to the surface. The deposition of mechanical energy there is, in my view, the primary cause of the expansion of the wind. It is only to be expected that such a wind will be irregular on a variety of timescales. The fluctuations in photospheric radius and in visual magnitude are an inevitable consequence. The formation of shells in the wind at some distance above the star is another.

Lamers: The high luminosities and low effective gravities of the LBV's suggest that radiation pressure may play a key role in their behavior. Several speakers at this colloquium have discussed the possible effects of radiation pressure for explaining the high and variable mass-loss rates of LBV's. I would like to make a few remarks about the connection between radiation pressure and the evolution of massive stars.

Stellar surface layers where $\kappa_F \geq 4\pi GMc/L$ (where κ_F is flux-mean opacity) will not be bound by gravity. The opacity depends on density, temperature, and composition. For Population I stars, κ_F in the photosphere has a maximum near $T_{eff} \approx 10000$ K and decreases for higher and lower T_{eff}. This implies that the luminosity in the H-R diagram where the photospheres become unbound will reach a minimum L at $T_{eff} \approx 10000$ K and will increase to higher or lower T_{eff}. Lamers and Fitzpatrick (1988, *Astrophys.J.* **324**, 279) suggested that this can explain the Humphreys-Davidson limit. [The idea was also described, *e.g.*, in *Science* **223**, 227 (1984).] In reality the outer layers may become unstable before L reaches the limit set by κ_F, because a loosely-bound atmosphere with a very small effective gravity may easily become unstable by secondary effects such as pulsation or turbulence. It may very well be that such secondary processes trigger the shell ejections in LBV's, but the major reason for their instability is still radiation pressure. I doubt that such secondary effects can trigger the large eruptions such as the η Car event, because the large amount of energy involved in such an event requires a more efficient mechanism operating deep below the photosphere.

If the LBV variability is due to triggering at small g_{eff}, then there are two ways to explain the co-existence of LBV's and normal supergiants in the same part of the H-R diagram. Either the more normal stars have higher masses and larger gravities than LBV's with similar luminosities, or the triggering is less efficient, or both. Lower masses for LBV's require that they have lost more mass than the normal stars. This agrees with their higher nitrogen abundances. Maybe the LBV's were more rapid rotators and lost more mass on the main sequence, but this is very difficult to test observationally. Or maybe they have lost more mass because they have suffered one or more large eruptions already. In that case, shouldn't we expect ring nebulae around most LBV's?

The question of triggering may have consequences for the formation of B[e] stars. These differ from LBV's in that they have most of their mass loss in equatorial (?) disks. This idea is similar to the classical Be stars, which are on the average fast rotators, but not fast enough to produce sufficient centrifugal forces. The equatorial ejection is supposed to be due to the combined effects of non-radial pulsation and radiation pressure. Rotation is important for creating non-radial pulsations with their highest amplitudes along the equator. Maybe a similar mechanism is operating in the B[e] stars with low g_{eff} and some triggering along the equator? As in the case of the Be stars, one does not need a large rotational velocity to eject in a disk-like fashion.

A critical test for the role of radiation pressure in the explanation of the LBV and B[e] phenomenon can be made by comparing their locations in the H-R diagram for galaxies with different metal abundances. Since κ_F will decrease with decreasing metallicity, one might expect that the LBV's in low-metallicity galaxies occur at higher luminosities than the galactic ones. One should keep in mind, however, that the luminosity of the LBV's will not only depend on κ_F but also on the M/L ratio, which may be different in low-metallicity galaxies.

General discussion:

Shore: In theoretical astrophysics various dimensionless numbers are useful or interesting. Here we have one--we might call it the Eddington number: $\Gamma < 1$ could play a role like (for instance) the Rayleigh number ($\gg 1$) that pertains to convection, with a phase transition. In the recombination mechanism we have a non-linear instability which may have 3 time-dependent equations connecting ionization, mass in the envelope, and velocity, all as functions of time. In dynamo theory, *e.g.*, that type of structure leads to a phase transition with lots of noise before it flips; it is the prescription for disaster because it's the signal of a Strange Attractor. We may be dealing with something that has a bistable character, flips back and forth, and does so because the instability that is driving it is intrinsically non-linear, buried fairly deep in the envelope but not in the core ... but if there is a strange attractor, if this is a bistable system, then we may never find what triggers it all; it's simply intrinsically unstable and the spectrum of flips may be unpredictable. I might also add that this could drive a close companion star over the edge of sanity, too.

Sreenivasan: ... Regarding the trigger for LBV explosions -- The facts that many modes are excited in non-radial pulsations and that their frequencies are close suggest mode-coupling. This gives rise to flickering at higher frequencies and a modulation at lower frequency. Occasionally, the amplitudes could add constructively and push the overlying material off at the escape velocity which has been lowered by radiation pressure, mass, and centrifugal force. Just when the trigger gets pulled cannot be predicted and this could well look like 'deterministic chaos' as Steve Shore remarked.

Davidson: Some of us have been wondering why non-linear behavior wasn't emphasized more at this meeting! As Hummer and Shore implied, even Herschel's and Clerke's choices of words 90 to 150 years ago ring chaotic, non-linear alarm bells when we read them today.

... About another matter: Henny Lamers referred to the effect of metallicity on an opacity-dependent instability. Here we have to be careful about what's conspicuous in an empirical H-R diagram, and what would not be. Suppose that the Upper Limit is lower for stars with higher heavy-element abundances. Then the Upper Limit that we **notice** in the diagram would really be an envelope representing the lowest metallicities that happen to be observationally accessible. Some stars of higher metallicity might be marginally unstable or might be LBV's well below that envelope. (Maybe this is pertinent to Katy Garmany's talk. Also, this should remind us that an Upper Limit, a locus of instability, and a locus of LBV behavior can logically be three different lines that need not coincide.) The stability boundary for stars with nearly primordial composition may be higher, but I Zw 18 isn't close enough to check!

Humphreys: We have reasonable data for just three galaxies: the Milky Way, LMC, and SMC. They do have different metallicities, but in the SMC, with the lowest metallicity, the sample size is so small that there are almost no very massive stars. Population statistics may dominate any effect due to metallicity differences. It would be nice if we could examine a big spiral galaxy with low metallicity.

Gallagher: With the possible exception of η Car, for each LBV we see evidence of only one really major ejection event. It would be good to look for additional faint outer shells; meanwhile, conceivably a major outburst is a one-time event. Are we perhaps being premature in assuming that LBV's live long enough to represent a significant mass-loss phase in the evolution of very massive stars?

Lamers: Indeed we don't know. Unfortunately, we know too few LBV's for a thorough statistical study. Major outbursts may occur on long timescales, possibly longer than 10^4 years.

Walborn: I am not sure that a distinction between normal and "Plinian" outbursts is well established observationally. Does the latter refer only to η Car, or does it include the 3-mag ranges of P Cyg and R 127? The difference may be only that η Car is more massive and hotter than the other objects, and thus has a larger bolometric correction.

De Groot: We speak of P Cyg's outburst as a 3-magnitude event, but we do not really know this. There are no observations from before the year 1600. The 3-mag amplitude is based on the star's behavior after that time, when an outburst was seen at maximum. The subsequent brightness of P Cyg may be at some intermediate level and might not represent the pre-outburst brightness.

Davidson: We've been told of many indications of bipolar structure around LBV's. Maybe the existence of an axis of symmetry implies that some sort of angular momentum is important. Any comments on bipolarity?

Gallagher: Of course bipolarity might be an indication of binarity. This kind of structure may be important this way because it is so easy for other evidence of close binaries to be hidden.

Appenzeller: A bipolar outflow geometry cannot prove the binary nature of its source, because rotation or a magnetic field can define a preferred axis in single stars also. In fact, the strikingly well-collimated outflows from T Tauri stars are now generally supposed to be single-star phenomena. It would be extremely interesting if one could learn more about the relatively sketchy indications of binarity, with observations like those described at this meeting by Paresce. Then one might start making use of this potential source of information.

Friedjung: I would like to make a comparison with classical novae. There is evidence for continued ejection driven by an object above the Eddington limit. In the nova case there may be an extra source of energy in outer layers which brings them above above the Eddington limit, but does not make the whole star fly away into space. Is something similar possible in LBV's, at least for the major outbursts?

Kahn: What you need is a layer of limited extent, deep down, where Γ exceeds unity. This layer is unstable but cannot lift off all the overlying material. It must therefore oscillate at finite amplitude, and so generate waves which ultimately reach the surface of the star and drive a wind from there.

Maeder: It has been said that the layers above the $\Gamma = L/L_{\rm Edd}$ level may be ejected. Indeed, the amount of ejecta can be larger, since the critical level may move inward during the episode of shell ejection. In view of wide differences between stars such as η Car and Var A, we may suspect that there is perhaps more than one unique mechanism helping to accelerate the exterior layers outward in very massive stars.

Hummer: During the past few days, many people have been discussing $\Gamma = g_{rad}/g_{grav}$ as though it were a constant parameter of each star. But Γ depends on radius and must be less than unity deep in the star. It increases with radius and if it reaches unity, that layer is unstable and starts to move outwards. Just where this happens depends on how the opacity and radiative flux vary with location and frequency. In every hot star, Γ reaches unity at the base of the stellar wind.

De Jager: For stars at the H-D limit, $\Gamma \approx 1$ at $\tau \approx 1$ for temperatures above 10^4 K. In some regions Γ can decrease outward. This situation (unstable inside, more stable outside) has interesting hydrodynamic consequences.

Davidson: If we don't care about the normal wind and look only at optical depths of the order of unity or somewhat larger (averaged appropriately over frequency, of course), then Γ in that region can characterize a star if we are careful with definitions. That was one point of an empirical remark I made earlier: In this sense $\Gamma = 1$ does not resemble the observed Upper Limit in the H-R diagram, but rather is too level; for a hint about why, see John Castor's comment following my contribution in the book from the Lunteren meeting. (When examined closely, Lamers and Fitzpatrick's results do not contradict this assertion.) But Γ = (some number around 0.8) in the same region does give a fit to the observed Limit. Why? -- Who knows!

Schmutz: Most people here seem to be happy with the instability found by the Munich group. I am not so happy, because I do not think it is a mechanism that can explain S Dor-type variations. There is no doubt that a change of iron ionization Fe III/Fe II is needed to explain the large difference between mass-loss rates at maximum and minimum (see also Leitherer *et al.*, these proceedings). But the result found by the Munich group is an instability only if the stellar parameters are close to the critical ones. The observed temperatures and mass-loss rates of R 71, AG Car, and R 127 in their minimum states are far from being critical for Fe III to recombine! The stars must change their radii by large amounts before the critical values are reached.

Appenzeller: Interpreting the new improved model atmosphere computations, one has to keep in mind that so far these results have been obtained for only a rather limited set of parameters. The very important 'unified' P Cyg models all assume the same fixed effective temperature (19000 K) and thus provide no direct information on the temperature dependences of such atmospheres. Such calculations for an extended set of parameters would obviously be useful.

Kudritzki: For most of this meeting we have been depending on just apparent brightnesses and temperatures estimated from energy distributions. We have almost no good hard numbers based on quantitative spectroscopy of LBV's; we know a little about the abundances in ejecta, but do not have much data on atmospheric or photospheric abundances or on gravities (except for one object). So I think that it is necessary to work out observational techniques to give good numbers. For example, high members of the Balmer series in absorption can be used to determine gravities; high-S/N absorption line measurements can give CNO abundances; *etc.* So far we have little to work on!

De Jager: I agree. It would also be very helpful to monitor radial velocities as well as the brightness of at least one selected LBV for at least half a year. That would make it possible to study its internal structure and to decide on its evolutionary status. For comparison one should also monitor a non-LBV star at the same location in the H-R diagram. Excellent data on one LBV is better than partial data on many.

Davidson: The need for more data is indisputable. Remember, though, that the paramount question about LBV's today is why they are so unsteady -- what triggers their eruptions and what supplies the energy for such events? It seems to me that our ignorance along these lines is due primarily to a lack of theoretical studies and only secondarily to a shortage of data (we have many observational clues even if they are not precise). At the same time, the unsteadiness of LBV's is in itself a significant constraint on their structure, if only we can understand the theory! What I'm trying to say is that this is a question that theorists can legitimately attack **now**, without having to wait for the improved observations.

Hillier: A comment to emphasize Roberta's point about low statistics. Here we have 5 panel members, and a 6th if we include Kris Davidson, the session chairman. After two had spoken I noticed a correlation, that their surnames begin with letters in the first half of the alphabet. More observations confirm that all 6 LBV's share this correlation. Is this significant? -- Selection effects may occur and we must examine the population from which they were drawn. A quick count by Allan Willis suggests that the theorists were drawn from a biased sample but observationalists were more uniformly distributed.

'Panel 1'

Kahn, Hummer

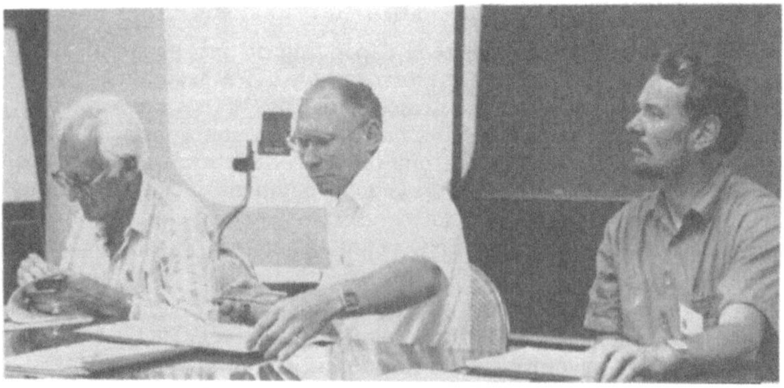

de Jager, Appenzeller, Lamers

SECOND PANEL DISCUSSION

[This panel and discussion was intended to concentrate on observational connections and future observational directions. Panel members were P. Conti, R.M. Humphreys, A. Maeder, O. Stahl, and N.R. Walborn, with A.F.J. Moffat acting as moderator or chairman. Remarks by panel members and other participants are given here in condensed form, partially reworded for brevity and clarity.]

Conti: Concern has been expressed throughout this meeting as to the definition of an LBV. While these letters originally meant Luminous Blue Variables, we have seen during these four days that some are not blue, some are not luminous, and some are not (yet) variable. As far as proposing a 'legal' definition, I am somewhat mindful of the judge who when asked to define pornography responded that he couldn't but "knew it when he saw it." In that spirit, I would like to suggest that LBV stars be considered Little Bit Violent stars. We would all agree that those objects showing this kind of behavior are LBV's. We probably also need to consider LBV **candidates**.

What is the role of LBV's in massive star evolution? They are similar to WN stars in having enhanced He and N abundances, comparable luminosities, masses, and mass-loss rates, and irregular variability, along with ejections of ring nebulae in several instances. While all LBV's have hydrogen, only some of the WN's have evidence of this element being present. It seems safe to conclude that **some** LBV's become WN stars although it is by no means clear that **all** do, nor that all W-R stars have gone through an LBV phase. It is premature to associate any W-R subtype with a predecessor LBV phase, even though a connection to WN8 stars is tempting. I would like to emphasize that there is no relation (yet) between any evolution model and any W-R subtype, especially since the latter have very heterogeneous properties.

I would like now to pose some questions: Are there different scenarios for 'bright' (R 127) and 'faint' (R 71) LBV's, above and below the red supergiant limit? Will all stars with anomalous emission lines eventually become LBV's? Does the **core** physics or the **surface** composition terminate the LBV phase? What role do binaries play in the LBV phases?

Humphreys: Not all luminous emission-line stars are LBV's. We have seen in this colloquium that there are actually very few known LBV's. Numbers presented here suggest that the duration of this stage is very short, only of the order of 20000 years. This is a short but physically important state in the evolution of very massive stars.

A variety of objects are near the luminosity/stability limit in the upper H-R diagram – LBV's, B[e] supergiants, unstable cool hypergiants, and 'normal' mass-losing O, B, and A-type supergiants. What are the relationships among these different objects, and what is the relation between the appearance of the upper H-R diagram and the cause of the instabilities?

K. Davidson et al. (eds.), Physics of Luminous Blue Variables, 249–254.

The temperature-dependent luminosity limit for hot stars and the upper luminosity boundary for the cool hypergiants are related. The observed limit for the hot stars is most likely due to radiation pressure and the opacity-dependent Eddington limit. The Kudritzki *et al.* 'hydrogen recombinator' (Kudritzki's word!) in the stellar atmosphere is a possible trigger. Where the opacity reaches a maximum at 8000--10000 K, the Eddington luminosity may be at its minimum. Below this luminosity, stars can evolve to become red supergiants or hypergiants. We also know from de Jager's studies of the atmospheres of cool hypergiants that these stars' surfaces can become unstable because of turbulent pressure. Thus the turbulent pressure gradient in cool atmospheres, as well as the radiation pressure obstacle to redward evolution of hotter stars, sets an upper limit to luminosities in the cool part of the H-R diagram. Based on observations of nearby galaxies, this limit is near a bolometric absolute magnitude of about −9.5. Three very luminous A-type supergiants with bolometric magnitudes around −10 thus occupy a critical place in the H-R diagram: Cyg OB2 #12, HD 33579, and B 324 in M 33. It is uncertain whether they can evolve to the red supergiant stage. Maeder's density inversion may play a role in stars like these.

The B[e] supergiants appear to be caused by a different type of instability. The observations show that they have relatively dense equatorial disks. This must be due to high mass loss at their equators, possibly caused by rotation or binary interaction. I doubt if there is any direct evolutionary connection with the LBV's.

One of our most important observational tasks is to identify and study more LBV's to improve our database on these very rare stars.

Stahl: Since we now apparently all know what an LBV is, we may turn to other stars.

In the upper H-R diagram we find not only LBV's, but also 'normal' stars and many other emission-line stars which have not (yet?) shown any outbursts and thus cannot be classified as LBV's. What is the relation of these stars to the LBV's? It is likely that that many of them are LBV's in quiescent phases. After all, R127 was detected behaving as an LBV only about five years ago; before that outburst it was a 'normal' Ofpe/WN9 star. This makes the Ofpe/WN9 stars especially promising LBV candidates, although some of them appear to be much fainter than R 127.

Nevertheless, it may be that some of these stars never show outbursts. The B[e] supergiants, for example, have been suggested to be a completely different class of emission-line stars, which do not show any outbursts. These stars may have disks formed by rapid rotation. However, observationally, it has yet to be shown that the B[e] supergiants do indeed rotate fast.

The role of binarity also still has to be clarified. So far only one binary, R 81, is known among the luminous emission-line stars. This star has not shown any outburst so far and it even has unusually small micro-variations. So the roles of rotation and binarity are unclear for LBV's and need to be attacked observationally.

Walborn: An interesting morphological challenge would be to devise criteria discriminating large red tomatoes from large red apples, while suggesting an evolutionary relationship between small green apples and large red ones, under the condition that one is not allowed to approach close enough to bite! The opposite might be more readily inferred. Moreover, small green tomatoes and large green apples also exist, not to mention crab apples! The problems of classifying and interpreting the amazing array of envelope phenomena that hot massive stars can produce are rather analogous. We are far from understanding the physical reasons for all this diversity, and the underlying generic relationships. I would like to highlight some aspects of three specific issues.

1. Eta Carinae: Usually we think of the LBV phase as intermediate between O and WR states, which may apply to the typical LBV. However, I suggest that η Car is a post-WN-A (narrow-line WN) object. This star is unique among known LBV's in terms of its location within a giant HII region associated with O3 and WN-A stars of initial masses at least 100 $M_{\odot}$, and in terms of the magnitude of its outburst amplitude (due to a higher effective temperature and bolometric correction?). The spectroscopic progression from O3 V through O3 If to WN-A is very smooth; it is difficult to imagine the chaotic η Car phase interpolated between O3 If and WN-A, when the relative appearance of these two kinds of spectra strongly suggests simply an increase in wind density between them. The WN-A stars have the highest hydrogen content and luminosities of all WR's, and it is plausible that they are produced by steady mass loss from the most massive O-type stars. It is noteworthy that no η Car counterpart is observed in 30 Doradus, despite the much larger early-O and WR populations there. Possibly this indicates a very short time-scale for the η Car phenomenon. If the suggestive similarities between the η Car nitrogen-rich ejected knots and the quasi-stationary flocculi in Cas A are significant, then η Car will become a supernova on a time-scale shorter than 10^4 years. The only caveat to the foregoing is the question whether the η Car phase is naturally experienced by all of the most massive stars, or corresponds to some pathological event such as a close binary interaction. If the former is the case, the suggestion is that the WR and LBV phases bear a fundamentally different relationship at the highest stellar masses, compared to their roles at somewhat lower masses.

2. Concerning the Ofpe/WN9 -- LBV *connection:* Are all Ofpe/WN9 objects quiescent LBV's? Conversely, are all LBV's (either in some mass-luminosity range or simply all) Ofpe/WN9 stars when they are at minimum? R71 definitely has a lower luminosity than other LBV's in the LMC, but some of the Ofpe/WN9's also have relatively low luminosities *e.g.*, BE 381). In this case P Cyg and S Dor would be in extended intermediate or maximum states, and their spectra at minimum have not yet been observed. The P Cyg light curve shown at this colloquium by de Groot indicates both brighter and fainter states during the 17th Century, with an intermediate light level since, consistent with the 'intermediate' B-type spectrum currently observed. Of course, the foregoing questions bear directly on the reality of the proposed LBV 'instability strip'; if all are actually Ofpe/WN9 at minimum, then the high-temperature limit of LBV excursions in the H-R diagram would be vertical as is the low-temperature one.

Some further tomato-vs.-apple questions: Does the behavior of HDE 269582 imply that some other Bpe objects in the LMC are really Ofpe/WN9's? If so, they are becoming remarkably numerous. Are some WN8, Ofpe/WN9, and LBV objects near or below the luminosity limit for red supergiants in the H-R diagram actually post-red-supergiants? Possible examples are P Cyg (see Lamers *et al.*, *Astron. Astrophys.* **123**, AG Car and R71 (McGregor *et al.*, *Astrophys. J.* **329**, 874), HD 96548 (Smith *et al.*, *M.N.R.A.S.* in press), and HDE 269227 (Ofpe/WN9--RSG spectrum composite, Schmutz *et al.* in these proceedings); in several cases the post-red-supergiant interpretation is suggested by the characteristic of the circumstellar nebulae. Finally, we need more observational evidence for the putative O → LBV → WNE/WC evolutionary sequence, such as physical associations among objects in these relatively short phases.

3. Disarray on the Late WN Sequence: For some time we have been used to thinking in terms of the 'WR phenomenon' in the context of massive objects and planetary nebula nuclei with superficially similar spectra. However, it may be necessary to emphasize this concept within the Population I sample itself: there may be qualitatively different physical origins for objects in adjacent subclasses and even within a given subclass. For instance, most WN6/7-A objects are in giant H II regions or young associations and have very high luminosities ($M_V \sim -6.5$ to -7.5); CQ Cep is not, but is a close binary, so it provides a poor paradigm for this subclass. Most WN8 objects are not in giant H II regions and have lower luminosities; some may be post-red-supergiants. The nature of the Ofpe/WN9 class in the LMC and its relation to the LBV's has been amply discussed here. Some galactic objects have been proposed as counterparts to the LMC Ofpe/WN9 class, and other isolated objects have been classified as WN10 or WN11 in the literature, but without a systematic morphological analysis relative to the existing WN sequence. A larger sample of such objects with homogeneous data and a systematic analysis are required in order to satisfactorily define the late WN sequence in the future. Several such objects have been presented at this meeting, a welcome development.

Maeder [points excerpted from a tape recording of his talk]: Can we understand LBV's without understanding evolution? I think the answer is *no*. Can we understand evolution without understanding LBV's? I think the answer is *no*. ... Here I wish to emphasize the relations between model ingredients, model outputs, and observations.
A model uses only a few equations but many ingredients: nuclear reactions and weak interaction parameters, opacities, boundary conditions, initial abundances, convection and overshooting, mass loss, diffusion, rotation, binarity ... When the model outputs agree with observations, everyone is happy but we do not make much further progress. When they disagree, people are not so happy, but future progress results. To make such progress, we must identify which particular features of the model are responsible for the disagreement; maybe the disagreement tells us that some assumed ingredient is wrong. But there is predictive power in models, too; for instance, LBV's may be gamma-ray sources because of ^{26}Al in their atmospheres and ejecta ...

Physical quantities represented in models and *observed* quantities are often different entities. To improve the connections, it would be good to obtain very detailed observations (photometry, abundances, pulse periods, etc.) of stars *within clusters*. Not all abundances constrain models equally well; *e.g.*, for WN stars the C/N ratio is more or less the same whatever the evolution, but C/H is a very constraining quantity. Finally, if everything could be predicted accurately from theory, then we would not have the pleasure of attending meetings on the topic!

General discussion:

Walborn: Roberta mentioned Cyg OB2 #12. I think that the association membership of this star is questionable; its late-B spectral type is incompatible with the Of character of the association. It might not be as distant or as luminous as generally assumed. We have to be especially careful about distances in Cygnus and Carina, where we're looking along spiral features.

Humphreys: #12 is the most highly reddened star in that association --

Walborn: But there is a strong gradient in the reddening in that direction.

Wolf: Many years ago I analyzed HD 33579, another of the luminous stars that Roberta mentioned. The methods would certainly not be considered state-of-the-art now, but the *log g* that I found indicated a low mass, of the order of 25 $M_\odot$. This indicates that HD 33579 is very evolved, it has lost much mass.

Cassinelli: With regard to rotation — (1) Is the observational situation regarding *v sin i* going to improve for the B[e] stars? R 126 is especially frustrating because there's evidence for rapid rotation but no value of *v sin i*. (2) Peter Conti mentioned that both LBV's and WR stars have ring nebulae. Did you mean to imply that they are spherically symmetric?

Walborn: Lacking good photospheric absorption lines, it is hard to see how we can observe *v sin i*. Realizing this, in my own mind there is some doubt about the rapid rotations.

Conti: I was not making any assertion about spherical nebulae. In most cases with hot central stars they are not symmetric. At least part of the reason could be rotation of the central object, and another cause may be the non-homogeneous interstellar environment into which expansion occurs.

Niemela: NGC 6888 has been mentioned in this connection. However, it is mostly swept-up material, maybe not directly comparable to the AG Car nebula.

Conti: Perhaps those two nebulae are not directly comparable. But what happens to an AG-Car-like object after the central star becomes a WN with a strong wind? Then, too, more material might be swept up.

Niemela: Just what is an anomalous O-emission-line star?

Conti: Two kinds of 'slash' stars were introduced by Walborn in the 1970's. Ofpe/WN stars have anomalous Si IV features and also other emission features not normally seen in Of-type stars. The other kind has features intermediate between types O and WN. R 127 was an example of the first class; maybe the other class can also lead to LBV's.

Walborn: R 127 was not only an example of the Ofpe/WN9 class, it was a **prototype** of that class. Still, the hypothesis that Ofpe/WN9's all become LBV's may be an extreme one.

Shore: But what other trick do we have for finding LBV's?

Walborn: We have to wait for their outbursts.

Shore: That'll take a lot of GSL's ('graduate-student-lifetimes').

Humphreys: The Hubble-Sandage variables offer hope that we can expand the sample of galaxies, especially if the Space Telescope becomes available. For the known H-S variables, older photometric data is available to supplement newer spectroscopic data.

Shore: Although we haven't had enough time for this point here, there is something very strange about the space distribution of the massive stars. They tend to appear in compact groups, groups much smaller than associations. Also, maybe we tend to look for LBV's where they are easy to find, say in fields that are not too crowded. are LBV's distributed uniformly in galactic disks? Abundance gradients may be relevant here.

Humphreys: In M33 the known LBV's are all on one side, and not very far out. Var A is farther out, in the outermost arm.

Moffat: The Upper Limit in the H-R diagram is fuzzy. How fuzzy?

Humphreys: Here we are definitely in the realm of small-number statistics, we don't have any way to know exactly where to draw the line, and no solution has been proposed during this colloquium. We can define it as the actual upper envelope, connecting just several of the most luminous stars. Or we can draw it about half a magnitude lower where there are more stars. A better definition requires both better statistics and a more definite theoretical background.

SECTION II: POSTER PAPERS, A BRIEF LIST OF CATALOGS, AND INDEX

Poster-discussion
chairmen:

van der Hucht

Willis

de Groot

P CYGNI: THE STAR THAT STARTED IT ALL

Mart de Groot,
Armagh Observatory, College Hill, Armagh, Northern Ireland, BT61 9DG.

P Cygni was discovered in 1600 AD during an outburst when it reached 3rd magnitude. After another outburst in 1655 the star is now showing irregular photometric variations with an amplitude in *V* of ~ 0.2 magnitude. Because of these variations and of its peculiar spectral characteristics, P Cygni has been classified as anything between supernova and W UMa system. However, its light curve is unlike that of any other star and P Cygni could well be considered the first discovered LBV.

Over the last three years an accurate photoelectric light curve has been put together. It is based on observations made by a number of amateur and professional observers around the world and shows the following features:

+ Variations are irregular with maxima and minima every 1 - 2 months;
+ The brightness variations seem to show more scatter when the star is faint.

There are few simultaneous spectroscopic and photometric observations of P Cygni. In one reasonably well documented case (Baliunas *et al.*, preprint, 1987) there is an increase in the star's brightness while the Hα emission intensity decreases.

Radial-velocity variations, both at visual wavelengths and in the UV, show ejections of shells at semi-regular intervals. (Van Gent and Lamers, *Astron. & Astrophys.*, **158**, 335; Markova, *Astron. & Astrophys.*, **162**, L3).

Polarimetric observations of P Cygni also reveal the presence of anisotropic mass flows which vary in both direction and time (Hayes, *Astrophys. J.*, **289**, 726).

The above observations can be explained as follows:

+ P Cygni ejects shells at about two-monthly intervals. These shells increase the opacity in the star's envelope so that the bright photosphere is obscured and the star's brightness decreases.
+ The shells are "clumpy" so that during their presence, at minimum brightness, there are larger photometric variations than during their absence, at maximum brightness.

K. Davidson et al. (eds.), Physics of Luminous Blue Variables, 257–258.

Note the following observed timescales of variations in P Cygni:

Radial velocities (UV)	60 - 75^d	Lamers *et al., Astron. & Astrophys.*, **149**, 29
Radial velocities (visual)	50 - 100^d	Van Gent & Lamers, *op. cit.* / Markova, *op. cit.*
photometry	25 - 60^d	*do.*/Percy *et al., Astron. & Astrophys.*, **191**, 248.
polarimetry	12 and/or 125^d	*do.*
rotation	$\approx$ 50^d	

These can be compared to the expected time scale for non-radial pulsation: ~ 28^d (Maeder, *Astron. & Astrophys.*, **90**, 311).

There is another star very similar to P Cygni: R81 in the LMC (Wolf *et al., Astron. & Astrophys.*, **99**, 351). Recently it was found to be an eclipsing binary with a period of 75^d (Stahl *et al., Astron. & Astrophys.*, **184**, 193). This is an excellent example of what can be achieved through long-term photometry of an interesting object.

P Cygni occupies a crucial position in the Hertzsprung-Russell diagram. A better understanding of this star will enable us to answer some very fundamental questions about the evolution of massive stars. E.g.: How does the mass-loss rate of a LBV vary with time? What mechanism is mainly responsible for the mass loss? Is it radiation pressure, binarity, non-radial pulsations, turbulent instability, or something different?

In order to identify the true (or most important) cause of P Cygni's mass loss we need more simultaneous photometry, spectroscopy and polarimetry.

This is a task for observers with regular access to these kinds of instrumentation. Many observers are already doing regular photoelectric UBV photometry. Those with regular access to spectroscopic equipment could obtain at least Hα profiles at a resolution of 0.25 Å or better. Polarimetry should be done as often as possible. In all cases, one good set of measurements per night will be sufficient to draw some very interesting conclusions.

So, let's have your observations, especially your spectroscopy, and your polarimetry. Contact the author at the above address for further details.

After nearly four centuries of discussing how interesting P Cygni is, it is time to become serious about more observations and unravel the mystery of its variations and find the clue to the secret of the evolution of massive stars !

THE WIND OF P CYGNI

J. Puls, A.W.A. Pauldrach, R.P. Kudritzki
Institut f. Astronomie & Astrophysik der Universität München
Scheinerstraße 1, D-8000 München 80, F.R.G.

The stationary features of the wind of P Cygni are considerably different from those of 'normal' supergiant winds with comparable luminosity. In contrast to such winds, which are generally accepted to be driven by radiation pressure, P Cygni's mass-loss rate is higher by a factor of 5, the terminal velocity is higher by a factor of 10, and the velocity law itself is much flatter than would be expected from a first glance at glance at typical scaling relations. However, these relations depend crucially and non-linearly on the star's distance from the Eddington limit, which for P Cyg is very small (see below). Here we investigate whether the acceleration mechanism of P Cygni's wind can also be explained by line pressure and to what extent self-consistent wind models represent the observed quantities (especially the IR energy distribution).

Stellar and wind parameters: The stellar parameters of P Cyg are fairly well known from an investigation by Lamers *et al.* (1983). Assuming $E(B-V) = 0.63 \pm 0.05$, they found $T_{eff} = (19300 \pm 2000)$ K, $R_* = (76 \pm 15)\,R_\odot$, $d = (1.8 \pm 0.1)$ kpc. According to the evolutionary tracks of Maeder and Meynet (1987), P Cyg is on its way back from the RGB with a mass of $23.5\ (+18,-9)\ M_\odot$, which results in $\log(g) = 2.05 \pm 0.25$ and $L/L_{Edd} = 0.73$.

The wind's terminal velocity is more uncertain. Cassatella *et al.* (1979) found a value of 300 km/s from the Mg II, Fe II lines, whereas Lamers *et al.* (1985) argued that the blue absorption edges are contaminated by turbulent motion, and that the true terminal velocity is 206 km/s.

The usual way to determine mass-loss rates is by the use of radio data. P Cygni is extremely well observed from 0.33 to 20 cm (Wendker 1987). A compilation of these data leads to a high correlation with the thermal emission model, from which we find (assuming $d = 1.8 \pm 0.1$ kpc): $\dot{M}(10^{-6} M_\odot/yr)/v_\infty(km/s) = 4.177\ (+1.45,-1.07) \times 10^{-2}$. A more difficult problem is the determination of the velocity law. Using the ground-based IR fluxes by Abbot *et al.* (1984, 0.98--20 μm) and the IRAS fluxes (20--100 μm), Waters *et al.* (1986) argued that the best fit could be obtained by a *linear* velocity law $v(r) = v_\infty\{0.1 + 0.057(r/R_* - 1)\}$, where the fluxes at 60 and 100 cm are due to enhanced emission by dust or originate from a shell at about $15\,R_*$. *This fit, however, is not unique.* A typical β-velocity law -- excluded by Waters *et al.* -- with $\beta = 4$ and a much smaller initial velocity ($v_0 = 4 \times 10^{-4}\, v_\infty$) reproduces the observed IR flux and the IR excess as well; both the linear and the $\beta = 4$ laws show the same density structure in the IR-emitting region.

The acceleration: The first quantitative analysis of the acceleration mechanism was carried out by Lamers (1986). On the basis of the linear velocity law he showed that neither turbulent pressure, nor dissipating sound waves, nor pure continuum pressure are able to provide sufficient acceleration to drive P Cygni's wind. He concluded that the acceleration must be due to line radiation pressure, probably by a large number of

K. Davidson et al. (eds.), Physics of Luminous Blue Variables, 259–260.

optically thin lines in the Balmer continuum. Simple algebra, however, shows that the radiative acceleration necessary to yield the linear velocity law should be due to optically thin lines only in the lower part of the wind ($r < 3R_*$, $v < 0.22 v_\infty$), whereas the outer part of the wind must be accelerated only by optically thick lines. In order to test these constraints, we calculated the line force for the linear velocity field, by applying the NLTE code of Pauldrach (1987) for the parameters of P Cyg. We found that the linear velocity law gives rise to an acceleration both by optically thick and optically thin lines (in ratio 0.61) throughout the wind, which is in strong contradiction to the applied input and shows that a linear velocity law cannot be obtained by radiative acceleration in the case of P Cygni. For the β-velocity law, however, the situation looks much better. The resulting NLTE line force (including line overlap, *cf.* Puls 1987) turns out to be nearly identical to that necessary for building up the β-law. Hence it is most probable that the acceleration mechanism is line pressure, resulting in a velocity field with the typical β-structure of radiatively driven winds, giving the IR excess.

Radiatively driven wind models: In order to examine how far self-consistent wind models are able to reproduce the observed features, we proceeded in two steps. First we calculated a stellar wind model with the given force-multiplier parameters $k = 0.088$, $\alpha = 0.524$, $\delta = 0.059$, and $\log(g) = 2.04$, resulting in $\dot{M} = 12.1 \times 10^{-6} M_\odot/yr$, $v_\infty = 213$ km/s, values that lie exactly in the expected region for P Cyg. *The observed IR excess was also reproduced.* The second step was (and is) the calculation of completely self-consistent models, including the simultaneous solution of hydrodynamics, NLTE rate equations and radiation transport with line overlap. This last topic is especially important for quantitative results, since the line overlap (strong line blocking) leads to a significant reduction of the line force. From the three tracks calculated by Pauldrach et al. (1988) we choose the $R = 76\ R_\odot$ track since this is the most probable radius from our investigation of IR and radio data. Up to now we have finished only our calculations for a model with $\log(g) = 2.0$, corresponding to model 3 by Pauldrach, on the upper track of the mass-loss function. Comparing our results to the results of the single-line approximation by Pauldrach, we find a reduction of both mass-loss rate and terminal velocity in the expected direction:

$\dot{M}/(10^{-6} M_\odot/yr)$ = 32 (single-line approx.) or 21 (with line overlap);
$v_\infty/(km/s)$ = 280 (single-line) or 228 (line overlap);
$10^7 \dot{M}/v_\infty$ = 1.17 (single-line) or 0.9 (line overlap).

Future work: The next steps in our fully self-consistent description of the wind of P Cygni, underway at present, are the calculation of a model grid around the discontinuity in the mass-loss function (see Pauldrach et al.) and the investigation of the ionization balance of metals (especially Fe II, III) in comparison to observations. From the work presented here, however, it is evident that the stationary features of P Cygni's wind can be explained with radiation pressure as the dominant acceleration mechanism, with stellar parameters $T_{eff} \approx 19300$ K, $\log(g) \approx 2.04$, $R_* \approx 76\ R_\odot$, *yielding a mass of* $\underline{23 M_\odot}$.

References: Abbott, Telesco, & Wolff 1984, *Astrophys.J.* **279**, 225; Cassatella *et al.* 1979, *Astron.Astrophys.* **79**, 223; Lamers, de Groot, & Cassatella 1983, *Astron. Astrophys.* **128**, 299; Lamers, Korevaar, & Cassatella 1985, *Astron.Astrophys.* **149**, 29; Lamers 1986, *Astron.Astrophys.* **159**, 90; Maeder & Meynet 1987, *Astron.Astrophys.* **182**, 243; Pauldrach 1987, *Astron.Astrophys.* **183**, 295; Pauldrach, Puls, & Kudritzki 1988, these proceedings; Puls 1987, *Astron.Astrophys.* **184**, 227; Waters & Wesselius 1986, *Astron.Astrophys.* **155**, 104; Wendker 1987, *Astron.Astrophys.Suppl.* **69**, 87.

A SOURCE FOR MASS-LOSS INSTABILITIES IN LBV'S – THE CASE OF P CYGNI

A.W.A. Pauldrach, J. Puls, R.P. Kudritzki
Institut f. Astronomie und Astrophysik der Universität München
Scheinerstr. 1, D-8000 München 80, F.R.G.

Since the basic mechanisms that produce photometric variations, shell ejections, and eruptions of LBV's are still unknown, it is worthwhile to investigate whether instabilities can occur when the improved self-consistent NLTE treatment of radiation-driven winds (see Pauldrach *et al.* 1986, Pauldrach 1987, Puls 1987, Pauldrach & Herrero 1988) is applied to objects lying in the LBV part of the H-R diagram. The motivation is obvious: LBV's have lost considerable fractions of their initial masses and hence have L/M ratios close to the Eddington limit. For such objects, radiation-driven wind theory predicts not only a strong dependence of the mass-loss rate on the self-consistently calculated parameters k, α, and δ, which result from the NLTE occupation numbers of the 133 ions contributing to the line force, but also on Γ (= L/L_{Edd}): $\dot{M} \sim k^{1/(\alpha-\delta)}\,(1-\Gamma)^{(\alpha-1)/(\alpha-\delta)}$. Here we investigate the dependences of $\dot{M}$ on M (through Γ) and on the physical environment of the atmosphere (through k, α, δ) separately. The calculations are performed over a large model grid of stellar parameters for P Cyg, a typical LBV (see Fig. 1).

To study the $(1-\Gamma)$-dependence in detail, we calculated k, α, δ for a low-mass model (# 5) and a high-mass model (# 1). Figure 2 shows the mass-loss rates computed using these two different sets and $\log(L/L_{\odot}) = 5.74$ over the range of $\Gamma(M)$. Apart from the strong dependence of $\dot{M}$ on $(1-\Gamma)$, the most significant feature of Fig. 2 is the large gap between the curves obtained with the two model sets. It is also remarkable that the observed $\dot{M}$ for P Cyg lies between these two curves. This means that, in the parameter range applicable to P Cyg, a physical process that has a dramatic influence on the line force is initiated in the atmosphere. In order to investigate this behavior in more detail we have carried out fully self-consistent calculations with a constant L (see Fig. 3). Surprisingly, rather than gradual transitions from the low to the high $\dot{M}$ curve of Fig. 2, these calculations show *strong discontinuities*, revealing an instability mechanism for P Cyg; the discontinuities occur in the observed $\dot{M}$ range and the corresponding models are close to the M-L relation predicted by Maeder and Meynet (1987). This mechanism can be activated by small changes of either $\Delta R/R < 1.5\%$ or $\Delta L/L < 3\%$. Although we do not understand how these photospheric changes occur, the

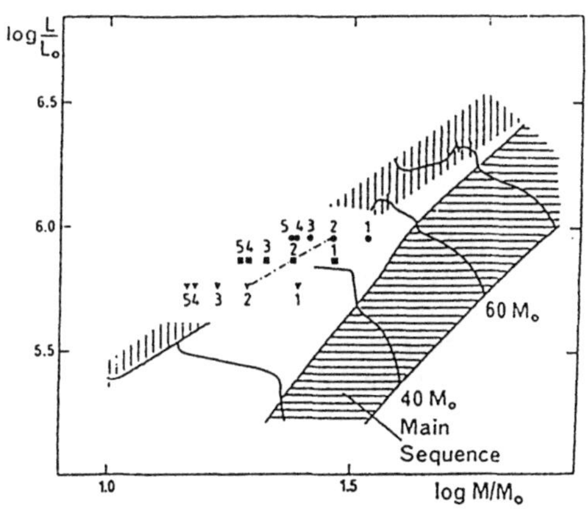

Fig. 1: Mass vs. luminosity. Models numbered 2 correspond to $\log L/L_{\odot}$ = 5.74, 5.86, 5.97 (Lamers *et al.* 1985) and the masses predicted by Maeder and Meynet's (1987) tracks for T_{eff} = 19300 K. Thus the dashed line is the M-L relation for post-RGB objects at the temperature of P Cygni. The other models tell us the parameters if we distrust the evolutionary theory and adopt different masses at T_{eff} = 19300 K for constant luminosities.

K. Davidson et al. (eds.), Physics of Luminous Blue Variables, 261–262.

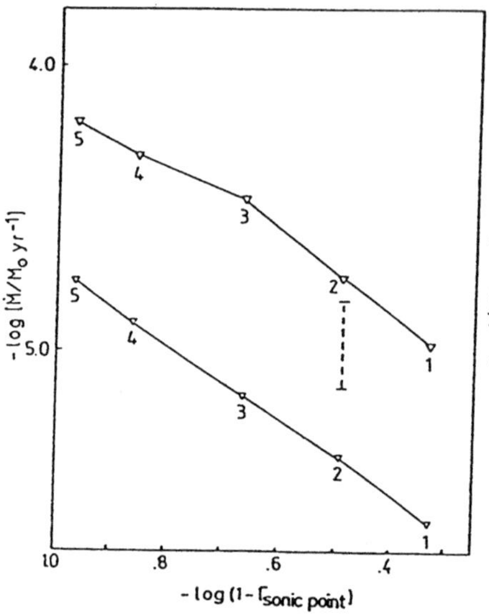

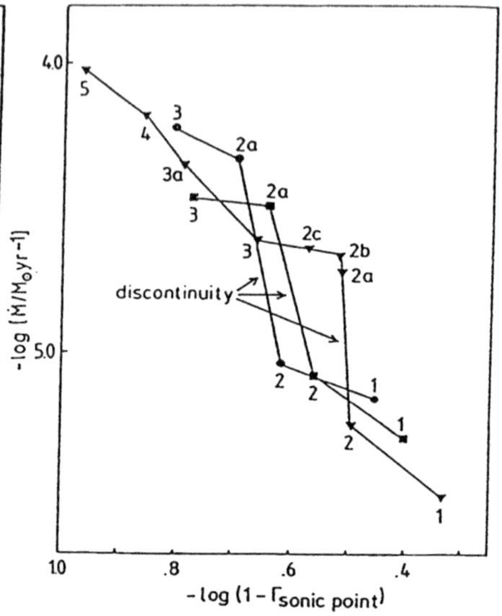

Fig. 2: Mass-loss rate vs. $(1 - \Gamma)$, where the approximate sets of k, α, δ are kept fixed. Vertical dashed line: rate observed for P Cyg. *Fig. 3*: Similar plot for the tracks of constant luminosity shown in Fig. 1. Models 2a, 2b, 2c, and 3a correspond to intermediate grid points. *(See text.)*

reason for the jump in $\dot{M}$ is clear from our calculations: the high-mass models with small Γ exhibit moderate mass-loss rates because the Lyman continuum is optically thin and, hence, due to the strong ionizing radiation field, higher ionization stages are preferred which are less important for the line force (see Pauldrach 1987). Moving toward lower masses and higher Γ, $\dot{M}$ increases according to the $(1 - \Gamma)$ dependence, until a critical density is reached at a certain critical Γ-value. At this critical density the Lyman continuum suddenly becomes optically thick, and consequently the reduced ionizing radiation field causes dramatic changes in the wind ionization and in the radiative line acceleration, producing the jump in mass-loss rate.

Conclusions: By means of our self-consistent NLTE calculations we have demonstrated that P Cygni's wind is highly unstable with respect to small changes in photospheric radius, $\Delta R/R < 1.5\%$, or in luminosity, $\Delta L/L < 3\%$. However, the present work is only a first step. The next steps will be:

-- a study of the mass loss in various phases of P Cyg;

-- an investigation of whether this instability mechanism is a common feature of all or most of the LBV's;

-- a study of whether the instability mechanism is self-initiating in the sense that the jump in $\dot{M}$ can explain the necessary small changes of the stellar photosphere.

References: H. Lamers, M. de Groot, & A. Cassatella 1983, *Astron. Astrophys.* **128**, 299; A. Maeder & G. Meynet 1987, *Astron. Astrophys.* **182**, 243; A.W.A. Pauldrach, J. Puls, & R.P. Kudritzki 1986, *Astron. Astrophys.* **164**, 86; A.W.A. Pauldrach 1987, *Astron. Astrophys.* **183**, 295; A.W.A. Pauldrach & A. Herrero 1988, *Astron. Astrophys.* **199**, 262; J. Puls 1987, *Astron. Astrophys.* **184**, 227.

THE IONIZATION AND THERMAL BALANCE IN P CYGNI'S WIND REVISITED

D.R.H. Johnson & J.E. Drew
Dept. of Astrophysics
Keble Rd.
Oxford OX1 3RH
England

ABSTRACT. The statistical and thermal equilibrium of P Cygni's wind as a function of radius is investigated using the model of Drew (1985) with an improved treatment of hydrogen. Provisional calculated hydrogen and helium line fluxes are compared with observation. The wind temperature is found to be cooler than observation implies, and a further heating source is tentatively identified.

1. The wind model and its aims

The basic structure and input parameters of the wind model were described by Drew (1985). One of the main aims of the current study is to update the treatment of hydrogen and helium in order to obtain reliable theoretical fluxes for the observed Hα, Hβ, Hγ, Brα and HeI 1μm, 2μm lines. So far, only the hydrogen treatment has been updated (to allow for 16 bound levels instead of just 6). The modelling of the neutral helium atom will ultimately be extended up to levels of principle quantum n=4, using the data of Berrington and Kingston (1987). It is hoped that determination of the He abundance in the wind of will then be possible, providing a valuable indicator of P Cygni's current evolutionary state.

Before this can be done, it is appropriate to consider the extent to which the solution obtained is sensitive to various model assumptions. One of these is the form of the input photospheric radiation field at the base of the wind. We check the sensitivity of the calculated wind structure by trying different low gravity Kurucz-Fitzpatrick atmospheres (Fitzpatrick, 1987) with effective temperatures falling in the range $T_{\star}$= 20000 - 25000 K. Another important assumption used by Drew (1985) was that the hydrogen Lyman lines are in radiative detailed balance. This is no longer imposed.

2. Results & Discussion

The temperature profiles obtained are displayed in Figure 1. The differences between the temperature profiles are relatively slight, although there is a definite correlation between effective temperature and wind temperature. In large measure the decrease in temperature (and consequent decrease in ionization) with respect to the results of Drew (1985) is due to the relaxation of the constraint of radiative detailed balance in the HI Lyman lines.

The computed hydrogen line fluxes (Hα, Hβ, Hγ and Brα) vary by a factor of 2 to 3 across the span of input atmospheres used, depending upon which line is considered, showing that the incident radiation field has to be chosen carefully. They are consistently lower than Johnson et al's

K. Davidson et al. (eds.), Physics of Luminous Blue Variables, 263–264.

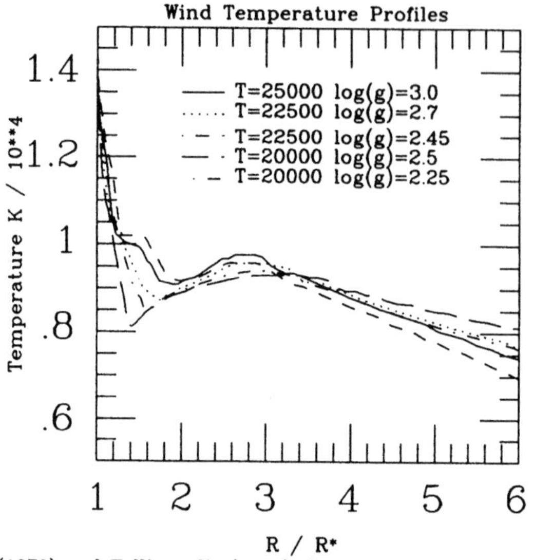

Figure 1. The computed temperature profiles labelled by their input Kurucz-Fitzpatrick atmosphere.

(1978) and Felli et al's (1984) observations by no more than a factor of 3. The computed HeI1μm and HeI 2μm fluxes fall well short of their observed values (Meisel et al (1982); A. Longmore and T. Geballe, private communication) and indeed switch to net absorption in two cases. This is undoubtedly due to the calculated wind temperatures being too low.

The near-neutrality of helium throughout the wind and the extremely low calculated HeI 1μm and 2μm line fluxes indicate a failure to populate the n=2 levels (Drew 1985 found that the main mechanism for ionization of H and He in P Cygni's wind was via n=2 excited states). We consider that this indicates the calculated wind temperature is too low since we expect the n=1 and n=2 HeI states to be roughly in Boltzmann equilibrium. Cassatella et al (1979) obtained a FeII excitation temperature of 12000 $\mp$ 900 K, well in excess of the 8000 K that prevails throughout the majority of the computed profiles. Clearly, it is necessary to establish what is responsible for the missing wind heating.

It has been noted that the UV spectrum of P Cygni is riddled with broadened FeII and FeIII absorption lines that are formed in the wind (Cassatella et al, 1979). It thus seems reasonable that these ions may play a large part in controlling the wind's thermal balance. At present, the only UV lines included in the thermal balance from these ions are FeII UV 1, 2 & 3 and some FeIII forbidden lines. We have thus just begun a pilot study of the Fe^{+}-Fe^{2+} equilibrium with a view to improving the thermal balance calculation.

References

Berrington, K.A., & Kingston, A.E., 1987. J. Phys. B. **20**, 6631.
Cassatella, A., et al., 1979. *Astr. Astrophys.*, **79**, 223.
Drew, J.E., 1985. *Mon. Not. R. astr. Soc.*, **217**, 867.
Fitzpatrick, E.L., 1987. *Astrophys. J.*, **312**, 596.
Felli, M., et al. 1983, ESLAB Symp.no.16, 'Galactic and Extragalactic Infrared Spectroscopy', p.13.
Johnson, H.L., Wisniewski, W.Z., Fay, T.D., 1978. *Rev. Mex. Astron. Astrof.*, **2**, 273.
Meisel, D.D., et al. 1982, *Astrophys. J.*, **263**, 759.

VARIABLE FLUX OF P CYGNI IN CONTINUUM AND SPECTRAL LINES

Indrek Kolka
W. Struve Astrophysical Observatory
202444 Tartu, Estonia, USSR

1. INTRODUCTION

Recently, cyclic variations have been discovered both photometrically in the continuum flux of P Cygni (Percy et al. 1988) and by position measurements of different components in spectral lines of this star (Kolka 1983, Markova, Kolka 1985, Lamers et al. 1985). Additionally, more stochastic polarimetric (Hayes 1985) and radio flux (van den Oord et al. 1985) variations were discovered, too. It has recurrently been stressed that simultaneous multiwavelength observations are badly needed. Fortunately, we had a possibility to complement our own observational data from 1982 with some additional data of the same year from astronomical literature to compare the B, V-filter flux with the behavior of spectral lines having two flux estimates for this period, too.

2. OBSERVATIONAL DATA

Our sample of observational data consists of

1) 14 spectrograms of P Cygni (from 9 June, 1982 to 5 November, 1982) on the Kodak IIaO plates covering a region 3600 Å - 3970 Å, photographed with the Cassegrain spectrograph ASP-32 attached to the 1.5 m telescope of Tartu Astrophysical Observatory giving dispersion of about 12 *Å/mm* in this region,

2) the results of radial velocity measurements and the profiles of selected lines, published by Markova (1986) and Markova, Kolka (1988),

3) the results of B, V-photometry by Percy and Welch (1983) with an additional B, V-estimate by the author, obtained on 6 April, 1982 parallel to the program of checking of the secular variations in the Tartu Observatory UBV-photometric system (Luud et al. 1977),

4) two radio flux estimates at 6 cm wavelength adopted from the compilation by van den Oord et al. (1985).

We present our data in Fig. 1. The left-hand part of this figure (Fig. 1a) shows the photometric behavior of P Cygni in B- and V-bands along with the results of two radio flux measurements (moments of radio observations estimated from Fig. 1 by van den Oord et al. 1985) and peak intensities relative to the continuum in emission components of three

K. Davidson et al. (eds.), Physics of Luminous Blue Variables, 265–267.

lines noted in the figure. The results obtained by the author are marked with filled symbols, open symbols denote the data from literature. We have preferred the peak intensities rather than the equivalent widths bearing in mind the difficulties with the rectification of P Cygni-type profiles (van Blerkom 1970). (The random error in these peak-intensities was estimated ±0.03 F_c (1σ) on the basis of continuum level fluctuations.)

The right-hand part of Fig. 1 allows a comparison of photometric variations in V-filter with the behavior of parameters, characterizing absorption components in spectral lines. Again, we have measured the "peak intensity", i.e. the flux relative to the continuum at the deepest point in the absorption component, and the position of this deepest point on the wavelength scale, giving the v_{rad} - picture in Fig 1b, based on four lines: H9 (circles), *He I* 3965 Å (quadrates), *He I* 3820 Å (triangles) and *O II* 3727 Å (rhombuses). As one can see the accordance of different measurements (open and filled symbols) is acceptable, allowing a reliable analysis of data.

3. CONCLUSIONS

It has been shown earlier (Percy et al. 1988, van Gent and Lamers 1986, Markova, Kolka 1985) that characteristic time-scales are about 40^d . . . 60^d for brightness variations and about 100^d . . . 120^d for radial velocity variations.

Here we obtained (cf. Fig. 1) for the first time a quite long cycle (200^d or more) in the velocity picture of the Balmer-line H9. In the behavior of other lines a cycle of 50^d - 60^d is possible (cf. the neighbourhood of JD 2445200!). Very interesting is the correlation of the maximum absorption depth with corresponding velocity-variations. This correlation in the 50^d . . . 60^d-cycle probably can be extended to the brightness oscillations considering additionally the general similarity of the photometric behavior at different epochs (cf. Fig. 1 and Percy et al. 1988).

As a possible explanation for the variations connected with the brightness-cycle we suggest the cyclically repeated high-density periods in spectrum-forming regions in the wind of P Cygni. Then the higher flux level in continuum (and, consequently, in related emission-peaks) is caused by a geometrical factor - a larger stellar disk, which is opaque at corresponding wavelengths. Greater absorption depths at these moments obviously indicate the higher opacity level in the line-forming region. The displacements on the velocity-scale can be accounted for by varying physical conditions, which are the result of density fluctuations. Longer periods (they may occur also in radio flux (cf. Fig. 1 and van den Oord et al. 1985)) of 100^d . . . 200^d may reflect the more smoothed (but cyclic!) changes in wind structure also related (directly or indirectly) to the density of the outflowing gas.

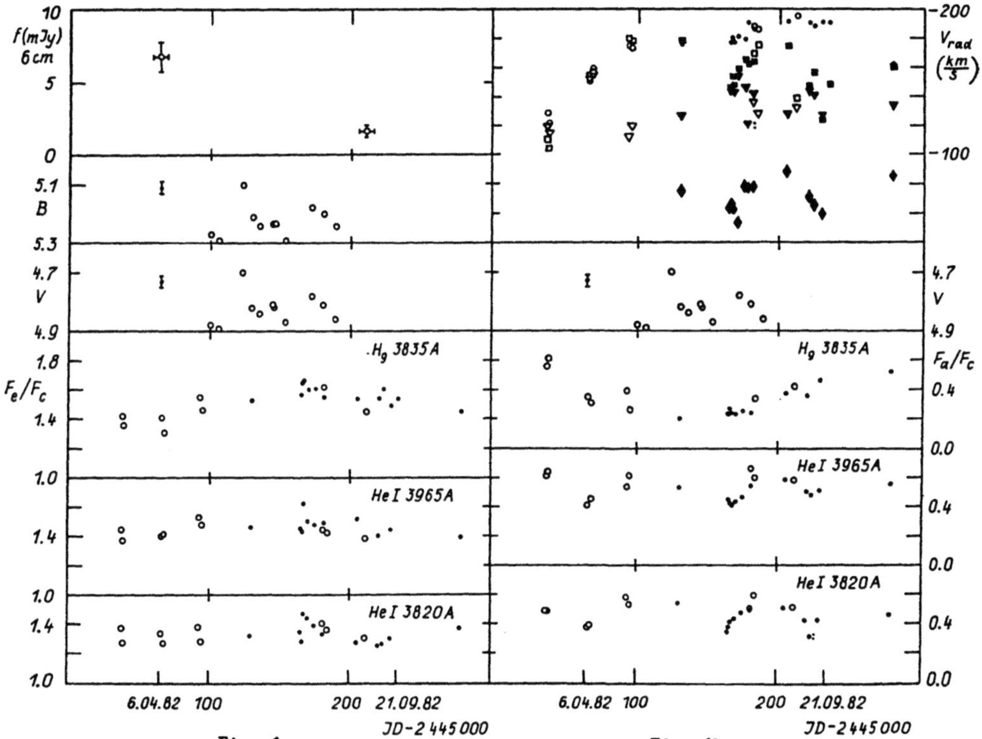

Fig. 1a Fig. 1b

(See text for explanations)

REFERENCES

Hayes D.P. 1985, *Astrophys. J.* **289**, 726.

Kolka I. 1983, *Publ. Academy of Sci. Estonian SSR* **32**, 51.

Lamers et al. 1985, *Astron. Astrophys.* **149**, 29.

Luud et al. 1977, *Publ. Tartu Astrophys. Obs.* **45**, 113.

Markova N. 1986, *Astrophys. Sp. Sci.* **123**, 5.

Markova N., Kolka I. 1985, *Astrofizika* **23**, 539.

Markova N., Kolka I. 1988, *Atlas of spectral lines of P Cygni in 1981-83* (in press).

Percy J.R., Welch D.L. 1983, *Publ. Astron. Soc. Pac.* **95**, 491.

Percy et al. 1988, *Astron. Astrophys.* **191**, 248.

van Blerkom D. 1970, *Monthly Not. Roy. Astron. Soc.* **149**, 53.

van den Oord et al. 1985, in *Radio Stars* (eds. R.M. Hjellming and D.M. Gibson), p. 111.

van Gent R.H., Lamers H.J.G.L.M. 1986, *Astron. Astrophys.* **158**, 335.

OPTICAL AND UV INVESTIGATION OF THE ENVELOPE OF AG CARINAE

ROBERTO VIOTTI
Istituto Astrofisica Spaziale, CNR, Frascati, Italy

ALDO ALTAMORE, CORINNE ROSSI
Istituto Astronomico, Universita' La Sapienza, Roma, Italy

ANGELO CASSATELLA
IUE Observatory, ESA, Madrid, Spain

The galactic LBV object AG Car is unique for assembling several interesting features: (i) a light curve typical of Hubble-Sandage variables, (ii) a largely variable P Cygni-type spectrum, (iii) a strong IR excess, and especially (iv) a small ring nebula. Caputo and Viotti (1970) found that during 1949-59 the AG Car spectrum varied between A1 and B0 equivalent spectral types. A dramatic evolution occurred in recent years, when the visual magnitude gradually faded from V=6 in 1981 to V=8 in 1985, followed by a rise of the line excitation from Aeq in 1981 (Wolf and Stahl 1982), to Beq in 1983 (Viotti et al. 1984), and to Ofpe/WN9 in January 1985 (Stahl 1986). This behaviour has represented a unique chance to follow a LBV in different spectral ranges during its major photometric and spectral changes. In the following we present the results of a long term monitoring of the star and of its nebula in the optical and UV. We propose that AG Car evolved in recent times from a red supergiant phase, and that the present LBV phase is characterized by a critical structure of the stellar atmosphere.

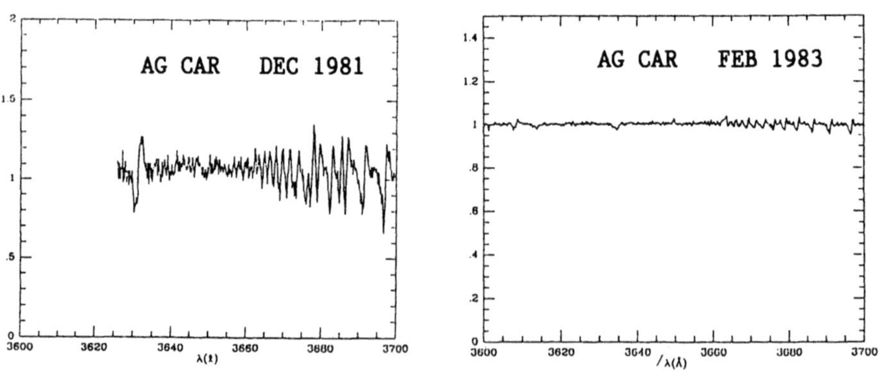

Fig.1. The spectrum of AG Car near the Balmer discontinuity in 1981-83.

K. Davidson et al. (eds.), Physics of Luminous Blue Variables, 268–270.

1. THE OPTICAL SPECTRUM. Coude' spectra of AG Car were obtained in November and December 1981, and during the fading phase in February 1983 (Fig.1). The 1981 spectra show strong Balmer and FeII emissions with P Cygni profile, and no HeI lines, while in February 1983 H and FeII faded and HeI appeared with weak P Cygni profile. In 1981 the Balmer series and the strong FeII lines of multiplet 42 show broad emission wings (Wolf and Stahl 1982). If produced by electron scattering they imply a Tel=9500-15000 K and τe=1.0 for the Balmer lines, and Tel=7000 K and τe=0.5 for FeII 5018 A. We have analyzed the 1981 FeII emission lines with the Self Absorption Curve method (Friedjung and Muratorio 1987), and derived a nearly Boltzmann-type FeII level population with an excitation temperature of 7200+/-350K. From a comparison with theoretical curves, we found that FeII is formed in a high velocity accelerated wind. Hα in Feb.1984 displayed a P Cygni absorption with a sharp edge at -138 km/s, and a narrow emission with wide wings (FWHM=27 A). At minimu in 1987 Hα appeared as a broad emission (Bandiera et al. 1988). The radial velocity of the i.s. CaII appears negative as found by Caputo and Viotti (1970), in agreement with the distance estimate of 2.5 kpc.

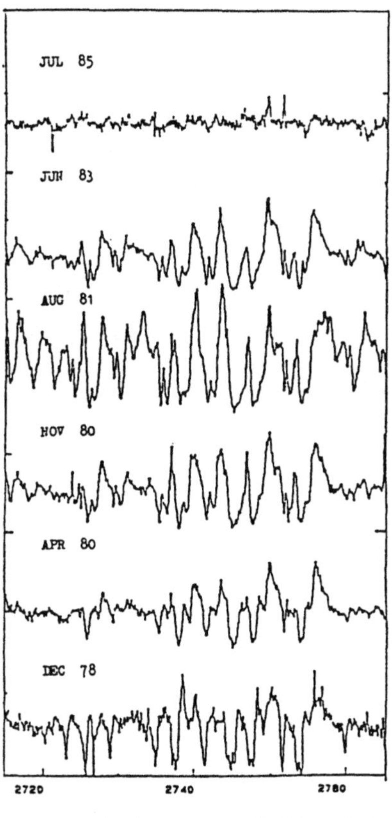

Figure 2. High resolution UV spectra of AG Car in 1978-85.

2. THE UV SPECTRUM. The long lifetime of IUE allowed the study of the UV spectrum of AG Car during its recent major variations. At high resolution the spectrum generally appears rich of FeII lines which strongly affect the energy distribution (Fig.2). The line intensity is variable according to the luminosity phase, and is larger at maximum. A double structure of the absorption component is clearly present during 1980-1983. From a study of UV to IR energy distribution, Viotti et al. (1984) and Viotti (1987) found that the large variations of AG Car occurred at almost constant bolometric magnitude, perhaps due to change of the envelope structure resulting in different energy distributions.

3. THE NEBULA. AG Car is surrounded by a ring nebula, closely resembling the classical Planetary Nebulae, except for the unusual fact that the central nucleus is a luminous P Cygni star! We have monitored the optical and UV spectrum of the nebula, and found at maximum a low excitation spectrum including FeII emissions. The Balmer decrement, when compared with case B, is in agreement with a reddening of about E(B-V)=0.6. Viotti et al. (1988) found that the nebula/star flux ratio is nearly constant. This result, and the

surface brightness distribution suggest that scattering by dust particles in the nebula should be the main source of the observed UV radiation of the nebula.

4. AG CAR AS LBV PROTOTYPE. The study of the recent light and spectroscopic history of AG Car can provide important constraints on the LBV phenomenon, as discussed in the following.

During recent times the effective temperature of AG Car (whatever be the physical meaning of this term) changed from less than 10000 K to more than 30000 K in a few years, whereas Mbol remained nearly the same. The effective radius of unit opacity has correspondently largely decreased. But this does not mean that the star has 'shrinked'! Viotti (1987) suggested that the structure of the atmosphere of AG Car is probably critical, so that small changes of the mass outflow could produce large opacity variations (for instance as a result of ionization of FeII to FeIII) followed by large changes of the 'effective radius' and energy distribution, but without large luminosity variations. Obviously the star has only apparently moved in the HR diagram.

The presence of dust in the nebula is putting severe constraints on the possible evolutionary models for AG Car. If the star is not binary, dust should be produced by AG Car itself. The easiest explanation is that some 10^3-10^4 years ago the star was a red supergiant with intense mass loss and dust formation in the wind. The star should thus have recently moved from this M-supergiant phase to the present LBV phase.

Finally, we would like to note that according to our estimate of the bolometric magnitude (about -8.5 mag) the star is underluminous with respect to most of the known LBVs. Although the distance estimate could be subject to large revision in future, we do not consider at present this result not acceptable, in view of the poor statistics of galactic (and extragalactic) LBVs, and the still poor knowledge of the origin of the variations. From our point of view AG Car is a peculiar luminous star which needs be studied individually, rather than as a member of a class of star. Thus we consider it non-physical to try to force the observations in order to have the 'correct' (i.e. model expected) luminosity.

REFERENCES.

Bandiera, R., et al. 1988, this volume.
Caputo, F., Viotti, R.: 1970, Astron. Astrophys. 7, 266.
Friedjung, M., Muratorio, G.: 1987, Astron. Astrophys. 188, 100.
Stahl, O.: 1986, Astron. Astrophys. 164, 321.
Viotti, R.: 1987, in Instabilities in Luminous Early Type Stars, H. Lamers and C. de Loore eds., Reidel, Dordrecht, p.257.
Viotti, R. et al.: 1984, NASA Conference Publ. 2349, p.231.
Viotti, R. et al.: 1988, Astron. Astrophys. 190, 333.
Wolf, B., Stahl, O.: 1982, Astron. Astrophys. 112, 111.

AG CAR: VARIABILITY, EXTINCTION, DISTANCE AND LUMINOSITY

Henny J.G.L.M. Lamers, Nick Hoekzema, Norman R. Trams
SRON Laboratory for Space Research, Utrecht, the Netherlands

Angelo Cassatella, Mike Barylak
ESA Satellite Tracking Station, Villafranca, Spain

I. INTRODUCTION

AG Car is one of the few Luminous Blue Variables (LBV's) in our galaxy. The star is a photometric variable which had a recent outburst in 1981 (Whitelock et al., 1983; Wolf and Stahl, 1982) and is surrounded by a resolved ring nebula (Thackeray, 1950; Paresce, 1989). These properties indicate that AG Car is a typical LBV. However, at its canonical distance of 2-2.5 kpc, derived from its assumed membership of the Car OB1 or OB2 association, AG Car has a bolometric luminosity of $M_{bol} \simeq -8.5$ to -9.0 and is located significantly below the Humphreys-Davidson limit, and also below the upperlimit for red supergiants (RSG). This implies that AG Car might have been a RSG and that the ring might be due to the interaction of the RSG wind and the LBV wind.

In this paper we present new evidence for a larger distance and a higher luminosity of AG Car. The star has not been a RSG and the ring nebula must be due to an LBV outburst.

II. THE EXTINCTION OF AG CAR

We have studied the UV and visual energy distribution of AG Car at nine different epochs when the star varied in brightness between V=6.75 and 8.42. For each epoch the energy distribution was compared with those of a set of standard supergiants ranging in spectral type between O9 Ia and B9.5 Ia. The energy distribution of AG Car, corrected for interstellar extinction with $0 \leq E(B-V) \leq 1\overset{m}{.}0$ in steps of $\Delta E(B-V) \simeq 0\overset{m}{.}05$ was compared with each standard star. This results for each epoch in a best-fit comparison star and a corresponding value of E(B-V) for AG Car. The result is listed in Table 1.

Although the energy distribution of AG Car varied drastically during the observation period, the resulting value of E(B-V) is consistently 0.63±0.02. The observed visual magnitudes were corrected for extinction with E(B-V) = 0.63 and the observed bolometric correction was applied to derive m_{bol} for five epochs. The values of m_{bol} are about constant with a mean value of $m_{bol} = 3.05 \pm 0.06$.

III. THE DISTANCE AND LUMINOSITY OF AG CAR

We derived an estimate of the distance of AG Car from the extinction versus distance relation for 27 normal stars with $9 < V < 14^m$ within 17' from AG Car. The stars were observed with the Walraven photometric system at ESO. The extinction-free colours were used to derive an estimate of T_{eff} (within ~ 1000

K. Davidson et al. (eds.), Physics of Luminous Blue Variables, 271–272.

TABLE 1

JD -2440000	V AG Car	Best-fit comparison star HD	Type	E(B-V) AG Car	V_o AG Car	BC obs	m_{bol}
3848	6.75	41117	B2Ia	0.65±0.05	4.80	1.70	3.10
4358	7.14	2905	B1Ia	0.65±0.05	5.19	2.23	2.96
4552	6.96	140379	B1.5Ia	0.65±0.05	5.01	1.96	3.05
4820	6.23	58350	B5Ia	0.65±0.05	4.28	1.20	3.08
5491	6.94	52138	B3Ia	0.60±0.05			
5846	7.37	190603	B1.5Ia	0.60±0.05			
6268	7.98	150898	B0.5Ia	0.67±0.05			
6384	7.92	150898	B0.5Ia	0.63±0.03	5.97	2.92	3.05
6606	8.04	150898	B0.5Ia	0.65±0.05			
				0.63±0.02			3.05±0.06

K) and of log g (within 0.1 for the brighter stars and within 0.4 for the faintest ones). The temperatures and gravities were compared with the predicted evolutionary tracks of Hejlesen (1980) and Maeder and Meynet (1987) to derive M_V and the distance. The value of E(B-V) was derived from the observed B-V and the one predicted by the Kurucz (1979) model atmospheres. The E(B-V)-d relation for stars near AG Car shows general increase from E(B-V) $\simeq$ 0.15 at d = 0.3 kpc to E(B-V) $\simeq$ 0.57 at d $\geq$ 6 kpc. The extinction of E(B-V) = 0.63 for AG Car indicates that the star is more distant than 5 kpc. This agrees with the kinematic distance of 7.2 kpc derived by Humphreys et al. (1989). It also agrees with the distance of d $\simeq$ 5±1 kpc derived from the empirical relation between the amplitude of the B-variations and the luminosity of LBV's by Wolf (1989). We adopt a distance of d $\simeq$ 6 ± 1 kpc. With this distance, the luminosity of AG Car is log L = 6.2±0.2 and M_{bol} = -10.8±0.4, which is well above the upper limit for red supergiants.

CONCLUSIONS

1. The extinction of AG Car at all phases is E(B-V) = 0.63±0.02 and the apparent bolometric magnitude at all phases is m_{bol} = 3.05±0.06 magn.
2. The distance derived from the E(B-V)-d relation of nearby stars and from the radial velocity of AG Car is 6±1 kpc.
3. The luminosity of AG Car is log L = 6.2±0.2 or M_{bol} = -10.8±0.4.
4. AG Car has not been a red supergiant. The ring must be the remnant of an LBV outburst. The diameter of the ring is 1.0 pc and its kinematic age is 1.10^4 yrs.

REFERENCES

Hejlesen, P.M. 1980, Astr.Ap.Suppl. 39, 347.
Humphreys, R.M., Lamers, H.J.G.L.M., Hoekzema, N. 1989, Ap.J. (in press).
Kurucz, R.L. 1979, Ap.J.Suppl. 40, 1.
Maeder, A., Meynet, G. 1987, Astr.Ap. 182, 243.
Paresce, F. 1989, These proceedings.
Thackeray, A.D. 1950, M.N.R.A.S. 110, 524.
Whitelock, P.A. et al. 1983, M.N.R.A.S. 205, 577.
Wolf, B. 1989, These proceedings.
Wolf, B., Stahl, O. 1982, Astr.Ap. 112, 111.

The optical micro variations of the two S Dor type stars AG Car and HR Car

A.M. van Genderen[1], P.S. Thé[2], M. Heemskerk[2], D. Heynderickx[3], I. Larsen[1], I. Wanders[1] and N. van Weeren[1].

1. Leiden Observatory, Postbus 9513, 2300 RA Leiden, The Netherlands
2. Astronomical Institute 'Anton Pannekoek', Roetersstraat 15, 1018 WB Amsterdam, The Netherlands
3. Astronomical Institute of the Catholic University of Leuven, Celestijnenlaan 200B, B-3030 Heverlee, Belgium

Abstract. VBLUW photometry of high precision of two S Dor type stars AG Car (in a quiescent stage) and HR Car was made in 1987 and 1988. Both stars show visual micro variations with amplitudes of $0^m.1 - 0^m.2$ with time scales $\bar{P} \sim 10^d$ for AG Car and $\bar{P} \sim 20^d$ for HR Car.

1. Introduction

The first series of VBLUW photometry of high precision for the S Dor type star AG Car in a quiescent stage, showed that micro variations occur of the same type as for R71 in the LMC (van Genderen et al. 1988, hereafter called Paper VI). The characteristic time scale appeared to be $\bar{P} = 13^d.8$. New observations were made between December 1987 and March 1988 together with the nearby S Dor type star HR Car.

2. Results

The observations were conducted with the 90-cm Dutch Telescope equipped with the Walraven VBLUW photometer and is located at the ESO, La Silla, Chile.
The observational technique is described in Paper VI. For both variables we used the comparison star HDE 303311 (B9V). The average mean error $(\frac{\sigma}{\sqrt{n}})$ per data point variable minus comparison star amounts to at most ± 0.001 log intensity scale.

Figure 1 shows the light curves of both variables relative to the comparison star as a function of Julian Date. The oscillations of AG Car show a time scale of $\sim 10^d$, which is not much different from the one found for the previous series of observations discussed in Paper VI. Even during the rise by $\sim 0^m.15$ in the last sequence, $\bar{P}$ stays the same. The reddening of the colours during this stage (not shown here), suggests that a new S Dor phase may be impending.

K. Davidson et al. (eds.), Physics of Luminous Blue Variables, 273–274.

HR Car oscillates with a time scale of ~ 20^d. The total range is 0.045 (log int.). It appears that the colour variations are much larger than for AG Car and R71. Since HR Car is not quite in its minimum stage (this should be somewhere near V_J = 8.7), it is possible that part of the colour variations originates in the circumstellar shell.

3. Conclusions

We conclude that also HR Car exhibits optical micro variations similar to AG Car, R71 and all other blue and yellow α Cygni variables (but non-S Dor type stars).

The oscillations of AG Car at the end of our sequence of observations occur at a level ~ $0^m.15$ brighter than for the previous series. The reddening of the colours during the sudden rise suggests that a new S Dor phase with enhanced mass loss is impending. However, the time scale of the oscillations did not change appreciably. One of the purposes to observe these variables regularly, is to see what happens with the oscillations if the star starts a new phase of excessive mass loss.

Reference

van Genderen, A.M., Thé, P.S., Augusteijn, Th., Engelsman, E.C. van der Grift, E.W., Prein, J.J., Remijn, R.L., Steeman, F.W.M., van Weeren, N. 1988, Astron. Astrophys. Suppl. 74, 453

Figure 1

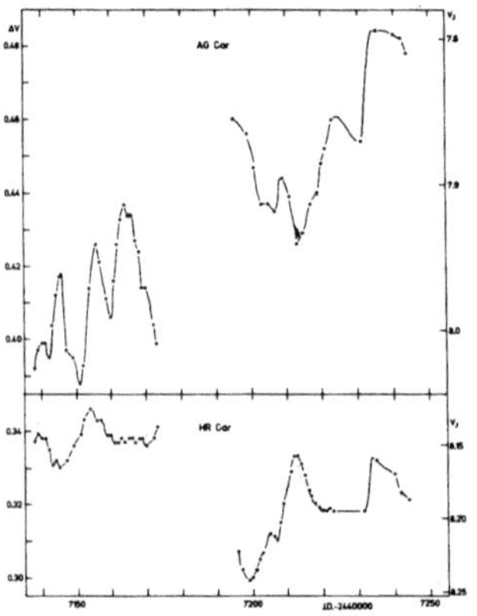

The light curves of the micro variations of AG Car and HR Car relative to the comparison star (in log intensity scale). For comparison purposes the V scale of the UBV system (with subscript J and in magnitude scale) is indicated at the right hand side.

ON THE PERIODICITY OF HUBBLE-SANDAGE OBJECTS IN M 31

G. R. IVANOV

Department of Astronomy, University of Sofia, 1126 Sofia, Bulgaria

The spectral and photometric characteristics of the Hubble-Sandage objects (HSO) are similar to those of the η Car and S Dor variables. They are actually referred to as S Dor-type in the General Catalogue of Variable Stars. Their light variations differ from star to star.

The aim of the present paper is to discuss a possible periodicity in HSOs in M 31. We used phtometry from Hubble(1929), Sharov(1973, 1981), Rosino and Bianchini(1973), and added some observations taken during 1980-87. The latter were obtained with the 2-m Ritchey-Chretien-Coude and 0.5-m Schmidt telescopes of the Bulgarian National Astronomical Observatory - Rozhen.

Table 1 containes the observational data on four HSOs. Columns 2-4 contain data from nine Schmidt-telescope plates. The absorption in the line of sight in the Galaxy is adopted to be E_{B-V}=0.08 and within the M31 associations it is up to 0.24 (Massey et al.,1986). Hence we decided to use E_{B-V}=0.08 for AE And and AF And. The variables VA-1 and V15 are located within the spiral arms near dust clouds and they are probably influenced by internal absorption in M31. For both of them we used E_{B-V}=0.24. In luminosity the HSOs are similar to the brightest blue supergiants. Their location on the colour-magnitude diagram suggests that they are young massive stars in a rapid evolutionary stage.

The HSOs are irregular variables. Still we attempted to look for periodicity using the method of Deeming(1975) - Fourier analysis. The periodicities we find range from 800 to 3700 days (Table 1). And indeed, Hubble(1929) noted a five-year cycle in AF And, which coincides with our value. The period-luminosity diagram of the M31 Cepheids is shown in Fig. 1. The four HSOs are marked with crosses. It appears that they lay well on the continuation of the period-luminosity relation.

Back in the 50s, Hubble and Sandage(1953) suggested that HSOs are appropriate as extragalactic distance indicators. However, their 5 to 10- year-long cycles make it inpractical for them to be used as such. Looking for possible explanations to Fig. 1, let us note that the equation $P\sqrt{\rho}$=Q could be transformed to:

$$\log P + 0.248\ M_{bol} + 3 \log T_{eff} - 12.50 = \log Q$$

We used the relations of deJager (1980) for converting the intrinsic colours B-V and M_B into T_{eff} and M_{bol}. The data from Table 1 give Q = 25.1 . This value is too large to explain the luminosity variations by pulsations.

The spectroscopic observations of Kenyon and Gallagher(1985) suggested that the HSOs might be massive binary systems. Variations in the properties of the stellar wind could be a possible cause for their variability.

K. Davidson et al. (eds.), Physics of Luminous Blue Variables, 275–276.

TABLE 1
HUBBLE-SANDAGE OBJECTS IN M 31

STAR	B	B-V	M_B	PERIODICITY (days)
AF And	16.6	-0.02	-7.8	1700
AE And	17.1	0.01	-7.3	1200
VA-1	16.8	0.20	-8.1	3670
V 15	18.1	0.40	-6.8	862

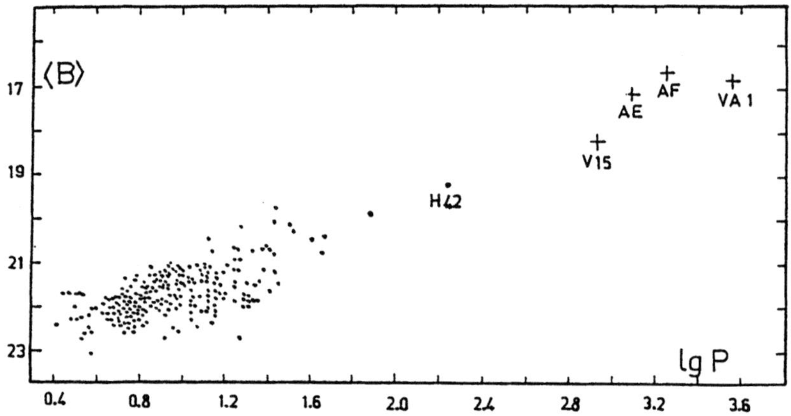

REFERENCES

Deeming, T.J. 1975, Ap. Space Sci., 36, 137.

De Jager, C. 1980, The Brightest Stars (D.Reidel: Dordrecht).

Hubble, E.P. 1929, Ap. J., 69, 103.

Hubble, E.P., and Sandage, A.R. 1953, Ap. J., 118, 353.

Kenyon, S.J., and Gallagher III,J.S. 1985, Ap. J., 290, 542.

Massey, P., Armandroff, T.E., and Conti, P.S. 1986, A.J., 92, 1303.

Rosino, L., and Bianchini, A. 1973, Astr. Ap., 22, 453.

Sharov, A.S. 1973, Peremenye Zvezdy, 19, 3.

Sharov, A.S. 1981, Peremenye Zvezdy, 21, 485.

ETA CARINAE: UV CONSTRAINTS ON POSSIBLE MODELS.
THE CLOSE BINARY HYPOTHESIS

GIOVANNI BATTISTA BARATTA
Osservatorio Astronomico, Viale del Parco Mellini 84
00136 Roma, Italy

ROBERTO VIOTTI
Istituto Astrofisica Spaziale, CNR
Via Enrico Fermi 21, 00044 Frascati, Italy

η Car is one of the brightest stars in our Galaxy. For its luminosity, huge mass loss, large variability can be considered as an ideal laboratory to study the LBV phenomenon, and to give constraints on possible models of LBVs. We propose that η Car is a - possibly binary - F-type hypergiant whose wind is heated by dissipation of mechanical energy.

Table 1. Basic data (and references) on η Car.

distance	2500	(7)	mass loss	0.075 M☉/yr	(3)
bol mag	0.0 mag	(3)	terminal vel	500-800 km/yr	(4)(5)
radiative power	$5x10^6$ L☉	(3)	E(B-V) interst	0.4 +/- 0.1	(7)(10)
mechan power	$7x10^6$ L☉	(3)	E(B-V) circum	0.7:	(3)(5)
eff temperature	7000-10000 K	(9)	Si/C abundance	0.3-2	(10)
eff radius	~1000 R☉	(10)	el density	10^9-10^{11} cm^{-3}	(10)

1. ULTRAVIOLET. With the aim of discussing possible models for η Car, we have analyzed (10) the UV spectrum of the star and compared with ground based observations. The simultaneous presence of both low and high ionization resonance lines with broad P Cygni profile suggests a wide ionization range throughout the whole wind, which is hard to explain with a simple photoionization model. The wind is most likely collisionally ionized. The most intense emission lines (NIII], SiIII], high excitation FeII and FeIII) display broad and narrow emission components, indicating that the expanding envelope of η Car is asymmetric with low and high velocity regions. The weakness or absence of CIII] 1909 A in emission and the strength of SiIII] (Fig.1) can be explained by a Si/C overabundance of a factor >3 to >20 (10). C and O are however present in the UV spectrum of η Car with saturated CII and OI resonance lines, whereas CIV is rather weak. Note also that OI is present with a strong IR emission (1)(8). We believe that at present no accurate abundance estimate is possible in the absence of a reliable wind model of η Car.

K. Davidson et al. (eds.), Physics of Luminous Blue Variables, 277–278.

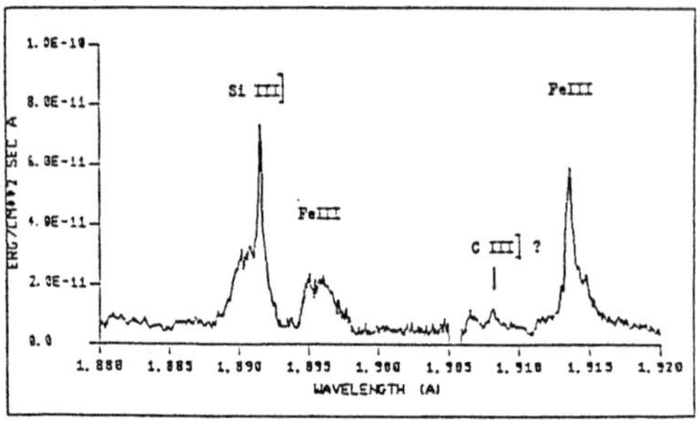

Fig.1 The UV spectrum of η Car near the SiIII and CIII lines.

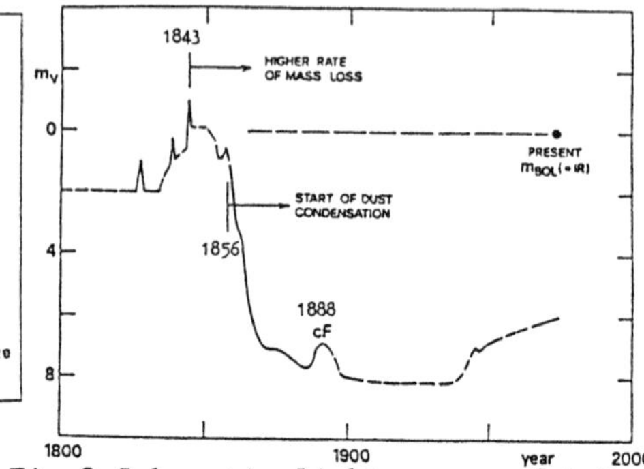

Fig.2 Schematic light curve of η Car

2. VARIABILITY. During the last century η Car underwent dramatic light variations (Fig.2). The large fading started in 1856 was due to the <u>start of dust condensation in the wind</u>, but the total power remained the same (3). The event could have been caused by an increase of the mass loss around 1843, followed by opacity enhancement in the wind, which favoured fast moleculae and dust grains formation (<u>dust catastrophe</u>)(6). The light history, luminosity and lifetime of η Car suggest that it is a massive object (>100 M☉), which is presently a rather <u>cool star</u> (Teff<10000K) in a rapid evolutionary stage (9). As in AG Car and other LBVs, the smaller light fluctuations and related spectral changes are likely caused by <u>small structural changes of the wind</u> followed by significant opacity variations, <u>not by shell ejection</u>.

3. BINARITY. Speckle observations revealed that η Car is a multiple system composed of 4 stellar objects, one much brighter than the others (11). Thus η Car is a <u>Trapezium-like system</u>, and the individual components may likely be close binary systems. Many of the peculiarities of η Car (e.g. the wide ionization range and the X-ray emission) could be caused or enhanced by interaction in a close binary system. Binarity is also suggested by the asymmetries of the emitting envelope and of the circumstellar nebula (the <u>homunculus</u>), suggesting preferential directions of mass ejection. Binary interaction might eventually affect the wind structure and be the origin the observed variations of η Car and possibly of other LBVs.

REFERENCES

Allen, D.A., Jones, T.J., Hyland, A.R.: 1985, Ap.J. 291, 280 (1).
Altamore, A. et al.: 1986, ESA SP-263, 303 (2).
Andriesse, C.D., Donn, B.D., Viotti, R.: 1978, MNRAS 185, 771 (3).
Bensammar, S. et al.: 1985, Astr. Ap. 146, L1 (4).
Pagel, B.E.J.: 1969, Nature 221, 325 (5).
Stencel, R.E.: 1986, in Circumstellar Matter, Reidel, p.529 (6).
Tapia, M. et al.: 1988, MNRAS 232, 661 (7).
Thackeray, A.D.: 1969, MNASSA 28, 37 (8).
Viotti, R., Andriesse, C.D.: 1981, The Most Massive Stars, ESO, 141 (10)
Viotti, R. et al.: 1988, in preparation (10).
Weigelt, G., Ebersberger, J.: 1986, Astr. Ap. 163, L5 (11).

HIGH RESOLUTION EMISSION LINE PROFILES IN BLUE LUMINOUS STARS

RINO BANDIERA
Osservatorio Astrofisico di Arcetri, Firenze, Italy

PAOLA FOCARDI
Istituto Astronomico, Universita' di Bologna, Italy

ALDO ALTAMORE, CORINNE ROSSI
Istituto Astronomico, Universita' La Sapienza, Roma

OTMAR STAHL
Landessternwarte, Heidelberg, Germany

Emission lines are often observed in high luminosity stars and provide evidence of the presence of extended stellar envelopes. Ha is the most frequently observed across the H-R diagram, but lines of HeI or FeII are also found in emission in these stars. They could be used as diagnostics of the structure of their outer atmospheres and winds. High resolution (1/dl $\sim 10^5$) high S/N profiles of Ha and HeI 5876 in the galactic LBVs η Car, AG and HR Car, and in the LMC star S22 have been obtained with the ESO CAT-CES during 1984-87, and are described in Figs.1-5. We find that these stars show a large variety of profiles with narrow and broad emissions, wide or multiple blue-shifted absorptions. The profiles are largely variable. Once, a kind of inverse P Cyg profile was observed in HR Car (Fig.4). These results indicate the presence of large scale phenomena and high velocity fields which are dramatically variable in time. Continuous HIRES monitoring of these stars is urgently needed.

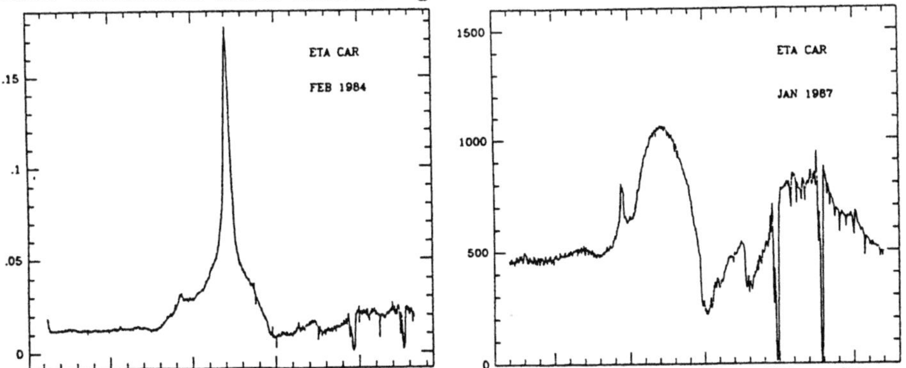

Figure 1. The large changes of the HeI 5876 A line in η Car.

K. Davidson et al. (eds.), Physics of Luminous Blue Variables, 279–280.

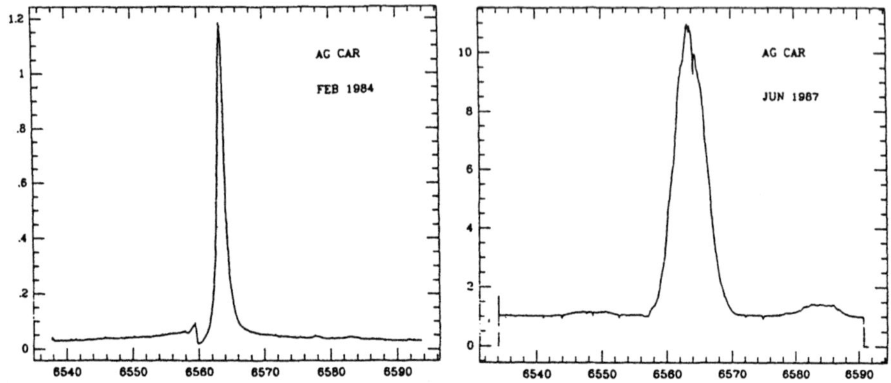

Figure 2. Variation of the Ha profile in AG Car.

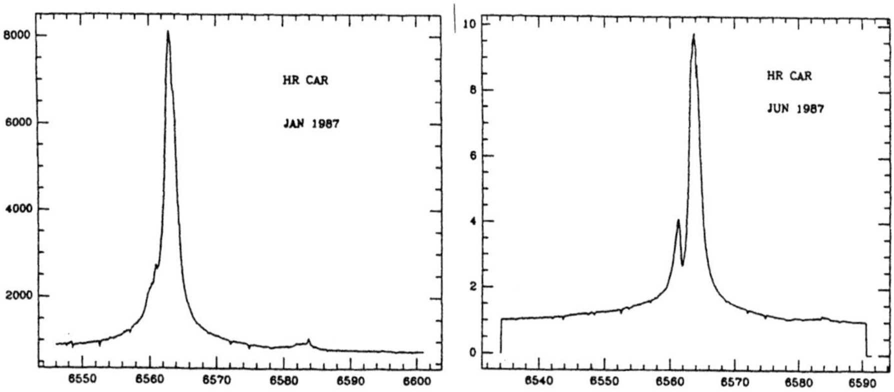

Figure 3. Variation of the Ha profile in HR Car.

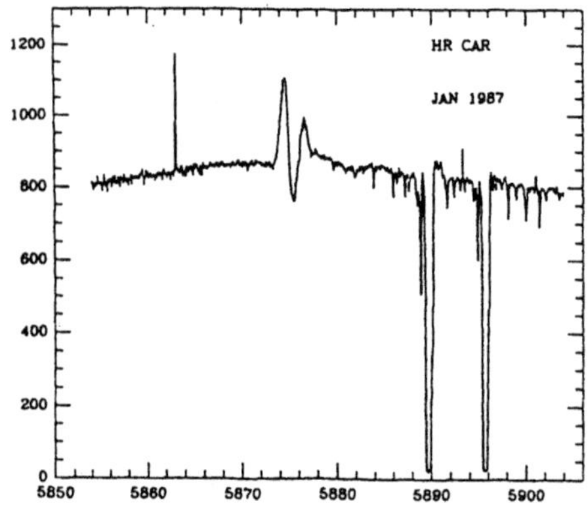

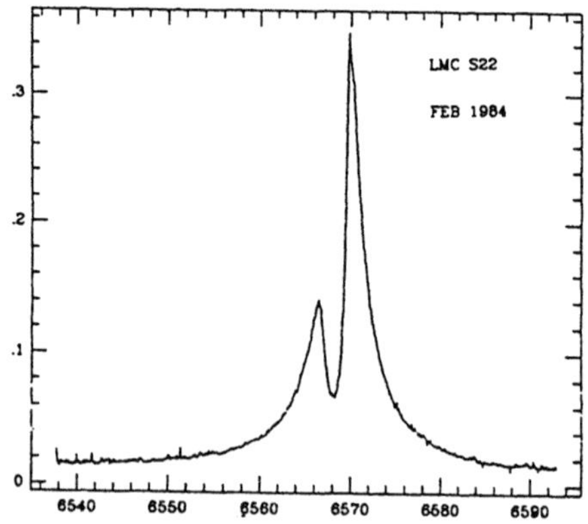

Fig.4. HeI 5876 A in HR Car.

Fig.5. Ha profile in S22.

ANALYSIS OF LBV CIRCUMSTELLAR ENVELOPES USING FeII LINES

CORINNE ROSSI
Istituto Astronomico, Universita' La Sapienza, Roma

GERARD MURATORIO
Observatoire de Marseille, Marseille, France

MICHAEL FRIEDJUNG
Institut d'Astrophysique, CNRS, Paris, France

ROBERTO VIOTTI
Istituto Astrofisica Spaziale, CNR, Frascati, Italy

Emission and absorption lines of FeII frequently dominate the optical and UV spectra of LBVs. Many luminous blue stars possess a rich optical FeII emission line spectrum, which can be used as diagnostic of their extended atmospheres (Viotti 1976a). Since the launch of IUE several LBVs were observed in UV, and in many cases (η Car, AG Car, P Cyg, Hubble Sandage variables, etc.) their spectra were found dominated by a large amount of FeII absorptions, presumably formed in cooler envelopes. Often these absorptions are so strong as to seriously affect the UV energy distribution. In order to study line formation outside LTE of such a complex ion, the FeII line intensities in luminous stars were analyzed using the Self Absorption Curve (SAC) method by Friedjung and Muratorio (1987). The fitting of the emission line fluxes in different multiplets to theoretical SACs obtained with different wind models, leads to the determination of the relative upper and also relative lower level populations. The overlapping of the individual multiplet curves gives the observational SAC which provides an estimate of the wind parameters (column density, acceleration parameter a, mass loss rate). This method of analysis was applied to the optical and UV high resolution spectra of several luminous stars. For instance we found that in AG Car the relative population of the FeII levels from 1.6 to 8.6 eV can be fitted to the same Boltzmann distribution with an excitation temperature of Tex=7200+/-350K, whereas for the VV Cep star KQ Pup the linearity extends from 0 to about 9 eV with Tex=6700 K (Viotti et al. 1988).

The results of the analysis were used to compute the theoretical UV spectra of hot stars with cool (=FeII) envelopes to be compared with low resolution UV observations of LBVs. To evaluate the effects of different stellar and wind parameters on the FeII absorption/emission, we have computed synthetic spectra in the range T*=10000 to 20000 K, and Mbol=-8 to -10. An example is given in Fig.1. We found that the energy distribution is strongly affected by the FeII transitions which are crowded in

K. Davidson et al. (eds.), Physics of Luminous Blue Variables, 281–282.

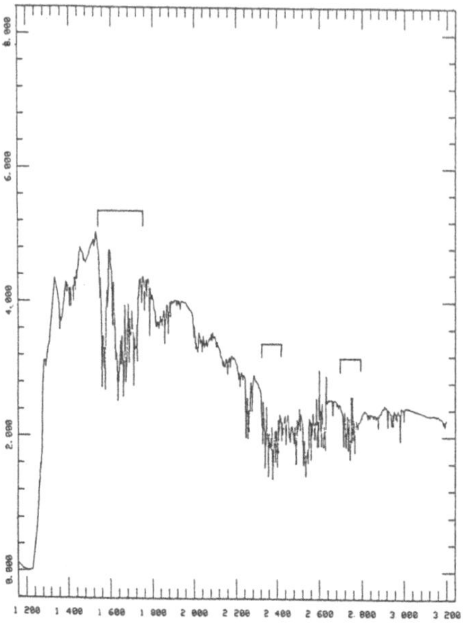

Fig.1. Synthetic spectrum of a hot star with a cool envelope. The selected FeII regions are indicated.

some characteristic broad regions (e.g. Viotti 1976b). This fact not only makes the identification of FeII in the low res spectra of faint objects easier, but also might allow a determination of the mass loss $\dot{M}$. To study this problem we derived for each model the FeII line absorption/emission in 3 ranges (Fig.1): 1540-1750A, 2330-2420A, and 2700-2800A. Some results for a characteristic wind velocity V of 300 km/s and for different values of Mbol, $\dot{M}$ and <u>a</u> are shown in Fig.2. We note that an increase of $\dot{M}$ produces an increase of the line blocking in the far-UV, while in the near-UV the absorption is filled-in by the prominent emission, especially from multiplets 62 and 63. The dependence on Mbol is less important, whereas in an accelerated wind (<u>a</u>=2.5) the near-UV line emission is smaller (Fig.2). These results indicate that the FeII UV features can be used as mass loss indicators of LBVs of intermediate temperature.

REFERENCES

Friedjung, M., Muratorio, G.: 1987, Astron. Astrophys.<u>188</u>, 100.
Viotti, R.: 1976a, Astrophys. J. <u>204</u>, 293.
Viotti, R.: 1976b, Mon. Not. R astr. Soc. <u>177</u>, 617.
Viotti, R., Rossi, C., Muratorio, G., Friedjung, F., Baratta, G.B.: 1988, 'A Decade of UV Astronomy with IUE', ESA SP-281, p.357.

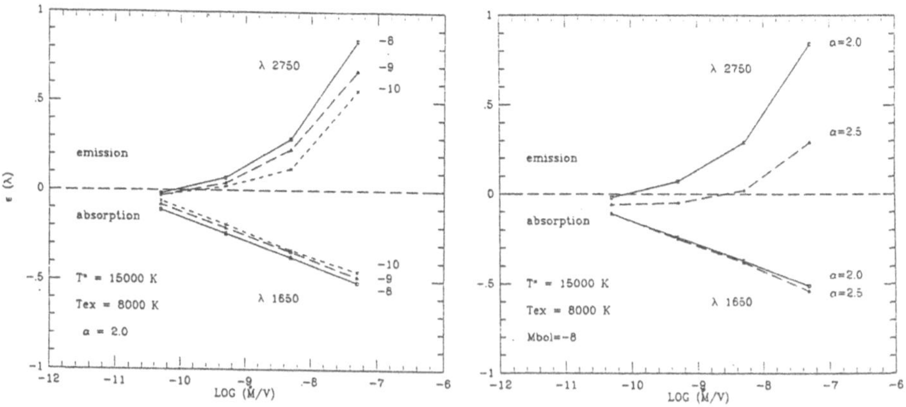

Fig.2. FeII line blocking/emission in the UV.

ON THE INTERPRETATION OF EMISSION WINGS OF BALMER LINES IN LBV'S

Ivan Hubeny and Claus Leitherer
Joint Institute for Laboratory Astrophysics
University of Colorado and National Bureau of Standards
Boulder, Colorado 80309-0440 USA

The optical spectra of Luminous Blue Variables (LBV's) are characterized by strong Balmer emission lines. In many cases, the narrow emission component (FWHM $\lesssim$ 300 km/sec) shows extended emission wings reaching up to ~1000 km/sec. These wings are generally attributed to electron scattering of line photons in the dense stellar wind (Bernat and Lambert 1978). In this paper we discuss an alternative mechanism which may partly contribute to the observed line wings.

Our computations are based on the model-atmosphere program TLUSTY (Hubeny 1988). These plane-parallel NLTE atmospheres assume hydrostatic and radiative equilibrium and include H and He only. For T_{eff} = 15,000 K (a typical value for LBV's), the lowest gravity, for which a stable model exists is log g = 1.65. We emphasize that we neglect the influence of sphericity but the effect discussed here originates in deep layers, where sphericity is very likely negligible.

The run of temperature vs. mass density shows the well-known NLTE behavior. The temperature assumes a minimum value around log m $\approx$ -1 and increases further out due to increased heating in the Balmer continuum. The location of the temperature minimum is sensitive to log g; the lower log g, the deeper the temperature minimum is shifted in the atmosphere. This behavior can be understood in terms of the lower gas pressure at a given log m resulting from lower gravity. A lower pressure is associated with lower number densities and thus lower opacity so that optical depth unity in the relevant transitions occurs at smaller column density. Abbott and Hummer (1985) discussed the analogous effect for O stars.

Figure 1 shows the resulting H_α profiles of the three NLTE models with log g = 2.5, 2.0, 1.65, and for comparison, of a log g = 1.65 LTE model. The lowest log g NLTE model shows pronounced emission wings extending a few hundred km/sec to each side of the line center. They are due to an NLTE coupling of the Balmer continuum and the Balmer lines. The wings are formed at the region of the temperature minimum where $S_\nu(H\alpha) > B_\nu$. We have $S_\nu(H\alpha) > B_\nu$ since the n = 2 level of hydrogen is depopulated relative to LTE by photoionizations in the Balmer continuum. The rate of photoionizations is higher deeper in

K. Davidson et al. (eds.), Physics of Luminous Blue Variables, 283–284.

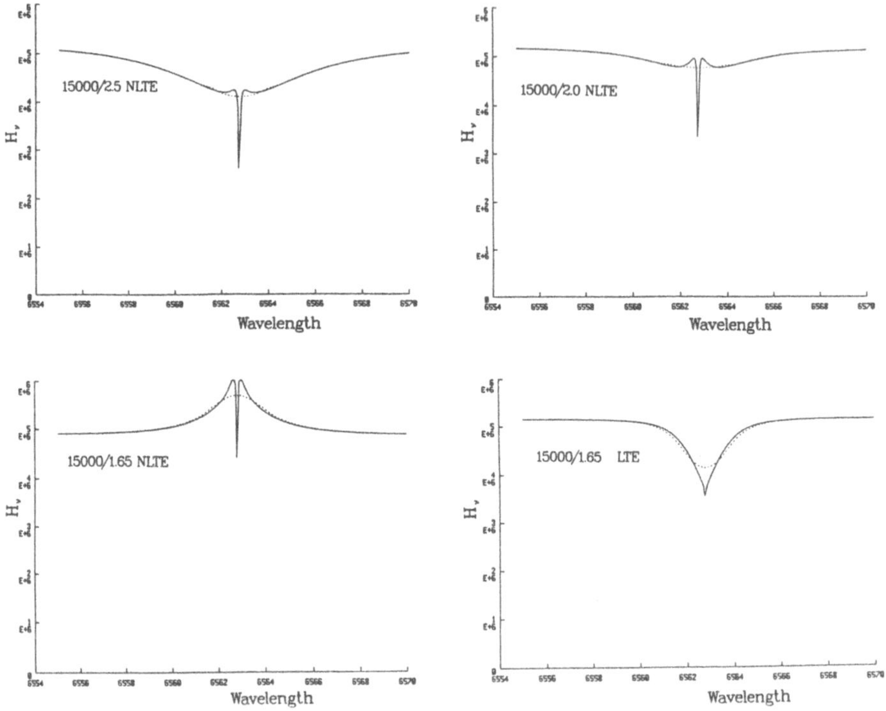

Fig. 1 NLTE H_α profiles for different values of log g. For comparison, the corresponding LTE profile is also shown. The dashed lines are rotationally broadened profiles with vsin i = 50 km/sec.

the atmosphere so that this effect is more pronounced if the wings are formed deeper in the atmosphere, i.e., if log g is lower.

The emission wings in H_α produced by a hydrostatic, plane-parallel, low-gravity NLTE atmosphere are in qualitative agreement with the observations although they are still weaker. For even lower log g values (which are derived in LBV's), however, the emission wings can be expected to strengthen considerably. In realistic models (including the photosphere and the stellar wind), the traditional electron-scattering mechanism and the effect discussed here will operate. A determination of wind properties assuming the wings are purely due to electron scattering may therefore be misleading.

ACKNOWLEDGMENTS. C. L. gratefully acknowledges support by a Feodor-Lynen-Fellowship of the Alexander-von-Humboldt Foundation and by NSF grant AST 85-20728 through the University of Colorado. I. H. is a 1987-88 Visiting Fellow at JILA.

REFERENCES

Abbott, D. C., and Hummer, D. G. 1985, Ap. J., **294**, 286.
Bernat, A. P., and Lambert, D. L. 1978, Pub. A. S. P., **90**, 520.
Hubeny, I. 1988, Computer Phys. Commun. (in press).

ATMOSPHERIC MODELS FOR LBV'S AT MINIMUM AND MAXIMUM STATES

Claus Leitherer,[1] Werner Schmutz,[1] David C. Abbott,[1]
Ana V. Torres-Dodgen,[1] Wolf-Rainer Hamann,[2]
and Ulf Wessolowski[2]
[1] Joint Institute for Laboratory Astrophysics, University of Colorado and National Bureau of Standards
[2] Institut fuer Theoretische Physik and Sternwarte der Universitaet, Universitaet Kiel

We study the relation between photospheric and wind parameters of LBV's on the basis of self-consistent atmospheric models. Our theoretical model consists of a spherically extended, dynamical NLTE atmosphere for the radiative transfer and a modified CAK code for the hydrodynamics. These codes are combined in an iterative scheme to achieve self-consistency.

The observed photometric variability is generally interpreted in terms of a variable photospheric temperature $T_{2/3}$ and radius $R_{2/3}$ with constant luminosity $L = 4\pi\ R^2(2/3)\ \sigma\ T^4(2/3)$. During maximum phase (when $T_{2/3}$ is around 8000 K), we find from our radiatively driven wind models that $\dot{M}$ increases strongly relative to the minimum phase. The increase in $\dot{M}$ is generated by a recombination of doubly to singly ionized iron-group elements. Our models predict mass-loss rates a factor of ~30 higher in the maximum than in the minimum and wind terminal velocities of about 200 km/sec for both states.

The derived mass-loss variations refer to the region of the atmosphere which is optically thin to continuum radiation. On the other hand, the observed photometric variations originate in photospheric regions where $\tau \simeq 1$. We investigate the possibility that the higher densities in the atmosphere due to higher $\dot{M}$ produce higher opacities and thus a larger photospheric radius. In such a scenario, the photospheric and wind variability could be accounted for self-consistently if this mechanism were self-initiating due to some instability mechanism (Appenzeller 1986).

T_{eff} is defined by the luminosity and the stellar radius $L = 4\pi\ R_*^2\ \sigma\ T_{eff}^4$ where R_* is chosen small enough to ensure sufficient Rosseland optical depth. According to this definition, T_{eff} and R_* are constant and remain unaffected by different mass-loss rates. Models with high $\dot{M}$ have $T_{2/3} > T_{eff}$ and $R_{2/3} > R_*$. Figure 1 illustrates the effect of increasing $\dot{M}$ on the energy distribution of R71. The models use T_{eff} = 16000 K, R_* = 81 $R_\odot$ (from the observed energy distribution at minimum), a velocity law with β = 1 (derived from the hydrogen line profiles), and log $\dot{M}$ = -6.2, -5.2, -4.2. Part of the UV flux is redistributed to longer wavelengths. However, the increase in UBV is $\lesssim$ 0.3 mag. Even using vastly different velocity laws, the

K. Davidson et al. (eds.), Physics of Luminous Blue Variables, 285–286.

observed variability could not be reproduced. The increase of $R_{2/3}$ cannot be explained by opacity effects with constant R*.

Davidson (1987) derived a relation between $\dot{M}$, the characteristic temperature of the photosphere T_o, and the velocity of the photosphere v_o. Our results imply that his diagnostic diagram can not be used to model an LBV at both maximum and minimum with one fixed index n of his velocity law $v(r) \sim r^n$. This is basically due to the enormous variation of v_o associated with a variation of $\dot{M}$. If log $\dot{M}$ = -6.2, continuum formation occurs deep in the subsonic region whereas v_o is of the order of the sound velocity for log $\dot{M}$ = -4.2. As a consequence, $R_{2/3}$ increases rather weakly even at high mass-loss rates if R* is constant.

We assume the core radius R* itself is variable between minimum and maximum. A converged model for R71 is derived from a fit to the observed energy distributions and the line profiles in minimum and maximum states. The results are summarized in the table below.

	Log L	M	T_{eff}	R*	$T_{2/3}$	$R_{2/3}$	log g_{eff}	log $\dot{M}$	v_∞	β
Minimum	5.5	12	16000	84	15500	88	1.3	-6.2	170	1.0
Maximum	5.5	12	10000	230	8100	332	0.0	-4.7	110	1.0

Notice the low gravity both in minimum and in maximum and the low mass. According to evolutionary tracks by Maeder and Meynet (1987), R71 is a post-RSG with an initial mass of ~25 $M_\odot$. The stellar wind properties derived from the line-profile analysis are in agreement with the prediction of the radiatively driven wind theory. However, we cannot self-consistently account for the observed radius variability. The mechanism responsible for the variable photospheric radius appears to have its origin in deeper, sub-photospheric layers.

This work has been supported by NSF grants AST88-02937 and AST88-06594 through the University of Colorado. C. L. acknowledges receipt of a Feodor-Lynen Fellowship of the Alexander-von-Humboldt Foundation.

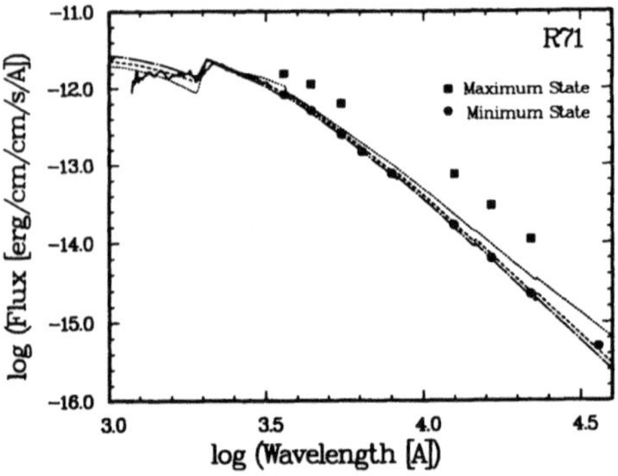

Fig. 1. Observed and calculated energy distributions of R71 at minimum and maximum. The IUE spectrum (solid line) refers to the minimum. The three calculated models have log $\dot{M}$ = -6.2 (dash-dotted), -5.2 (dashed), and -4.2 (dotted) assuming a constant core radius.

Appenzeller, I. (1986) in IAU Symp. 116, "Luminous Stars and Associations in Galaxies," Dordrecht: Reidel, p. 139.
Davidson, K (1987) Ap. J., 317, 760.
Maeder, A., and Meynet, G. (1987) Astr. Ap., 182, 243.

ATMOSPHERIC PARAMETERS AND ACCELERATIONS IN THE OUTER PARTS OF LUMINOUS HOT STARS

Hans Nieuwenhuijzen and Cornelis de Jager
Astronomical Institute and Laboratory for Space Research
Beneluxlaan 21, 3527 HS Utrecht, the Netherlands

1. OUTLINE

In the atmospheres of the most extreme luminous stars, close to the Humphreys-Davidson limit, the inward gravitational acceleration is for a great part compensated by outward accelerations due to radiation pressure, turbulent pressure and dynamic pressure of the stellar winds. As a result the effective acceleration is very small, resulting in blown-up atmospheres that can no longer be considered plane-parallel or in hydrostatic equilibrium.

In order to estimate the physical conditions in the outer parts of the atmospheres of extreme stars an iteration procedure was developed based on the equations

$$g_{eff} - \frac{k_R}{\tau_R} P = 0 \quad (1)$$

$$k_R = k_R(P,T) \quad (2)$$

$$-g_{eff} = -g_{grav} + g_{grav}\Gamma_{Edd} + \frac{1}{2}\frac{d(v^2)}{dr} - \frac{\psi}{2\rho}\frac{d}{dr}(\rho s^2) \quad (3)$$

$$T(\tau_R) = f(T_{eff})$$

Here τ_R is the Rosseland mean optical depth of the level considered,
k_R the Rosseland mean absorption coefficient at that level.

The functions (2) and (4) are read from Kurucz's tables and models, while the function ψ is an empirical ratio between the squared microturbulent velocity component $<\xi_\tau^2>$ and the squared velocity of sound, s^2. The three last terms at the right-hand side of eq.(3) are respectively the radiative acceleration, and the accelerations resulting from dynamic wind pressure and turbulence.

The system of equations is solved iteratively with $g_{eff} = 0.5\ g_{grav}$ as a starting value. The calculations presented here refer to the level $\tau_R = 0.1$.

K. Davidson et al. (eds.), Physics of Luminous Blue Variables, 287–288.

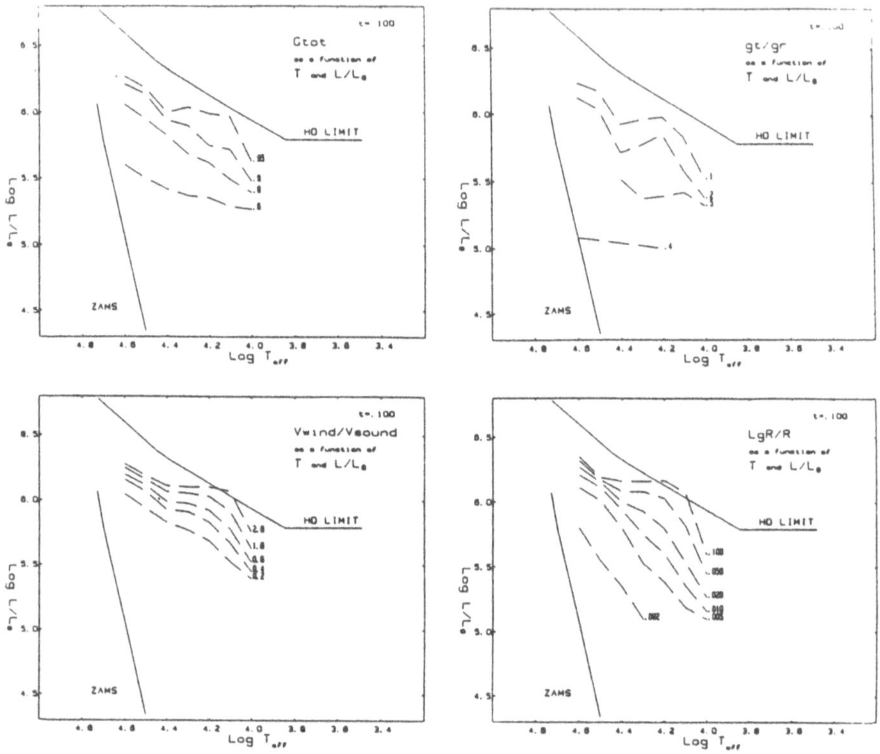

2. RESULTS

The figures give, for the upper part of the Hertzsprung-Russell diagram respectively: $-g_{out}/g_{grav}$ (G_{tot}); g_{turb}/g_{rad} (gt/gr); v_w/s and log $R/R_\star$ (LgR/R), where v_w is the wind velocity and $R-R_\star$ is the geometric distance between the levels τ_R=0.1 and τ_R=2/3, the latter assumed to refer to the star's radius.

The diagrams show that at the HD limit $g_{out} \approx -g_{grav}$, and that the stars' atmospheres become quite extended there.

Crucial for the understanding of the LBV phenomena seems that the upper boundary of the LBV area in the Hertzsprung-Russell diagram coincides with the line $v_w \approx s$ and where the ratio g_{turb}/g_{rad} drops rapidly to a small value. Inside the LBV-area the ratio g_{turb}/g_{rad} reaches large values, close to 0.5 and even higher. This shows that dynamic effects may be predominant for explaining the LBV phenomenon.

The final version of this paper has been submitted to Astr. Astrophys.

NLTE ANALYSIS OF R84

W. Schmutz[1], C. Leitherer[1], A.V. Torres-Dodgen[1],
M. Vogel[2], P.S. Conti[1], W.-R. Hamann[3], U. Wessolowski[3]
(1) Joint Institute for Laboratory Astrophysics
University of Colorado and National Bureau of Standards
Boulder, Colorado 80309-0440, USA
(2) Institut für Astronomie
ETH Zentrum, 8092 Zurich, Switzerland
(3) Institut für Theoretische Physik und Sternwarte
Universität Kiel, 2300 Kiel, Germany

The LMC star R84 (=HDE 269227 =BR 18) belongs to the group of Ofpe/WN stars believed to be closely related to the Luminous Blue Variables. Support for such a relation comes from the spectral resemblance of these stars to AG Car during visual minimum, and from the observed outburst of the Ofpe/WN star R127.

The spectral analysis of R84 presented here is based on model calculations with the NLTE comoving-frame code described by Wessolowski et al. (1988) and references therein. The helium model atom was represented by 28 levels and hydrogen by 9 levels. The free model parameters were varied until the observed line profiles and the absolutely calibrated and dereddened continuum flux were reproduced. The comparison of the theoretical continuum flux distribution with the observed one yields a reddening of $E_{B-V} = 0.1$.

It was possible to determine the stellar temperature within an intrinsic error of only a few 100 K. However, systematic effects, e.g. from the adopted velocity law, are expected to be important. Therefore, the accuracy of the derived parameters can be estimated only. We believe the error in temperature to be of order 2000-3000 K. For the other parameters (Table 1), we believe the luminosity and the mass-loss rate to be accurate to within a factor of two, and the helium abundance uncertain by ±0.1.

Comparing the stellar luminosity and the helium abundance with the predictions of evolutionary models (Maeder and Meynet, 1987), we conclude that R84 is in the Post-Red-Supergiant phase and that its initial and current masses are 25 $M_\odot$ and 15 $M_\odot$, respectively.

Table 1. The parameters of the model that reproduced best the observed spectrum of R84 at January 13, 1982:

log $L/L_\odot$	R_*	T_*/K	$R(\tau=2/3)$	$T(R_{2/3})$	log $\dot{M}/(M_\odot/yr$	N[He]
5.5	24 $R_\odot$	28500 K	29 /$R_\odot$	26000 K	-4.75	0.63

K. Davidson et al. (eds.), Physics of Luminous Blue Variables, 289–290.

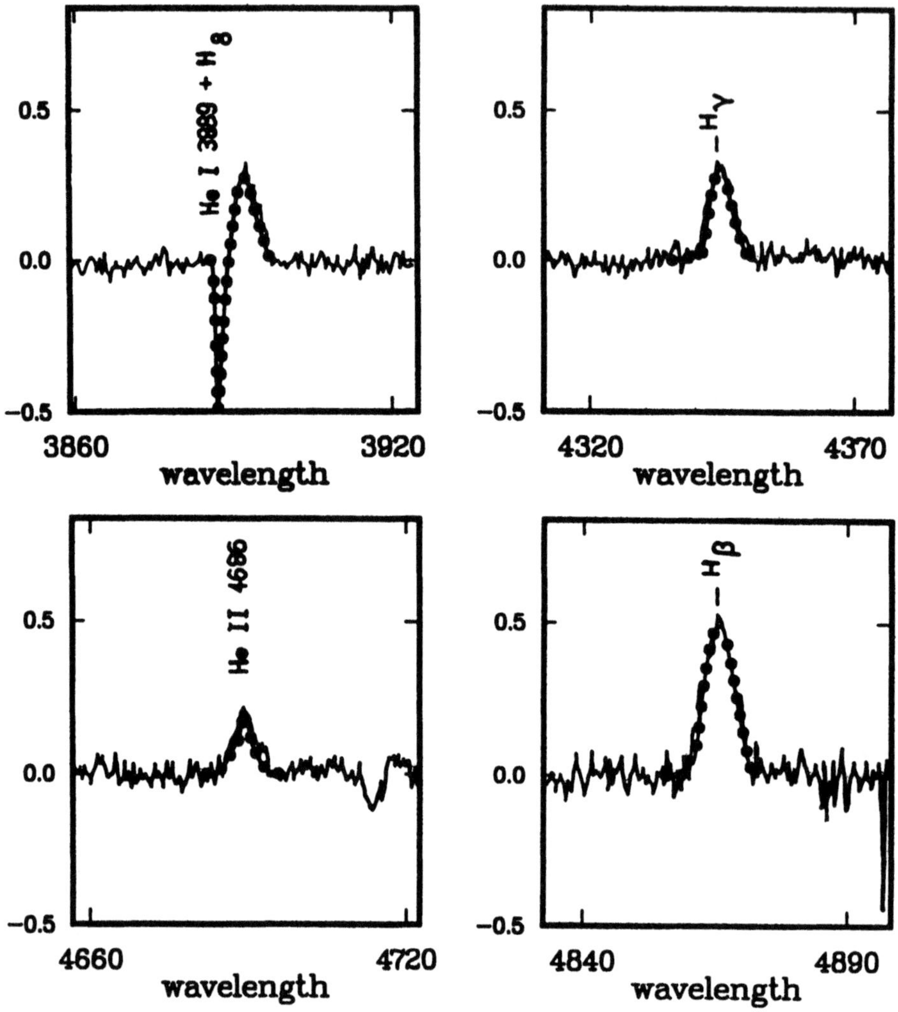

Figure 1. Observed line profiles of R84 at Jan. 13, 1982 (Stahl et al. 1982). Superimposed are the synthetic profiles calculated with the parameters given in the text. Note, that the normalized fluxes are plotted logarithmically.

References

Maeder,A., Meynet,G.: 1987, Astron. Astrophys. 182, 243

Stahl,O., Wolf,B., de Groot,M., Leitherer,C.: 1985, Astron. Astrophys. Suppl. 61, 237

Wessolowski,U., Schmutz,W., Hamann,W.-R.: 1988, A. & A. 194, 160

Acknowledgements. This work was supported by the Swiss National Foundation and by the NSF Grant AST-8802937. C.L. acknowledges receipt of a Feodor-Lynen Fellowship from the Alexander-von-Humboldt Foundation.

Physics of Instabilities in Radiatively Driven Stellar Winds

S.P. Owocki, Bartol Research Institute/ University of Delaware
J.I. Castor, Lawrence Livermore National Laboratory
G.B. Rybicki, Harvard-Smithsonian Center for Astrophysics

One type of variability that is ubiquitous among luminous blue stars is that associated with their massive stellar winds. Ultraviolet spectral lines show variable narrow absorption components that may arise from dense shells or clumps in the wind; the observed soft X-ray and non-thermal radio emission from these stars may be produced in embedded wind shocks associated with these shells. The line-driving of such winds is known to be strongly unstable, and recent numerical simulations of the dynamical, nonlinear evolution of such instabilities indicate that such shocks and dense shells do indeed form (Owocki, Castor, and Rybicki 1988; hereafter OCR). However, the structure computed in these dynamical simulations is quite different, and in some sense opposite, what was anticipated in earlier heuristic shock models (e.g. Lucy 1984, Krolik and Raymond 1985, Abbott 1988); the highest speed material is very *rarefied*, not dense, and the strongest shocks that form are of the *reverse*, not forward, type. In subsequent simulations we have found that the detailed character of this structure can depend on many variables, but this tendency always to form high speed rarefied waves and reverse shocks is quite robust. It is thus important that we have a clear understanding of the reasons for this tendency, and it is on this point that we focus our comments here.

As pointed out in OCR, this formation of high speed rarefactions is a natural consequence of the fact that the inward-propagating (relative to the fluid) radiative-acoustic mode, for which the density and velocity variations are nearly opposite in phase, is much more unstable than the outward mode, for which density and velocity are nearly in phase. In a formal, mathematical sense, this last result was already clear from the linear stability analysis of Owocki and Rybicki (1984), but we now offer the following brief explanation of its physical basis:

For ordinary acoustic waves, the velocity and density perturbations are exactly in phase for the outward mode, but opposite in phase for the inward mode. Addition of a perturbed line force with a component in phase with the velocity causes work to be done on the wave, thereby tending to amplify it. However, during this amplification the velocity and density can no longer remain exactly in or exactly out of phase. This is because mass continuity requires that the associated intrinsic temporal decrease or increase in the perturbed density must be balanced respectively by a positive or negative divergence (i.e.

K. Davidson et al. (eds.), Physics of Luminous Blue Variables, 291–292.

gradient) in the perturbed velocity. The sense of the required phase shift is such that the perturbed pressure force (which varies as the negative of the perturbed density gradient) now has a component that opposes the line-force for the outward mode, but reinforces it for the inward mode. The result is that the outward mode is only weakly unstable, whereas the inward mode is strongly unstable. Amplification of such unstable inward mode waves naturally leads to wind structure in which the highest velocity material is very rarefied, and kinematic steepening of this structure naturally leads to the formation of the reverse shocks.

Note that the only ***nonlinear*** effect here is the simple kinematic steepening of the large amplitude waves into shocks. Relatively complicated, ***dynamical***, nonlinear effects like line shadowing, which played such a prominent role in earlier heuristic shock models, play a rather minor role here, only affecting various details of the structures. Thus this very robust property of the nonlinear wind structure is really a straightforward consequence of the properties of the ***linear*** instability. If we had only been clever enough, we could have predicted it! Of course, we usually aren't so clever, and it is the great value of numerical simulations that, when carefully analyzed, they can teach us basic physics that we could have known without ever doing them.

We thank J. Cassinelli for thoughtful questions that helped stimulate us to develop a sounder physical understanding of our numerical results. The computations described here were performed at the San Diego Supercomputer Center and the Lawrence Livermore National Laboratory. This work was performed in part under the auspices of the U.S. Department of Energy by the Lawrence Livermore National Laboratory under Contract No. W-7405-ENG-48. S.P.O. acknowledges partial support from NSF grant AST 86-11824.

References

Abbott, D.C. 1988, in *Solar Wind VI*, V.J. Pizzo, ed., NASA Conference Proceeding, in press.

Krolik, J. and Raymond, J.C. 1985, *Ap. J.*, **298**, 660.

Lucy, L.B. 1982, *Ap. J.*, **255**, 286.

Owocki, S. P. and Rybicki, G. B. 1984, *Ap. J.*, **284**, 337.

Owocki, S. P., Castor, J. I., and Rybicki, G. B. 1988, *Ap. J.*, **335**, in press. (OCR).

FAR INFRARED CIRCUMSTELLAR "DEBRIS" SHELLS: CLUES TO THE EVOLUTION OF MASSIVE STARS?

R.E. STENCEL and C.D. GARMANY
Center for Astrophysics and Space Astronomy
University of Colorado
Boulder, Colorado 80309-0391 U.S.A.

ABSTRACT: A survey of the infrared properties of late–type supergiant stars, using the IRAS database, reveals that about 25% of the stars possess resolved, extended circumstellar shells. These shells are typically several arc minutes in apparent size and therefore on the scale of parsecs at the source. Furthermore, among the resolved sources, there is an inverse correlation between physical size of the infrared shell and B–V color, suggesting that these shells are formed while the objects are red supergiants, but continue to expand ballistically, while the star evolves blueward from that extreme. These shells may be the material swept up into ring nebulae when the central star develops a fast wind.

1. Introduction

Despite the comparatively coarse angular resolution of the *Infrared Astronomy Satellite* (IRAS), it is now known that it resolved a number of stellar sources. Perhaps the most extreme example is the peculiar G supergiant variable R CrB, where the 18 arc-minute far infrared shell was estimated to contain between 0.2 and 6 $M_\odot$ of material (Gillett et al. 1986).

The survey of the infrared properties of late–type supergiants in the catalog of Humphreys (1978) by Stencel, Pesce and Bauer (1988, 1989) found that many late–type supergiants show evidence for dust in their infrared flux distribution, and that about 25% are resolved at 60μm. More importantly, among the resolved sources, the physical extent of the multi–arc-minute infrared shells appear to be largest among the supergiant stars with smallest B–V color.

2. Discussion

We argue that this observation is significant for understanding the evolution of massive stars, perhaps including the LBV stars, for the following reasons. The LRS data indicate that the dust forming phase is most efficient while the star is at the reddest extreme of its evolutionary track, based on the presence of the 10μm silicate feature. This and the inverse correlation with B–V color suggest that the resolved, infrared shell is the debris of that phase of evolution when dust forming occurred, and the subsequent expansion is simply ballistic. The multi–parsec sized shells among the G and F supergiants require $\sim 10^5$ years

K. Davidson et al. (eds.), Physics of Luminous Blue Variables, 293–294.

to expand at 20 km/sec, the typical red supergiant wind velocity.

As the star continues its blueward evolution, ultimately the radiation field of the star should begin to alter the infrared characteristics of the debris shell, in the sense of first dissociating, then ionizing and shocking the inner material as a function of the surface temperature and stellar wind characteristics of the central star.

Van Buren and McCray (1988) have discussed the infrared shells of various early-type stars, and described several bow wave situations where the wind and stellar motion are subsonic (e.g. δ Per, B5 III) and the resoved infrared shell shows a bow wave shape. An expanded survey of A and F type supergiants may reveal similar structures. At the extreme, the galactic Wolf–Rayet ring nebulae (Chu et al. 1983), which share similarities in physical size and material content with the debris shells of late–type stars, could be the evolutionary consequence of the continued evolution of the central, massive star within such a structure. The analogy with lower mass stars that form planetary nebulae, is difficult to resist.

McGregor et al. (1988) in their study of the infrared emission from the ring surrounding the galactic S Doradus variable AG Car conclude that the ring structure is most likely due to material that was lost by AG Car during a prior evolutionary phase and subsequently swept up bv the present hot superwind phase. Similarly, the optical and ultraviolet light echoes of Supernova 1987A are beginning to reveal the evolutionary history of the stellar wind in that former star. In principle, therefore, a more complete survey of the optical and infrared structures of intermediate spectral type objects may reveal an important evolutionary link between red and blue phases for 30 to 60 $M_{\odot}$ stars, perhaps along the lines of suggestions by Lamers et al. (1983). Such infrared inventories of OB Associations are underway by Stencel, MacConnell and others.

We are grateful for partial support of this research from NASA grants JPL 957632 and NAG5-816 to the University of Colorado.

REFERENCES

Chu, Y., Treffers, C. and Kwitter, K. 1983 *Ap.J.Suppl* **53**, 939.
Gillett, F., Backman, D., Beichman, C. and Neugebauer, G. 1986 *Ap.J.* **310**, 842.
Humphreys, R. 1978 *Ap.J.Suppl.* **38**, 309.
Lamers, H., deGroot, M. and Cassatella, A. 1983 *Astron. Astrophys.* **123**, L8.
McGregor, P., Finlaysson, K., Hyland, A., Joy, M., Harvey, P. and Lester, D. 1988 *Ap.J.* **329**, 874.
Stencel, R., Pesce, J. and Bauer, W. 1988 *A.J.* **95**, 141.
Stencel, R., Pesce, J. and Bauer, W. 1989 *A.J.* submitted.
Van Buren, D. and McCray, R. 1988 *Ap.J.* **329**, L93.

HIGH-DISPERSION SPECTROSCOPY OF THE B[e] SUPERGIANT S 111

O. STAHL
Landessternwarte Heidelberg, Königstuhl
J. SMOLINSKI
N. Copernicus Astronomical Center, Torun
B. WOLF and F.-J. ZICKGRAF
Landessternwarte Heidelberg, Königstuhl

ABSTRACT. We studied the luminous emission-line star S 111 in the LMC with high-dispersion spectroscopy and classified it as a new member of the class of B[e] supergiants.

S 111 is a little studied emission-line star in a dense cluster in the LMC which is known as HDE 269599. A finding chart with the identification of the star in the cluster is given by Appenzeller et al. (1984). Stahl and Wolf (1986) discuss the evidence for a circumstellar shell around the star. More recently McGregor et al. (1988) found from photometry and spectroscopy in the near-IR that the star (their HDE 269599s) has an infrared excess due to hot dust. We obtained high-dispersion and high S/N-ratio spectra in the wavelength ranges 3900-4950 Å and 5750-6800 Å with CASPEC and one LWP-IUE spectrogram.

The optical spectrum of S 111 is dominated by the Balmer lines which are strongly in emission. These lines have P Cygni structure and the absorption component is clearly split in two components. The terminal velocity of the wind as measured from the Balmer lines is 645 km sec^{-1}. The He I lines also have P Cyg structure with a very similar edge velocity (640 km sec^{-1}). In addition to the Balmer lines and the He I lines, the optical spectrum of S111 shows many narrow emission lines of FeII, TiII, CrII, VII and NaI. Forbidden lines of [FeII], [NII], [SII], [NiII] and [OI] are also present. A section of the red spectrogram of S 111 is shown in Figure 1. Gaussian profiles have been fitted to the profiles of selected, unblended emission lines. Both the allowed and forbidden lines give an emission line velocity of 266 km sec^{-1}, which we use as the systemic velocity. The FWHM of the emission lines (uncorrected for the instrumental width) are 75 km sec^{-1} for the lines from allowed transitions (mainly FeII lines) and 40 km sec^{-1} for the forbidden lines. The instrumental width is about 15 km sec^{-1}, so the lines are well resolved.

The IUE UV spectrum shows many multiplets of FeII and the resonance doublet of MgII. The FeII lines in the UV show P Cyg structure with a smaller terminal velocity than the P Cyg lines in the optical ($\approx$ 250 km sec^{-1}). MgIIλ2800 is in emission. A P Cyg absorption component seems to be present, but in this case most of the absorption is probably of interstellar origin. The observations clearly show that S111 should be classified as a new B[e] supergiant. The characteristics of these

K. Davidson et al. (eds.), Physics of Luminous Blue Variables, 295–296.

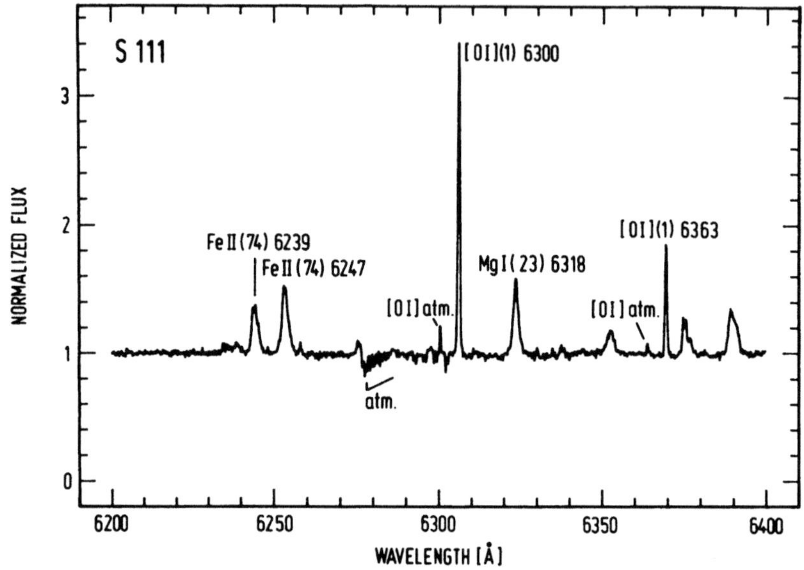

Figure 1: Section of the red spectrogram of S111 around the [OI] lines at 6300, 6363 Å. The [OI] lines are clearly much sharper than the lines of FeII and MgI.

stars are (see e.g. Zickgraf et al. 1986) strong Balmer emission lines and lines of FeII, [FeII] and other forbidden emission lines, most strongly usually [OI]λλ6300, 6363. Likewise the LWP-spectrum with the numerous FeII-lines is typical for B[e]-stars (cf. e.g. the case of R66 described by Stahl et al. 1983). All these stars also have a strong IR excess indicative of hot circumstellar dust. In all known cases the forbidden emission lines are much narrower than the stellar wind velocity, thus suggesting a different zone for the origin of these lines. This region most likely is a circumstellar disk.

References

Appenzeller, I., Klare, G., Stahl, O., Wolf, B., Zickgraf, F.-J.: 1984, *ESO Messenger* **38**, 28

McGregor, P.J., Hillier, D.J., Hyland, A.R.: 1988, *Astrophys. J.*, in press

Stahl, O., Wolf, B., Zickgraf, F.-J., Bastian, U., de Groot, M.J.H., Leitherer, C.: 1983, *Astron. Astrophys.* **120**, 287

Stahl, O., Wolf, B.: 1986, *Astron. Astrophys.* **158**, 371

Zickgraf, F.-J., Wolf, B., Stahl, O., Leitherer, C., Appenzeller, I.: 1986, *Astron. Astrophys.* **163**, 119

SLASH STARS: TWO TYPES, TWO SCENARIOS ?

P. S. Conti and Bruce Bohannan
Joint Institute for Laboratory Astrophysics
Astrophysical, Planetary and Atmospheric Sciences
University of Colorado, Boulder, Colorado, U.S.A.

Of/WN, or as we will call them here, "slash '/'" stars were first described by Walborn (1977; 1982a; 1982b) as hot luminous stars that have spectroscopic properties of both O and low-ionization WN spectral classes. Walborn called attention to Sk -67° 22, an O3If*/WN6-A star that had strong O-type absorption lines, but also a *broad* emission feature at HeII λ4686Å. The '/' symbol was used to indicate that there was difficulty in assigning either the Of or the WN type unambiguously. He labled other LMC stars as Ofpe/WN9 not only because of a similar ambiguity in the classification but also the presence of narrow "nebular" emission lines associated with the star. Conti and Garmany (1983) pointed out the similarity of Sk -71° 34 to Sk -67° 22. A total of ten stars with Ofpe/WN9 character have now been identified in the LMC (Bohannan and Walborn 1989) : HDE 269927C, 269445, 269582, 269687, 269858, 269227; Sk -66 040, -67 266, -69 297; BE 381.

There has also been a long standing classification problem among hot luminous stars in distinguishing between the "most extreme Of star" (HDE 313846, Hutchings 1979) and the "least extreme WN star" (HD 93162, Conti 1976). Leep (1979) has investigated the spectra of known strong (emission) line Galactic Of stars and weak (emission) line WN stars and was able to classify them both to the O and to the WN type. We asked ourselves whether all Of stars such as these should be considered as '/' stars and concluded that this would be pushing the classification too far. We thus adopted as an arbitrary condition that the emission line strength of HeII λ4686Å had to be at least 1A equivalent width. With this criterion about half of the known Of stars (disregarding the Walborn (f) and ((f)) types) and a very few WN objects might be considered as '/' stars. These candidates are listed in table below, where we have generally adopted the Of and WN subtypes of Leep (1979) but also re-examined their spectra. The designation before the '/' is the leading classification; that is, if the emission spectrum dominates over the absorption one, the WN comes first, and conversely.

It will be noticed that the Of/WN stars in the LMC and the Galactic candidates contain mostly relatively late WN subtypes with a range of Of classes up to O4; a few stars combine both an early WN and an O3f or O4f subtype. It is these combinations we propose as Of/WNL "(/L)" or Of/WNE "(/E)" subtypes.

The quantitative spectroscopic properties of Of stars may pass smoothly through the '/' objects to pure WN characteristics. In proceeding from Of to WN types, the stellar winds increase in strength; it is possible that the composition also alters gradually in favor of enhanced helium and nitrogen line strengths and abundances (Voels *et al.* 1989; McGregor *et al.* 1989). We would imagine that this transition would further

K. Davidson et al. (eds.), Physics of Luminous Blue Variables, 297–298.

involve weakening and violet shifting of the absorption spectrum, with a concomitant strengthening of the emission line spectrum, as the wind becomes more optically thick. Alternatively, perhaps the Of/WN stars are binaries, with the Of and WN spectra coming from the individual stellar components? Although it is impossible to assert that a given star cannot be a binary, studies of several objects in our list have not revealed a single case of a double-lined system (*e.g.* Conti, *et al.* 1977; Conti *et al.* 1979).

Table of Candidate '/' Stars in the Galaxy

Star	*Walborn*	Slash '/'	Star	*Walborn*	Slash'/'
HD14947	O5If+	O5f/WN8	HD93131	WN6-A	WN6/O4f
HD15570	O4If+	O4f/WN8	HD93162	WN6-A	WN6/O4f
HD16691	O4If+	O4f/WN8	HD150958	O6.5Ia(n)f+	O6.5f/WN9
ζ Pup	O4I(n)f	O4f/WN8	HD152386	O6:Iafpe	O5.5f/WN9
HD92740	WN7-A	WN7/O5f	HD152408	O8:Iafpe	O8f/WN9
HD93129	O3If*	O3f/WN4.5	HDE313846	O7:Iafpe	O7f/WN9

The slash '/' stars may be a transition group between Of and WN in their spectroscopic properties. Are they also a set of stars in transition for their evolution? Do the early and late subtype combinations foretell a somewhat different origin, or fate? One of the original Walborn (/L) stars, HDE 269898 = R127, had a recent violent outburst (Stahl *et al.* 1983) which now with a spectrum similar to S Doradus at maximum light (Wolf *et al.* 1988) qualifies it to be a prominent member of the Luminous Blue Variable (LBV) family. Will others of these '/' stars also become variable at some time in the near future? Further observational and theoretical work is critical to a clearer physical understanding of these stars which appear to play significant—but still poorly understood—roles in the evolution of massive stars.

This research was supported in part by grants from the National Science Foundation to PSC and to BB. Support of early stages of this work by a Small Research Grant from the American Astronomical Society to BB is gratefully appreciated.

REFERENCES

Bohannan, B. and Walborn, N.R. 1989 submitted to *ApJ.*

Conti. P.S. 1976 in *Proc. 20th. Colloq. Int. Ap. Liege*, p. 193.

Conti, P.S., Garmany, C.D., and Hutchings, J.B. 1977, *ApJ.* **215**, 561.

Conti, P.S., Niemela V.S., and Walborn, N.R. 1979 *ApJ.* **228**, 206.

Conti, P.S. and Garmany, C.D. 1983 *Pub. Astron. Soc. Pac.* **95**, 411.

Conti, P.S. and Massey, P. 1989 *ApJ.* in press.

Hutchings, J.B. 1979 *Pub. Astron. Soc. Pac.* **91**, 361.

Leep, E.M. 1979, in *Mass Loss and Evolution of O-Type Stars (IAU Symp. No.83),* eds. P.S. Conti and C.W.H. DeLoore (Dordrecht:D. Reidel), p. 471.

McGregor, P.J., Hillier, D.J., and Hyland, A.R. 1989 submitted to *ApJ.*

Stahl, O., Wolf, B., Klare, G., Cassatella, A., Krautter, J., Persi, P., Ferrari-Toniolo, M. 1983, *Astron. & Astrophys.* **127**, 49.

Voels, S.A., Bohannan, B., Abbott, D., Hummer, D.G. 1989 *ApJ.,* in press.

Walborn, N.R. 1977 *ApJ.* **215**, 53.

Walborn, N.R. 1982a *ApJ. Letters* **254**, L15.

Walborn, N.R. 1982b *ApJ.* **256**, 452.

Wolf, B. Stahl, O., Smolinskik, J., and Cassatella, A. 1988 Astr. Ap. Suppl.. **74**, 239.

Blobs in the Winds of Wolf-Rayet Stars

C. Robert, L. Drissen and A. F. J. Moffat,
Département de Physique, Université de Montréal.

Abstract. Concentrations of material have been observed to propagate in the winds of many W-R stars. These are short-lived, stochastic features of unknown origin.

1) High precision polarimetry

Thomson scattering is the main process responsible for the linear polarization in the dense and highly ionized winds of W-R stars. A net, non-zero level of polarization results as a consequence of an asymmetric distribution of the scatterers (electrons) or light source(s).

Linear polarization data (with an average rms error of $\pm$ 0.015%) have been obtained since 1984 for a sample of about 30 W-R stars (St.-Louis *et al.* 1987, 1988; Drissen *et al.* 1987; Robert *et al.* 1988a, b). All these objects display random polarimetric variations with a typical timescale of a day or less (after subtracting off the periodic modulation in the case of a binary system). Figure 1 shows a clear correlation between the polarization scatter and the spectral subtype especially if one separates WN and WC. Cooler, slow-wind stars of late subtypes (especially WN8 and WC9) show the strongest variations, while hotter, fast-wind stars are quieter.

Blobs of dense material are proposed to be propagating outwards in the winds of all W-R stars. These blobs are formed less frequently, are partially destroyed or have less time to grow in the case of the faster stellar winds (Drissen *et al.* 1987; Robert *et al.* 1988a).

2) High resolution spectroscopy

High signal-to-noise (300-500) and high resolution (0.1 Å) spectra have revealed small bumps particularly in the CIII 5696 Å line of the WC stars WR 135, 137 and 140, and in the HeII 5411 Å line of the WN stars WR 134 and WR 136 (Moffat *et al.* 1988).

Typically, the bumps appear randomly in time, at different velocities and fade away on a timescale of several hours. Difference spectra of WR 135 (Figure 2) show that the bumps that are red-shifted relative to the line center become even more red-shifted with time, while blue-shifted bumps become more blue-shifted with time.

These small structures are interpreted as blobs (as seen independantly in polarization?) of emitting wind material accelerated outwards along with the general wind.

3) Origin

The origin of the blobs is not known. Non-radial pulsations are not believed to develop in W-R stars (Matthews and Beech 1987; Cox and Cahn 1988; however see Noels and Scuflaire 1987). The theory of vibrational instabilities (Maeder 1985) predicts a correlation of the dispersion with spectral subtype which is opposite to that observed in Figure 1. A phenomenon related to fast rotation is unlikely, according

K. Davidson et al. (eds.), Physics of Luminous Blue Variables, 299–300.

to the spectroscopic behavior of the blobs. Radiatively driven instabilties appear to be the most promising explanation. However, a simple model based on the growth rate of the amplitude of the perturbations (Owocki, Castor and Rybicki 1988) in the context of this theory fails to reproduce the trend in Figure 1, which is also seen in continuum light variations (Lamontagne and Moffat 1987).

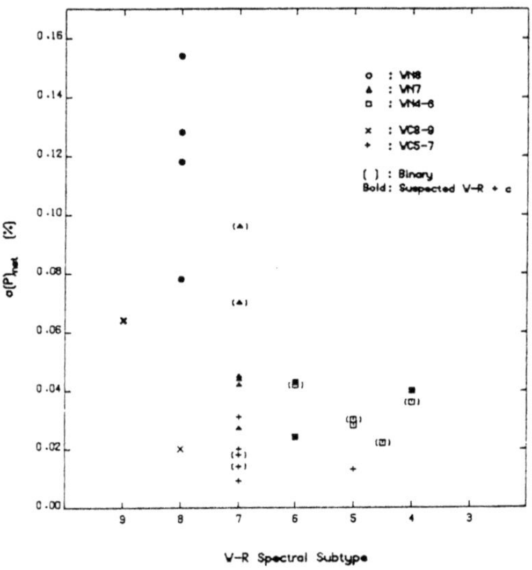

Figure 1. Polarization random scatter plotted as a function of spectral subtype (from Robert *et al.* 1988a). The polarization scatter is the rms deviation from the average polarization for single stars or from a model curve through the phased data for binaries.

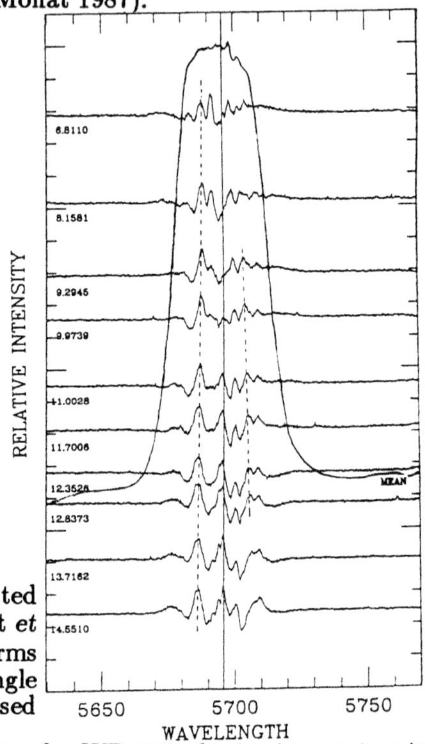

Figure 2. Difference from the mean of spectra for WR 135 obtained on July 6/7, 1988 at Canada France Hawaii Telescope. The mean spectrum (same scale) is the average of all spectra collected during the four night run. The UT time is given for each observation. The vertical line indicates the mean central wavelength. Dashed lines show the radially outwards acceleration of two sample blobs. Note that the dips are not real; they are an artifact of the subtraction method.

References:

Cox, A. N., and Cahn, J. H. 1988, *Ap. J.*, **326**, 804.
Drissen, L., St.-Louis, N., Moffat, A. F. J., and Bastien, P. 1987, *Ap. J.*, **322**, 888.
Lamontagne, R., and Moffat, A. F. J. 1987, *A. J.*, **94**, 1008.
Maeder, A. 1985, *Astr. Ap.*, **147**, 300.
Matthews, J. M., and Beech, M. 1987, *Ap. J.*, **313**, L25.
Moffat, A. F. J., Drissen, L., Lamontagne, R., and Robert, C. 1988, *Ap. J.*, **334**, 1038.
Noels, A., and Scuflaire, R. 1987, in *Instabilities in Luminous Early Type Stars*, ed Lamers and de Loore (Dordrecht: Reidel), p. 213.
Owocki, S. P., Castor, J. I., and Rybicki, G. B. 1988, submitted to *Ap. J.*
Robert, C., Moffat, A. F. J., Bastien, P., Drissen, L., and St.-Louis, N. 1988a, submitted to *Ap. J.*
Robert, C., Moffat, A. F. J., Bastien, P., St.-Louis, N., and Drissen, L. 1988b, submitted to *Ap. J.*
St.-Louis, N., Drissen, L., Moffat, A. F. J., Bastien, P., and Tapia, S. 1987, *Ap. J.*, **322**, 870.
St.-Louis, N., Moffat, A. F. J., Drissen, L., Bastien, P., and Robert, C. 1988, *Ap. J.*, **330**, 286.

LSS4005 AND NaStl, TWO GALACTIC Ofpe/WN or B[e] OBJECTS*

K.A. van der Hucht, SRON Space Research Utrecht, The Netherlands
P.M. Williams, Royal Observatory, Edinburgh, Scotland, U.K.
A.M. van Genderen, Leiden Observatory, The Netherlands
P. Mulder, Kapteyn Astronomical Institute, Groningen, The Netherlands
W.J. Zealey, Department of Physics, University of Wollongong, Australia

Optical and IR spectroscopy of two galactic objects previously considered to be extreme late WN stars (van der Hucht et al., 1984), LSS4005 (formerly tentatively listed as WR85a) and NaStl (WR122), show both objects not to be WR stars, but Ofpe/WN- or B[e]-like objects.

A CASPEC spectrum of LSS4005 obtained in Spring 1985 displays numerous narrow ($\Delta\lambda \approx 20$ km.s^{-1}) emission lines of NII, [NII], NIII, FeII, [FeII], FeIII, [FeIII] and strong emission of HeI and HeII. Except for SiIVλ4089, no absorption lines are present in the 3800-6500Å wavelength range and, in particular, no P Cygni absorption components. This indicates that LSS4005 does not have the dense stellar wind characteristics of WR stars. A medium-resolution CCD spectrum of NaStl shows strong similarities with LSS4005. IR spectra of NaStl in the J, H and K bands show HeI, HeII and NII emission lines.

Walraven photometry of both stars shows Balmer jumps indicative of effective temperatures of the order of 12000 K. Such temperatures are rather low for objects showing HeII emission lines.

Walraven and ground-based IR photometry betray both objects to be light variables ($\Delta m \approx 0.2$ mag for LSS4005, $\Delta m \approx 0.6$ mag for NaStl), and to have heated ($T \approx 900$ K) circumstellar dust envelopes, showing wider ranges of temperature than those associated with WR stars (Williams et al., 1987).

The low-ionization emission lines and the circumstellar dust shells point to a B[e] classification, but the HeII emission is indicative of an O[e] class, while the NIII emission lines argue for a Ofpe/WN status. The composite appearance of both LSS4005 and NaStl could mean a double shell structure or binarity.

In the literature the objects have no equals in the Galaxy nor in the LMC. More observations are needed to understand these enigmatic objects.

References:

van der Hucht, K.A., Williams, P.M., Thé, P.S.: 1984, in: A. Maeder & A. Renzini (eds.), Observational Tests of the Stellar Evolution Theory, Proc. IAU Symp. **105** (Dordrecht: Reidel), p. 273

Williams, P.M., van der Hucht, K.A., Thé, P.S.: 1987, Astron. Astrophys. **182**, 91

---

*) Based on observations collected at the European Southern Observatory (La Silla, Chile), the United Kingdom Infrared Telescope (Mauna Kea, Hawaii, USA) and the Anglo-Australian Observatory (Epping, NSW, Australia)

K. Davidson et al. (eds.), Physics of Luminous Blue Variables, 301.

Investigating turbulence: Colloquium participants on a white-water rafting excursion

Some members of the Local Organizing Committee: Andre Grandchamps, Tony Moffat, Louise Cloutier, Madeleine Bergevin

BRIEF ABSTRACTS OF ADDITIONAL POSTER PAPERS PRESENTED AT IAU COLLOQUIUM 113

AGES OF LUMINOUS BLUE VARIABLES FROM THEIR NEBULAR AND STELLAR ENVIRONMENT, *M.-C. Lortet.*
A study of the spatial distribution of different kinds of young objects in the Magellanic Clouds shows that S Dor-like stars (in a broad sense) are not the very youngest ones. A few specific ideas are developed in a longer article in these proceedings. Maps will be produced and discussed elsewhere.

CURRENT OBSERVATIONS OF THE HUBBLE-SANDAGE VARIABLES IN M 31 AND M 33, *R.M. Humphreys, K. Davidson, O. Stahl, B. Wolf, and F.-J. Zickgraf.*
Recent moderate- and high-resolution spectra of H-S variables covering 4000--9000 Å have been obtained, giving estimates of mass-loss rates and electron densities in the ejected shells. From [N II] and [Fe III] emission in the red spectrum of AE And, we estimate the total mass of the [N II]-emitting gas to be at least 0.006 $M_\odot$. This is very likely the mass ejected during its last outburst, and is additional evidence that the mean mass-loss rate during the total LBV phase is greater than the typical 10^{-4} to 10^{-5} $M_\odot$/yr.

QUANTITATIVE SPECTROSCOPY OF B-SUPERGIANTS IN THE SMC, LMC, AND GALAXY, *H.G. Groth, R.P, Kudritzki, K. Butler, D. Husfeld, S. Becker, F. Eber, & E. Fitzpatrick.*
High-resolution, high-S/N spectra of nine Ia supergiants (B0 to B2) taken with the ESO 3.6m telescope and CASPEC are analyzed by NLTE methods. All objects are found to have helium-enriched atmospheres, which points to enhanced mixing of CN-burned material during the early evolution of massive stars. This is confirmed by the NLTE abundance determinations for two of the objects, which both show clear nitrogen enhancement.

VARIABILITY OF THE O8Ib(f)p STAR HD 167971 IN THE INFRARED, *C. Chavarria-K., C. Leitherer, L. Terranegra, and E. de Lara.*
The triple-star system HD 167971 was observed during nine photometric nights in the summers of 1987 and 1988. The mean photometric values are: $K = 5.18 \pm 0.02$, $J - H = 0.28 \pm 0.01$, $H - K = 0.14 \pm 0.01$, $H - L = 0.25 \pm 0.04$, $K - M = 0.41 \pm 0.06$; the errors reflect the observed variability. Variability in K follows the same pattern as in V, except that the amplitude in K is smaller by a factor of two. The flux excesses in the L and M bandpasses are larger than those found by Leitherer *et al.* (1987, *Astron. Astrophys.* **185**, 121), suggesting that the mass-loss rate is variable on time-scales of one year or less.

WINDS IN CLOSE BINARY STARS: UW CMa (O7 Iaf + OB), *M.F. Corcoran & S.R. Heap.*
UV spectra from the IUE and X-ray images from the Einstein Observatory reveal that: (1) The primary is the source of the observed photospheric and wind spectrum; (2) The secondary periodically occults portions of the wind; (3) The mass-loss rate has maxima at the primary's tidal bulges; (4) Material from the primary's outer tidal bulge expands more slowly than does the bulk wind; (5) $L_X = 2 \times 10^{32}$ erg/s, independent of orbital phase.

K. Davidson et al. (eds.), Physics of Luminous Blue Variables, 303–308.

THE INFLUENCE OF CLOSE BINARY EVOLUTION ON THE DISTRIBUTION OF MASSIVE STARS IN THE H-R DIAGRAM, *D. Vanbeveren.*

A statistical study is carried out on the distribution of different kinds of massive star classes in the H-R diagram, using recent calculations of massive single-star and massive close-binary evolution. The influence of the mass tranfer process during the critical Roche lobe overflow phase of a primary component on the relative frequencies of different classes is outlined. It is shown that without an exact knowledge of how mass transfer takes place, the meaning of an initial mass function determined by classical methods is unclear, while a direct comparison of the observed and theoretically predicted blue/red star ratios is meaningless.

PULSATING B STARS IN THE MAGELLANIC CLOUDS: FIRST RESULTS FOR THE LMC, *C. Sterken & M. Jerzykiewicz.*

We have undertaken a search program for detecting β Cephei stars in the LMC. Three program stars are found to exhibit short-term brightness variations. There is one promising β Cep candidate: star 257 in the catalog of Brunet *et al.* (1975) shows a sinusoidal light curve with an amplitude (half-range) of 0.0070 mag and a period of about 5 hours.

RAPID VARIABILITY IN O STAR WINDS, *H. Henrichs, L. Kaper, & R. Zwarthoed.*

Rapid variability is seen in the Si IV P Cygni profiles in high-resolution IUE spectra of three O stars: λ Cep, O6 If; 68 Cyg, O7.5 III; and 19 Cep, O9.5 Ib. The variability occurs in the form of discrete absorption features, which march from low to high velocity within a day to a week (depending on the star), indicative of rapid changes in the stellar wind structure. The origin of this variability is not clear, but we confirm earlier suggestions that the beginning of a new 'episode' might be correlated with the rotation period of the star. There is growing evidence that the above behavior is in fact representative of all O stars.

TRANSIENT FEATURES IN THE He I P CYGNI PROFILES OF TWO EXTREME Of STARS, *A.W. Fullerton, D.R. Gies, & C.T. Bolton.*

Variable absorption features have been detected in the absorption trough of the He I λ5876 P Cygni profiles of two luminous Of stars, HD 151804 and HD 152408. These features propagate from red to blue over several days. They resemble the discrete opacity enhancements observed in UV wind lines of O stars, and are likely caused by the same mechanism.

THE WINDS OF CHEMICALLY EVOLVED O STARS, *I.D. Howarth & R.K. Prinja.*

Results of a differential analysis to investigate the surface chemistry of ONC, supergiant, and Of stars are presented, based on a quantitative ultraviolet survey of the winds of 203 galactic O stars (Howarth & Prinja 1989, *Astrophys. J. Suppl.*). The enhanced surface nitrogen abundance in ON stars is confirmed, while nitrogen appears to be depleted (by a factor of 5) in the OC stars. O supergiants do not exhibit non-solar atmospheric abundances, but Of stars appear to have mild ($< 2 \times$) surface nitrogen enhancements and may reveal CNO-processed material.

ON THE USE OF Hα AND He II λ4686 AS MASS-LOSS INDICATORS IN LUMINOUS BLUE STARS, *A. Gabler, R. Gabler, & R.P. Kudritzki.*

Our 'unified' model atmosphere code, which avoids the artificial division between photosphere and wind, is used to calculate correct theoretical Hα and He II λ4686 profiles by varying the wind parameters for a given model (T_{eff}, g, R fixed). The results can be used directly to derive mass-loss rates from observed profiles or equivalent widths.

RATES OF MASS LOSS IN THE UPPER H-R DIAGRAM, *H. Nieuwenhuijzen & C. de Jager.*

We have used 258 observed values for the rate of stellar mass loss in the upper part of the H-R diagram, in conjunction with masses derived from evolutionary calculations, to search for simple parametrizations of $log\ \dot{M}$. While a previously derived 20-term expression gave an overall O-C standard deviation of $\sigma = 0.49$ in $log\ \dot{M}$, we now find σ only 20% larger for the expression: $log\ \dot{M} = 8.78 \times 10^{-15}\ L^{1.24}\ M^{0.18}\ R^{0.81}$, where L, M, and R are in solar units and $log\ \dot{M}$ is in solar masses per year.

UNIFIED NLTE MODEL ATMOSPHERES INCLUDING STELLAR WINDS AND SPHERICAL EXTENSION, *R. Gabler, A. Gabler, A. Pauldrach, J. Puls, & R.P. Kudritzki.*

Self-consistent model atmospheres are presented which use only g and L as free parameters. The model atmospheres -- calculated for hot Of-supergiants and O-giants -- produce energy distributions and typical 'wind lines' (He II λ4686, He II λ1640, Pα, Bα) together with normal photospheric lines, as observed in stars of this type.

RADIATION DRIVEN WINDS OF HOT LUMINOUS STARS -- THE LUMINOSITY DEPENDENCE OF STELLAR WIND LINES, *A. Pauldrach, R.P. Kudritzki, & J. Puls.*

A grid of wind models along evolutionary tracks of massive O stars is calculated using the improved radiation-driven wind theory, in order to investigate changes of spectral morphology during the course of stellar evolution. The results show that the improved stellar wind models can describe the significantly different observed behavior of the profiles of the Si IV line with respect to those of C IV and N V as a function of spectral type and luminosity class. Agreement with observations is also found for the computed mass-loss rates and the ratios of calculated terminal wind velocities and escape velocities. For the wind efficiencies, a strong dependence on mass was found, together with a pronounced decrease from main sequence stars to supergiants and a dramatic increase toward LBV's.

ON THE STABILITY OF RADIAL PULSATIONS OF MASSIVE STARS, *J. Klapp & M. Corona.*

We investigate the radial stability of massive stars using the classical Ledoux quasi-adiabatic approximation, the Castor quasi-adiabatic approximation, and a fully non-adiabatic calculation. We find that the three methods of calculation give similar results provided that a small region in the outer layers of the star is neglected for the classical approximation. The critical mass for stability of massive stars is found to be in agreement with previous work. We explain the reason for the discrepancy between this and previous work by one of the authors.

LUMINOSITY AND RADIUS RELATIONS FOR WOLF-RAYET STARS, *N. Langer.*

On the basis of a comprehensive set of numerical models for W-R stars, we demonstrate that all observable properties of WNE/WC/WO-type stars depend solely on their mass and surface chemical composition. This enables one to derive very accurate relations for luminosities and radii as a function of these two parameters for various W-R subtypes (*cf.* N. Langer, *Astron. Astrophys.*, in press).

WC STARS: HOT STARS WITH COOL WINDS AND A STRATIFIED IONIZATION STRUCTURE, *D.J. Hillier.*

Detailed modeling of WC4-6 stars confirms a C/He abundance of > 0.1 (by number). The importance of dielectronic recombination for C III λ2296 can explain the discrepancies between the abundances derived from UV lines and those based on analyses of optical recombination lines. Dielectronic recombination is also very important for the formation of the strong optical C III lines. Because of the high C/He abundance, line cooling is very efficient. Typically we find T_e = 9000 K at $N_e = 10^9$ cm^{-3} and T_e = 20000 K at $N_e = 10^{11}$ cm^{-3}. The ionization state of the gas varies with radius.

SPECTRAL ANALYSIS OF WN STARS IN THE LMC, *W. Schmutz & P.S. Conti.*

Besides the sophisticated modeling of a particular W-R stellar spectrum (*e.g.* Hamann *et al.* 1988, *Astron. Astrophys.* **194**, 190), a relatively simple way has been introduced of inferring W-R atmospheric parameters using model grids. From a few observed quantities, the stellar parameters of the W-R stars can be determined. Schmutz *et al.* (1988, *Astron. Astrophys.*, in press) have published the stellar parameters for 30 galactic W-R stars. Here, the same procedure is applied to WN stars in the LMC. Based on the preliminary analysis presented in this poster, no systematic differences are found between galactic and LMC W-R stars.

THE EVOLUTIONARY STATUS OF HD 62910 (WN6-C4), *A.J. Willis & D.J. Stickland.*

A detailed analysis of high-resolution UV (IUE) and optical spectra of HD 62910 confirms its WN6-C4 composite classification, and shows that both its WN and WC features are formed in the same stellar wind. Abundances derived from a non-LTE Sobolev analysis indicate a zero H-content and He/C/N ratios intermediate between those of WNE and WC stars. We conclude that HD 62910 is on its way from the WN6 to the WCE stage.

TWELVE NIGHTS OF LINE PROFILE VARIATIONS OF HD 191765, *S.R. McCandliss & B. Bohannan.*

HD 191765 (WN6) was observed for twelve nights over essentially the entire visible wavelength region to search for correlated line profile variations. No convincing periodic variations were found in either the line flux or line centroid measures. Periodograms of the flux variations have more power on time scales of a few hours, a time scale characteristic of the wind flow. Line profile variations of the Pickering series of HeII indicate that lines formed farther out in the wind suffer more variation than those formed deeper. Our results suggest that no single physical mechanism is responsible for the line profile variations.

CCD OBSERVATIONS OF EMISSION LINE PROFILE VARIATIONS IN WOLF-RAYET STARS, *G. Koenigsberger & P. Massey.*

We report preliminary results of the observations of HD 164270 (WC9), HD 177230 (WN8), HD 192163 (WN6), and HD 193077 (WN6) in the wavelength range 4600--4720 Å, at a resolution of 0.3 Å. Variations on a day-to-day timescale are detected in the four systems. The variations in the two WN6 systems occur within approximately -600 km/s and +500 km/s of the line center, with the shortest timescale for variability being greater than 2 hours.

HIGH-RESOLUTION IUE VARIABILITY IN V444 CYGNI (WN+O6), *N. St.-Louis, A.J. Willis, & L.J. Smith.*

The ultraviolet wind eclipse spectrum of V444 Cygni (ratio of the spectra at phases 0.0 and 0.5) shows selective absorption in C IV λ1549.0, He II λ1640.4, and N IV λ1718.55 and numerous absorption features forming a large depression at λλ < 1500 Å Using experimental wavelengths and intensities of Fe V and Fe VI transitions with each line broadened to a gaussian profile (FWHM = 1000 km/s) and added linearly, we show that the λλ < 1500 Å absorption is largely due to Fe V, Fe VI. However, to achieve a good fit required shifting the resulting spectrum by -1100 km/s.

SEARCH FOR ULTRAVIOLET VARIABILITY IN HD 151932 (WR 78), *N. St.-Louis, L.J. Smith, & A.J. Willis.*

A series of 16 IUE spectra of HD 151932 obtained over 4 consecutive days in May 1988 shows no significant variability, but the Si IV λλ1393,1402 P Cygni profiles in these spectra show substantial differences compared to an archival spectrum secured in 1980. The 1988 Si IV lines show an excess absorption compared to 1980, extending a further ~ 350 km/s. These differences are similar in character to the Si IV profile variations previously reported for HD 96548.

INTENSIVE PHOTOMETRY OF SOUTHERN WOLF-RAYET STARS, *L.A. Balona & J. Egan..*

Intensive photometry of 17 W-R stars shows no evidence for p-mode pulsations. In WR 6 and WR 40, two simultaneous periods consistent with g-mode pulsations are found. Many W-R stars show single-period coherent light variations, but in some cases it is unlikely that these can be attributed to binary effects.

NEW RESULTS ON THE PHOTOMETRIC VARIABILITY OF TWO WN8 STARS, *E. Gosset, J.-M. Vreux, J. Manfroid, C. Sterken, & M. Remy.*

We present recent results on the variability of WR 40 and WR 16. For WR 40, a detailed analysis is in press (*M.N.R.A.S.*). For WR 16, we present the first extensive, reliable data set and show that the variability of WR 16 is not as simple as usually thought. The power spectrum is dominated by a peak corresponding to a period of about 10 days, but other significant peaks cannot be neglected. We suggest that the two periodicities discovered by Manfroid *et al.* (1987, *Astron. Astrophys.* **185**, L17) in their differential photometry could be attributed to the comparison star HD 86000.

RECURRENT DUST FORMATION BY WOLF-RAYET STARS,
P.M. Williams, K.A. van der Hucht, & P.S. Thé.

We present new data in our continuing study of Wolf-Rayet stars which show variations in their infrared fluxes attributable to variations in the rates of dust formation in their winds. There are four established examples:

WR 140 = HD 193793 (WC7+O4-5): variation periodic with orbit (2900 d)
WR 137 = HD 192641 (WC7+O): last maximum 1984.5, interval ~ 12 y
WR 48a (WC8): last maximum 1979, interval > 9 y
WR 70 = HD 137603 (WC8+B0): perhaps periodic (~ 400 d?)

The variation of WR 140 is well linked to its orbital motion, the infrared outbursts occurring near the time of periastron passage; WR 70 and WR 137 are also binaries. There are other candidates whose variation is still being established.

DISPLACED He I ABSORPTION LINES IN WOLF-RAYET STARS: REVISIONS TO V_∞, *P.M. Williams & P.R.J. Eenens.*

We present new spectra of six WC 7--9 stars showing the 2.058 μm He I 2s—2p absorption line displaced by velocities around 0.7 of the stars' terminal velocities derived conventionally. It is argued that these lines are formed in the outermost regions of the winds and give more accurate representations of the wind terminal velocities than the P Cygni profiles of saturated CIV lines observed in the UV. Consequently, mass-loss rates derived for these stars must be reduced in proportion.

HD 5980: A WR+WR BINARY,
V.S. Niemela.

The radial velocities of emission and absorption lines in the spectrum of HD 5980, the brightest W-R star in the SMC, imply that this eclipsing binary has two WN-type components, and probably a line-of-sight O-type neighbor. Both components of the WN4.5+WN3 binary are more massive than 40 $M_\odot$.

HUNTING FOR FAINT NEW GALACTIC WOLF-RAYET STARS,
M.M. Shara, M. Potter, A.F.J. Moffat, & L.F. Smith.

Broadband-narrowband He II λ4686 UK Schmidt direct plate pairs of a nine-square-degree region in Carina have been digitized and searched for faint new Wolf-Rayet stars. A ctalog of one million stars located on each plate yielded one hundred candidates. Eight new W-R stars were spectroscopically confirmed: six are WN and two are WC.

WOLF-RAYET STARS AND THE LUMINOUS STAR CONTENT OF GIANT H II REGIONS IN M 33, *L. Drissen, A.F.J. Moffat, & M.M Shara.*

CCD images of the two largest giant HII regions in M 33, obtained with a set of narrow-band filters sensitive to strong Wolf-Rayet emission lines, reveal eight W-R candidates in NGC 604 and nine in NGC 595. The flux in their emission lines is similar to that of normal Galactic W-R stars. The WR/O number ratio in NGC 604 ($\approx$ 0.1) is similar to the general Galactic ratio, but seems higher in NGC 595.

A FEW CATALOGS AND REFERENCES THAT INCLUDE MOST KNOWN LBV'S AND RELATED STARS:

BE: Bohannon & Epps, *Astr. Astrophys. Suppl.* **18**, 47 (1974). Stars in LMC.

HD: "The Henry Draper Catalog," *Harvard Annals* **91--100** and **112** (1918--1936). Properly "HDE" beyond star number 225 300. Many additional HD numbers can be found in the auto registration office in Heidelberg.

He 3: Henize, *Astrophys. J. Suppl.* **30**, 491 (1976).

HR: Yale Bright Star Catalog, most recent (4th, 1982) edition by Hoffleit & Jaschek. "HR" originally meant "Harvard Revised Photometry", *c.* 1908.

Hubble-Sandage Variables in M 31 and 33: For positions and refs. see Humphreys *et al.*, *Astrophys. J.* **278**, 124 (1984).

HV: "Harvard Variables" in the Magellanic Clouds; see, *e.g.*, Payne-Gaposchkin, *Smithsonian Contrib. to Astrophys.* **9** (SMC, 1966) and Number 13 (LMC, 1971)

MWC: "Mt. Wilson Catalog"; Merrill & Burwell, *Astrophys. J.* **78**, 87 (MWC 1--416, 1933); **98**, 153 (417--666, 1943); and **110**, 387 (667--1088, 1949).

R: "Radcliffe Catalog" of stars in Magellanic Clouds; Feast, Thackeray, & Wesselink, *M.N.R.A.S.* **121**, 337 (1960). Stars R 1--50 are in SMC, R 51--158 are in LMC.

S: Henize, *Astrophys. J. Suppl.* **2**, 315 (1956); stars in Magellanic Clouds. Properly, should have prefixes LHα 115 - in SMC and LHα 120 - in LMC, *e.g.* LHα 120 - S 128. S merely means star (as opposed to N for nebula).

Sk: Sanduleak's stars in Magellanic Clouds. In LMC, *Cerro Tololo Inter-American Observatory Contr. 89* (1969); in SMC, *Astron. J.* **73**, 246 (1968) and **74**, 877 (1969).

Alternative designations for some relevant stars are also listed by Carlson & Henize, *Vistas in Astron.* **18**, 47 (1979) and by Shore & Sanduleak, *Astrophys. J. Suppl.* **55**, 1 (1984).

INDEX TO STARS AND GALAXIES MENTIONED IN THIS BOOK, WITH MANY CROSS-REFERENCES TO ALTERNATIVE DESIGNATIONS:

(*The cross-references are probably not complete. Also, the arrangement may seem inconsistent in some places, but this should be harmless in practice. --* KD)

<u>GPSR Compliance</u>

The European Union's (EU) General Product Safety Regulation (GPSR) is a set of rules that requires consumer products to be safe and our obligations to ensure this.

If you have any concerns about our products, you can contact us on ProductSafety@springernature.com

In case Publisher is established outside the EU, the EU authorized representative is:

Springer Nature Customer Service Center GmbH
Europaplatz 3
69115 Heidelberg, Germany

Batch number: 10607222

Printed by Printforce, the Netherlands